Transfer of Technology

Sustainable and Unsustainable Experience from Israel

Edited by
A. Paul Hare

ISBN 0-9785565-9-3
ISBN 978-0-9785565-9-4

Camera-ready manuscript prepared by
Aaron Leeper
Words and Designs
P. O. Box 3064, Beer Sheva 84130 Israel

Biblio Books
an In Other Words company

Contents

Preface

This may be the last book I write or edit as a member of the Blaustein Institutes for Desert Research (BIDR), Ben-Gurion University of the Negev, where I have been a faculty member since 1984. The previous books I have authored or edited that relate to the experience of living and working in the desert are:

- *The Hebrew Israelite Community*. Hare, A. P. (ed.) (1998). Lanham, MD: University Press of America.
- *Desert Regions: Population, Migration, and Environment*. Portnov, B. A., and A. P. Hare (eds.) (1999). Heidelberg: Springer.
- *Foreign Experts and Unsustainable Development: Transferring Israeli Technology to Zambia, Nigeria and Nepal*. Schwartz, M., and A. P. Hare (2000). Aldershot, UK: Ashgate.
- *Israel as Center Stage: A Setting for Social and Religious Enactments*. Hare, A. P., and G. M. Kressel (eds.) (2001). Westport, CT: Bergin and Garvey.
- *The Desert Experience in Israel: Communities, Arts, Science, and Education in the Negev*. Hare, A. P., and G. M. Kressel (eds.) (2007). Lanham, MD. University Press of America.

Short introductions to the material found in each of the 17 chapters of this book are provided in Chapter 1, entitled "Transferring Technology." Half of the authors of these chapters I first knew as neighbors during the time I lived with my family at Midreshet Ben-Gurion. As for the other researchers at the BIDR, I only discovered what they had been doing there after I sent an e-mail to each of them asking if they had written anything about transfer of technology they had developed in their various disciplines, which they considered worthy of inclusion in this book. I was pleasantly surprised by their responses. Much creativity has been taking place at the BIDR, as you will discover, and I didn't know the half of it. Robert Gilmour, a chemical safety specialist and the author of Chapter 15, is the lone exception to the abovementioned connection to the BIDR.

synonymous terms, since research results have value as knowledge only when they are put to effective use. Transferring research results into practical applications is what technology transfer is all about. In the USA, the federal government is the largest source of funding for basic research. The new means of disseminating information rapidly and broadly (e.g., the Internet) now provides accessibility and ease-of-use that were lacking in 1997 when Gresshoff edited his volume.

Hawthorne (1971) summarizes the proceedings of a conference of Ministers held in Istanbul—with participants from Greece, Portugal, Spain, Turkey, and Yugoslavia—to explore possible measures these countries might take to facilitate or improve the transfer of knowledge they required. They concluded that the effective transfer of knowledge is not the mere importation of a productive process, but rather the adaptation and absorption of key elements of that technology into the industrial life of the host society. The main problems discussed included: licensing, the role of Research and Development, infrastructural activities, and institutional mechanisms and planning.

A book entitled *Technology Transfer Practice of International Firms* (Bradbury, 1978: 299) records a set of guidelines for the transfer of science and technology, wherein enterprises should:

- endeavor to ensure that their activities fit satisfactorily into the scientific and technological policies and plans of the countries in which they operate, and contribute to the development of national and scientific capacities, including, as far as appropriate, the establishment and improvement in host countries of their capacity to innovate;
- adopt practices in the course of their business activities that permit the rapid diffusion of technologies to the fullest extent practicable, with due regard to the protection of industrial and intellectual property rights;
- set reasonable terms and conditions when granting licenses for the use of industrial property rights or when otherwise transferring technology.

Allen (1977) calls attention to the similarities and differences between science and technology in a book entitled *Managing the Flow of Technology*, which focuses on Research and Development organization. For science, both the input and the output of the system are information in the form of human language. Technology is also an ardent consumer of information; however, the output is in the form of products or processes. The scientist's principal goal is to publish a paper. The technologist's goal is to produce a physical change in the world. For a scientist, the output of one phase can be the input for the next phase. For a technologist, because outputs are usually in a different form, they generally cannot serve directly as inputs for the next phase. However, as a by-product, technology may provide verbally encoded information in the form of documentation. Examples of technology transfer and innovation are provided by Mogavero and Shane (1982) with an example from the National Aeronautics and Space Administration of the USA. The overlaps between international business and technology transfer are reported in a volume edited by Agmon and Von Glinow (1991). Most of the articles in the present volume deal with the transfer of technology.

Reporting data primarily from three countries—the United Kingdom, Spain, and Greece—Charles and Howells (1992) describe the technology in Europe through public and private networks, based mainly in firms, higher education institutions and public research establishments. There are three interrelated processes: hardware transfer, information transfer, and knowledge transfer.

With a closer look at institutions of higher education, Matkin (1990) compares the transfer of technology at the University of California, Berkeley, at the Massachusetts Institute of Technology, at Stanford University, and at Pennsylvania State University.

One of the features of technological change—especially in developing areas—is the extent of foreign direct investment (Cantwell, 1999a, Vol. I and 1999b, Vol. II). In the collection of articles in Volume I the focus is on the firm or business rather than the level of the country or product. For multi-national corporations the focus is on science and engineering knowledge. The collection in Volume II deals with research and development.

Social technology transfer is considered to be the ensemble of methods used to change or maintain the behavior or thinking of individuals, small groups, communities, or organizations in order to reach desired ends (Chatterjee, 1990). Some examples are diagnostic casework, Gestalt therapy and program planning and budgeting systems. The transfer of social technology from one context to another depends on their similarities. An example of social technology transfer is provided in the present volume, in Chapter 17 on conflict resolution.

Some of the literature on social technology appears under the heading of "organizational capabilities," defined as the ability of an organization to perform and extend its output actions (Dosi, Nelson, and Winter, 2000).

Case studies of the transfer of technology from the USA to Latin America reveal that major economic and political variables made the transfer difficult. In Mexico there was illegal immigration to the USA, in Brazil a confrontation over nuclear energy, and in Columbia the attempt by the local government to gain more control over the management of energy and mineral resources (Baranson, 1981). Examples of similar problems in the present volume are given in Chapter 2 on settlements.

In a monograph covering the period from the 1960s through the 1970s, Eckaus (1977) notes that science and technology were increasingly recognized as powerful influences in economic growth processes in developing countries. However there was a growing concern about the apparent incongruities between the goals of developing countries, their labor supply conditions, and other resource endowments and the technologies these countries were importing. Advanced technologies were characterized by the use of less labor per unit of output, meaning that they not only failed to expand employment as fast as output but, in some instances, may have actually displaced labor and contributed to already substantial unemployment. Large plants and equipment were required, placing a strain on countries where capital was scarce. The technology strategies were measured against criteria of appropriateness. The criterion most commonly applied was net output maximization or cost minimization. However the only standard for deciding the "appropriateness" of technological decisions is in reference to the general goals of development.

A study of the adoption and diffusion of imported technologies in Korea was conducted in the late 1970s by Enos and Park (1988). The typical process went through several phases: planning, negotiation with foreign suppliers, design, construction, start-up and operation of equipment, securing improvements, developing new products and techniques, absorption of imported technology, and diffusion of imported technology.

Additional case studies are provided: Elmusa (1994) indicates the plusses and minuses of introducing drip irrigation and greenhouses to farms in the Jordan Valley; Granstrand (1999) addresses the economics and management of intellectual property versus natural resources in Japan; Cantwell (2004), in a collection of studies, demonstrates the advantages of "clusters" of firms by providing a pool of skill workers, local suppliers for necessary inputs, easy transmission of new ideas, and access to a university if the firm is science based; Dunning and Narula (2004) list the problems in catching up with the developed countries faced in developing countries that are part of multinationals and have foreign direct investment, with examples from Korea and Taiwan; Romanos and Auffrey (2002) describe the factors involved in sustainable development in intermediate-sized cities in a growth region of Thailand; Muir (1997) records the technology transfer system in the United States, with particular reference to licensing; The United Nations Industrial Development Organization (1979) provides a system of methods and mechanisms for diversified financial and technical cooperation, geared to the special and changing requirements of international industrial co-operation, as well as a set of guidelines for bilateral industrial co-operation; Turpin et al. (2002) give an account of innovation knowledge systems and regional development in China and Australia; Contractor (1985), on the subject of licensing in international strategy, outlines a guide for planning and negotiations.

Dunn (2005) suggests that it helps to have an "enabling adopter," who is a member of a work group who identifies a collaborative technology that could enhance the group's effectiveness and then works proactively to encourage the group to adopt the technology successfully. She found that the use of the "enabling adaptor" role had four phases that differed from the classic technology-transfer model in three ways: a stronger outcome orientation, greater attention paid to the coordination of users, and a shift from a linear to a circular model, reflecting the assumption that once adopted, the technology changes the work itself.

Contents of this book

In Chapter 1, on transferring technology, which you have just finished reading, A. Paul Hare summarizes the literature on the transfer of technology with particular reference to desert technologies.

In Chapter 2 on settlements, Moshe Schwartz and A. Paul Hare follow up—in the 1990s—the sustainable and unsustainable features of settlements and forms of agriculture introduced by Israeli agriculturalists and community planners in the 1960s in Zambia, Nigeria, and Nepal.

In Chapter 3, on building maintenance, Isaac Meir and Simon L. Hare describe the problems of maintaining various forms of buildings and homes built to follow solar principles, mainly due to lack of experience with solar construction on the part of tenants.

In Chapter 4, on solar construction, Yair Etzion describes a set of projects having to do with solar housing that failed to transfer.

In Chapter 5, on infrared and solar energy, A. Paul Hare summarizes the way that Jeff Gordon used infrared light to reduce the condensation of moisture on the windows of a racetrack in Singapore, innovative optics to serve the photovoltaic revolution, and the concentration of desert solar energy for laser surgery.

In Chapter 6, on remote sensing, Arnon Karnieli describes his use of remote sensing to compare the reliability of vegetation degradation and drought indices in Mongolia and assessing land degradation around water points in Kazakhstan.

In Chapter 7, on gully incision, Yoav Avni notes that geomorphic changes leading to natural desertification versus anthropogenic land conservation and gully incision are key factors in desertification in the arid environment of the Negev Highlands in Israel

In Chapter 8, on nomads and sheep, Allan Degen combines reports of Bedouin households and sheep production in Israel, the rejuvenation of forests with fodder trees and shrubs to sustain goat production in the hills of Nepal, browse selection by Karakul sheep in relation to plant composition and estimated metabolizable energy content, and asks the question "can pastoral nomads return to their traditional livelihood of raising livestock in the Registan desert of southern Afghanistan?"

In Chapter 9, on fish in the desert, Samuel Appelbaum reports that the technology for raising fish—for food and as ornamental fish—in brackish desert water at the Blaustein Institute has been transferred to kibbutzim in Israel.

In Chapter 10, on dust on solar collectors, Sergey Biryukov suggests an alternative method for cleaning dust from solar collector surfaces in desert and semi-desert areas, where water for cleaning is too expensive.

In Chapter 11, on grey water, Amit Gross uses advanced technologies for grey water treatment and reuse and nutrient recycling from organic waste for the production of organic fertilizers and protection of the environment.

In Chapter 12, on ground water pollution, Ofer Dahan describes water infiltration and pollutant transport through the deep vadose zone.

In Chapter 13, on ectoparasite population dynamics, Boris Krasnov describes the immune response to fleas in a wild desert rodent in southern Israel; the effects of parasite species, parasite burden, sex of host, and host parasitological experience—and the fact that the "mean abundance" predicts the prevalence of the distribution of fleas among small mammals in Slovakia.

In Chapter 14, on crust and dust, Eli Zaady applies methods for biogenic soil crust analysis and airborne particle study in desert environments to enhance runoff production in the northern Negev desert.

In Chapter 15, on the transfer of vital safety information in the chemical industry, Robert Gilmour outlines the procedures, based on experience in the Negev, that are used in the chemical industry in other countries.

In Chapter 16, on nomads, Gideon M. Kressel and Hendrik J. Bruins discuss the problems involved in integrated drought contingency planning in Israel, Kenya, and India.

In Chapter 17, on conflict resolution, A. Paul Hare reports on applications of methods of nonviolent protest and conflict resolution in Curacao, Cyprus, and South Africa.

A sample of work at the Blaustein Institutes by Bertrand Boeken

As a sample of the work being done to transfer technology to forestry in Israel, Bertrand Boeken provides the following account of interactions with organizations responsible for practical ecosystem management.

Since 1991, Moshe Shachak, Eli Zaady, and I (Bertrand Boeken), with a large number of research students and technical assistants, have been coordinating efforts with Avi Perevolotsky and Eugene Ungar from the Agricultural Research Organization - Volcani Center, with Menachem Sachs and Itzik Moshe and others from the Jewish National Fund (JNF) Forestry Division, Southern District and with Ramy Gerty and Shmuel Arbel from the Ruppin Research Station for Erosion and Runoff Research, Ministry of Agriculture. Since the late 1990s, we have also been working closely with Yoshua Shkedy and others in the Nature and National Parks Protection Authority (NNPPA).

The JNF—which has made great strides in the reforestation of Israel—and the NNPPA, represent the stakeholders for applying scientific understanding to land management and nature conservation, each with its own goals and policies. Our approach to technology transfer has not been based solely on telling them what we think, but rather on cooperation and mutual exchange of ideas.

From the beginning of our work with the JNF (starting with the Savannization Project, which has continued in various other long- and short-term projects), we have been looking at ecological problems from both scientific and practical management points of view. More specifically, the exchange of ideas has been based on:

- Incorporating commonly used management techniques in the design of field experiments,
- Helping the JNF update their basic management approach to reflect current ecological understanding, and
- Conveying research results and recommendations to management.

Through our work with the NNPPA, we have also been instrumental in their adoption of new ecological approaches for nature conservation management. Along with the practical and intellectual cooperation we share with our colleagues in the JNF and the NNPPA, we also organize and participate in meetings and workshops with JNF- and NNPPA-field managers and supervisors as a means of exchanging information with them at that level.

Beyond the mutual transfer of concepts and technology and the cooperation and intellectual interactions among us, an unexpected dividend has come from the many Bedouin who work as JNF field workers and who assist in construction and maintenance of cooperative field experiments: these Bedouin have learned to apply the ideas and techniques they observe to their own agricultural practices.

Content-wise, exchange of concepts and transfer of techniques are based on our systems approach to ecology, which entails cutting-edge insight into how ecosystems function, how their function is related to their structure, and how humans influence these. These insights stem from theoretical and empirical research on cooperative projects that were partly based on management. The following are important highlights of these insights:

Exchange of ecological concepts;

- Ecological systems are determined by interactions among organisms, landscape structure and resources;
- Human activities are part of ecological systems (by land use for grazing and agriculture, and by disturbance, degradation and restoration of landscapes);
- Ecological sustainability of human activities (from land use by grazing, crop cultivation and recreation to conservation of biodiversity) which depends on understanding the ecological basis and consequences of human land use;
- Most small-scale human activities in dryland primarily and initially impact landscape structure, which cascades to produce major changes in resource dynamics and biological production and diversity;
- Usually there are feedback mechanisms involved that lead to unexpected non-linear and non-equilibrium system changes, to thresholds and irreversibility in state transitions.

Recommendations for management

- Spatial landscape heterogeneity (of scattered shrub/tree patches in an open matrix) is crucial for primary production, plant, animal and microbial biodiversity, resource preservation and sustainable land use;
- Crust cover of the inter-shrub matrix determines runoff (+) and soil erosion (-);
- Negative grazing effects can be mitigated by protection of shrub patches, and adjustment of stocking rates;
- Variability of relationships and state-transitions along the Israeli rainfall gradient requires variable management (stocking rates, preservation of vegetated patches in more arid regions and of open patches in more humid zones).

Recommendations for restoration

- Ecosystem functions in degraded landscapes can be rehabilitated by altering landscape structure;
- Addition of patches (contour dykes, slope mini-catchments) is efficient in situations where shrub patches are small, sparse and/or absent.

In spite of our productive cooperation with the JNF (and the NNPPA) land managers in various field projects and intellectual exchange, there remain differences of opinion, approach and implementation due to differing aims, financial and organizational constraints, and sometimes, misconceptions. An example is the restoration of degraded watersheds, which should be implemented with spatial variation from non-interference to extensive construction of contour dykes, in order to create large-scale heterogeneity and to allow natural processes of recovery and adjustment to take place. In addition, the transfer of ideas and techniques to the level of national and regional policy seems to be very slow to set in motion, as they are limited by current attitudes, opinions and regional politics and often by discrepancies in the language and terminology used. The same seems to be the case for transfer of ideas about dryland land use and degradation to international organizations (UN, NGOs), and to associated social science communities.

References

Agmon, T., & Von Glinow, M. A. (1991) *Technology transfer in international business.* New York: Oxford University Press.

Allen, T. J. (1977) *Managing the flow of technology: Technology transfer and the dissemination of technological information within the R & D organization.* Cambridge, MA: The MIT Press.

Bradbury, F. R. (Ed.) (1978) *Technology transfer practice of international firms.* Alphen Aan Den Rijn, The Netherlands: Sijthoff & Noordhoff.

Baranson, J. (1981) *North-South Technology Transfer: Financing and Institution Building.* Mt. Airy, MD: Lomond Publications.

Cantwell, J. (Ed.) (1999a) *Foreign direct investment and technological change: Volume I: Theories of technological change.* Cheltenham, UK: Edward Elgar Publishing.

Cantwell, J. (Ed.) (1999b) *Foreign direct investment and technological change: Volume II.* Cheltenham, UK: Edward Elgar Publishing.

Cantwell, J. (2004) *Globalization and the location of firms.* Cheltenham, UK: An Elgar Reference Collection.

Charles, D., & Howells, J. (1992) *Technology transfer in Europe: Public and private networks.* London: Belhaven Press.

Chatterjee, P. (1990) *The transferability of social technology: Explorations in the knowledge structures of the helping professions and their transfer.* Lewiston, NY: The Edwin Mellen Press.

Contractor, F. J. (1985) *Licensing in international strategy: A guide for planning and negotiations.* Westport, CT: Quorum Books.

Doheny-Farina, S. (1992) *Rhetoric, innovation, technology: Case studies of technical communication in technology transfers.* Cambridge, MA: The MIT Press.

Dosi, G., Nelson, R. R., & Winter, S. G. (2000) *The nature and dynamics of organizational capabilities.* Oxford, UK: Oxford University Press.

Dunn, S. P. (2005) The enabling adopter: An emerging role in the diffusion of complex technologies. *Dissertation Abstracts International: Section B: The Sciences and Engineering,* 66 (3-B).

Dunning, J. H., & Narula, R. (2004) *Multinationals and industrial competitiveness: A new agenda.* Cheltenham, UK: Edward Elgar.

Eckaus, R. S. (1977) *Appropriate technologies for developing countries.* Washington, DC: National Academy of Sciences.

Elmusa, S. S. (1994) *A harvest of technology: The super-green revolution in the Jordan Valley.* Washington, DC: Georgetown University, Center for Contemporary Arab Studies.

Enos, J. L., & Park, W. H. (1988) *The adoption and diffusion of imported technology: The case of Korea.* London: Croom Helm.

Granstrand. O. (1999) *The economics and management of intellectual property: Towards intellectual capitalism.* Cheltenham, UK: Edward Elgar.

Ghatak, S., Deadman, D., & Eadie, C. (1981) *Technology transfer to developing countries: The case of the fertilizer industry.* Greenwich, CT: JAI Press.

Gresshoff, P. M. (Ed.) (1997) *Technology transfer of plant biotechnology.* Boca Raton, FL: CRC Press.

Hawthorne, E. P. (1971) *The transfer of technology.* Paris: Organization for Economic Co-operation and Development.

Hsu, C. W., & Chen, H. H. (2003) The Taiwan innovation system. In L. V. Shavinina (Ed.), *The international handbook on innovation* (pp. 976–997). New York: Elsevier Science.

Matkin, G. W. (1990) *Technology transfer and the university*. New York: Macmillan Publishing Company.

Mogavero, L. N., & Shane, R. S. (1982) *What every engineer should know about technology transfer and innovation*. New York: Marcel Dekker, Inc.

Muir, A. E. (1997) *The technology transfer system: Inventions: marketing, licensing, patenting, setting, practice, management, policy*. Latham, NY: Latham Book Publishing.

Romanos, M., & Auffrey, C. (Eds.) (2002) *Managing intermediate size cities: Sustainable development in a growth region of Thailand.* Dordrecht, The Netherlands: Kluwer Academic Publishers.

Turpin, T., Xielin, L., Garrett-Jones, S., & Burns, P. (Eds.) (2002) *Innovation, technology policy and regional development: Evidence from China and Australia*. Cheltenham, UK: Edward Elgar.

United Nations Industrial Development Organization. (1979) *Industry 2000—New perspectives*. New York: United Nations.

Wilhelm, B. E. (2003) Innovation process in Switzerland. In L. V. Shavinina (Ed.), *The international handbook on innovation* (pp. 915–944). New York: Elsevier Science.

Chapter 2

Settlement Assistance to Zambia, Nigeria and Nepal

Moshe Schwartz and A. Paul Hare*

General information

The study

The remarkable agricultural achievements of *kibbutzim* (communal settlements) and *moshavim* (pl. of *moshav*: smallholder cooperative settlements) soon after the establishment of the State of Israel led a number of less developed countries (LDCs) to approach it, immediately upon their political independence, and request technical assistance to develop cooperative agricultural settlements. Following such assistance, a number of integrated projects adopting elements of the *moshav* system were implemented in the late 1960s and early 1970s in these countries. Usually, the villages were established in clusters. Examples of such projects are the Farm Settlement Scheme in Eastern Nigeria, the Nawalpur and Khajura projects in Nepal and the Kafubu and Kafulafuta projects in Zambia's "Western Province" (the Copperbelt). Project objectives were similar and included:

---

* With Oliver S. Saasa, I. Eugene Nwana, Krishna Devkota, Bart Peperkamp, and Mirjam A. F. Ros-Tonen, who drafted this account.

Reprinted, with some deletions, with permission from the NIRP Research for Policy Series 13, by Moshe Schwartz et al. Published by: Royal Tropical Institute, KIT Publishers, P. O. Box 95001, 1090 HA Amsterdam, The Netherlands.

- increasing food production;
- establishing cooperatives to develop agriculture;
- developing the rural areas of the countries involved;
- generating employment opportunities in rural communities.

This study was designed in the early 1990s to find out what happened to the cooperative agricultural settlements established in LDCs in the 1960s with the assistance of Israeli planners and agriculturalists. The sample consisted of two projects in Zambia and Nepal each and three of the fourteen projects established in Nigeria.[1]

In two of the three countries, a change in the political climate caused the Israelis to leave before the settlements were fully established. In Nigeria, the Civil War (Nigeria-Biafra) of 1967–70 caused the Israeli planners to depart, while the termination of diplomatic relations between Israel and the African countries of the Organization of African Unity compelled the Israelis to withdraw abruptly from Zambia in 1973. Following these events, the projects in both countries dwindled and, in the case of Nigeria, some projects collapsed entirely, while others went on. The involvement of the Israelis in Nepal continued in the 1970s and 1980s. However, from the mid-1970s on they restricted it to agricultural technology transfer, ceasing their participation in the management of the Nawalpur project.

Hypotheses and methodology

That cooperatives could make an important contribution to agricultural growth in less developed countries (LDCs) was an article of faith for many scholars, especially in the 1950s and 1960s of the 20th century. The belief was especially strong with regard to Africa, some of whose ethnic groups had a tradition of cooperation, though they lacked formal cooperatives. Many African leaders shared the faith in cooperatives and used them on a massive scale as "development machines." However, such state sponsorship of cooperatives involved channeling large sums of money through them, fostering corruption among their leaders and members, a risk always present when cooperative associations are used as "development tools." Moreover, the effect of individual entrepreneurship seems to pull against the cooperative association. In these days of private enterprise ascendancy, one needs to redefine the factors likely to sustain the cooperative spirit at the community level. It seems that where people trust each other—while states are cautious in their use of cooperatives—cooperation is willingly accepted, not only as a means of gaining access to state funds, but also because cooperative associations provide economies of a scale (for marketing or input purchase), that also facilitates some activities (e.g., access to irrigation or agricultural machinery) that farmers might otherwise not be able to afford.

To test this hypothesis, and assuming that the agricultural cooperatives established in the three countries were of the *moshav* type, the principal researchers in each of the three countries conducted case studies of sample settlements, interviewed administrators and key persons in the communities and government, and reviewed background information, which helped shape a frame for discussing the successes or failures of the settlement activity. In Nigeria, two settlements were located in the East (Ibo) region and one in the West (Yoruba) region. However, differences between countries were generally greater than differences between settlements.

Theoretical orientation

International context: assistance to Third World in 1950s and 1960s

When the former colonies and protectorates in Asia and Africa gained independence and turned into "new states," they faced fundamental problems. Among these were: establishing the authority of central government; ensuring law and order, setting up effective educational structures, improving infrastructures, meeting the challenges of public health, sponsoring industrialization, and modernizing agriculture (Shils, 1963). To cope with these problems, technical assistance was needed, especially in agriculture.

Many developed countries were willing to offer assistance to LDCs. Helping the new states to overcome their problems, maintaining good relations with them, or soothing one's colonial guilt feelings, were not the sole motives for doing so. The assistance was also intended to promote business interests, gaining access to strategic commodities, such as copper (Zambia) and oil (Nigeria). Finally, some countries were considered strategic because of their location (Nepal, between India and Communist China, or Zambia, close to South Africa), or because of their leadership standing (Nigeria, the largest and most populated country of Black Africa, as well as that with the biggest military). The Cold War spurred industrial nations to offer the new states increasingly attractive assistance packages—technical and financial. Some never-colonized Asian and African states—Ethiopia, Liberia, Nepal and Thailand—also gained access to the same opportunities.

International and national organizations such as the Food and Agriculture Organization (FAO), the Ford Foundation and the Rockefeller Foundation were also involved in technical-agricultural assistance. Demand and supply of assistance focused on food production and agricultural modernization. Research was conducted to develop "wonder" seeds for wheat, rice and maize. Agricultural extension and training were organized and demonstration farms set up to promote the use of selected grains, chemical fertilizer and pressure irrigation, and to improve animal husbandry.

In the new states, demand for new settlements was not as brisk as that for food and agricultural modernization. Where resettlement was required on a large scale, it often was for settling army veterans, resettling refugees or increasing density in sparsely populated areas. On a smaller scale, resettlement served demonstration purposes, i.e., was used to spread novel agricultural technologies and modes of organization. Israel was one of the few states having resettlement expertise.

Israel as a source of agricultural and settlement assistance

Israel appeared qualified to provide both agricultural and settlement assistance. No other country had accommodated in such a short time so many settlers, most of them illiterate, and almost all with no farming experience, turning them into modern farmers.

Israel was also famous for achievements in specific agricultural branches: citrus, for one, of which it was a major exporter; and also vegetables, in which Israel had attained western yield levels, as well as animal husbandry (poultry and especially dairy) in which it had set world productivity records. Finally, Israel had proved skillful in the use of specific agricultural technologies, such as pressure irrigation, mostly using sprinklers (drippers were to enter mass use later on).

In addition to all that, Israel was a developing country, presumably able to understand other members of the "species" as well as show empathy for them and provide them with effective assistance. Also, Israel's climate was closer to that of more LDCs than the climate prevailing in Europe or most of North America. Western states and assistance organizations (such as the Ford Foundation and the FAO) considered Israel as a promising source of agricultural and settlement assistance (FAO study group, 1956; United Nations, 1954).

Besides the expectation that Israeli assistance would indeed help the agriculture of Third World states to "take off," the West saw that assistance as a useful socialistic alternative to Soviet Bloc influence, an important consideration in the Cold War years. The West was well aware that what Israel most wanted from LDCs was their votes in the UN General Assembly, to prevent the Arab states from passing resolutions against it. It was felt that even if Israel did receive those votes, this would entail no threat to the West.

In African countries, Israel ran more projects, and employed more experts than in all other LDCs put together. For example, in 1966, out of 23 agricultural projects ten were located in African countries, as opposed to five in Asian and eight in Latin American countries.

Israel believed that its efforts in Africa would not be in vain. For most African leaders, however, maintaining bilateral relations with Israel was one thing and supporting it in international forums, another. Bilateral relations were maintained even though Arab states did their utmost to undermine them. In international forums, however, the picture for Israel was not encouraging.

The new states may have had their own motives for preferring Israeli assistance to that of some other nation. They may have perceived that it was less threatening to their newly gained independence to be indebted to a small state like Israel, than to owe favors to a superpower, or even to some other large industrial state. Also, for new states with socialistic leanings, Israel offered experience in cooperative-socialistic forms of organization, without the strings usually attached to assistance from the Soviet Bloc. While the social-democratic countries of Western Europe also had such experience, they had never applied it on a large scale to Third World farmers, as Israel had done.

After successful Israeli assistance to Ghana and other West African states, several Francophone African countries—Dahomey, Guinea, Ivory Coast, Upper Volta and the Central African Republic (Congo-Kinshasa)—and some Anglophone countries had also started receiving Israeli assistance. By the mid 1960s, Israel was heavily involved in providing medical assistance, establishing military youth organizations, and setting up agricultural settlement projects.

The operational philosophy of Israeli assistance organizations

Three Israeli organizations were the most involved in providing agricultural assistance to developing countries. The Ministry of Foreign Affairs, through MASHAV (acronym for the Center for International Cooperation), was in charge of policy and of budgeting the assistance. Its field personnel kept in touch with decision makers in the countries receiving assistance, helping to implement projects and relaying information to headquarters. Agricultural assistance operations were actually run by the Ministry of Agriculture. Its commercial daughter company, the Agricultural Development Company, AGRIDEV, played an important role in foreign assistance operations. Last, but not least, the Center for International Agricultural Development Cooperation, CINADCO, in Rehovot/Beit Dagan played an advisory role, holding prestige and influence, especially in matters involving resettlement assistance.[2]

The operational philosophy of Israeli settlement assistance was grounded on three formative experiences. The first, after Israel's War of Independence, was the setting up of immigrant *moshavim* where vegetables had been introduced and, later on, poultry, to boost farmer incomes. Another one was the Lahish region settlement project in the years 1954–55, which included new planning elements, such as setting up clusters of 5 or 6 settlements grouped around a rural service center, where they received health, education and municipal services. Additional elements were the blending of mixed farming with industrial agriculture—oil seeds, sugar beets and cotton (Ben Arie, 1970; Arnon and Margulies, 1984). The third formative experience was the establishment of the regional purchasing cooperatives of the new *moshavim*, between the years 1956–59, and their rapid development. These purchasing cooperatives helped obtain a regular supply of agricultural inputs on credit and at reduced prices and to mobilize credit for development, while also functioning as "political machines" as well as agribusiness conglomerates.

These formative experiences were reflected in the physical plans of the resettlement projects. In all three countries they were of the Lahish type (locally adapted), while the houses were exact copies of *moshav*-type houses in the late 1950s (49 square meter two-room houses). Usually, the planners also relied on Israeli technologies (in fields such as irrigation and animal husbandry), while the plans reflected the cost-consciousness of Israeli settlement projects in the late 1940s and in the 1950s.

Field personnel stationed in the developing countries and desk personnel in Israel sometimes differed in their operational philosophy. The latter tended to uphold general planning principles and to make the lessons of Israeli experience into the yardstick of proper planning and organization (Box 2.1). Conversely, field personnel—who often spent 3–4 years in the country where they operated—tended to emphasize its unique needs. They demanded professional autonomy, feeling that they, not their advisors, would pay for failures. Also field

Box 2.1 Proverbs of settlement

The social and professional background of planners and experts affected what they did in subtle ways. For example, the Israeli experience in the 1950s had generated a set of standard recommendations on how to avoid social, organizational, or political problems, or how to cope with them once they appeared; a kind of common wisdom for planners and experts. Paraphrasing Herbert Simon, Schwartz and Hare (2000) labeled those recommendations "proverbs of settlement." Some of them are:

- Prefer educated farmers, they learn faster;
- Reject sick and old candidates for settlement, they are trouble;
- Engineer common formative experiences, it strengthens the settlers' mutual bonds;
- To improve motivation, put on work clothes and work shoulder to shoulder with the settlers, but also provide them with monetary incentives;
- If settlers are financially disappointed, try poultry; it gets cash rapidly;
- If you cannot neutralize the scheming of local politicians, co-opt them into your organizational set-up.
- If you run a cooperative, accurate and up-to-date accounting is a must.

personnel were eager to get things done before leaving "their" developing country, so as to be able to take credit for them. Field personnel often tried to present a united front against desk people, who may sometimes have been too confident in their ability to manage projects at a distance (Schwartz, 1996).

An important feature of Israel's operational philosophy was its reluctance to fund development assistance. While the State of Israel's growing security burden was one motive, Israeli experts were also convinced that if a country funded its own projects, rather than using other countries' money, it would try harder. If the developing country was unable to provide funding, alternative sources were American and UN organizations (thus, the Ford Foundation paid for the planning by Israel of a rural development project in Nigeria). Israel's effort to minimize its own share in funding development assistance was successful and unique. More than half of Israel's assistance was paid for from non-Israeli sources, thus generating a sizable technical assistance program with little capital investment (Laufer, 1967: 52).

Elaboration of the research

The results of the three case studies were published in a series of "Field Notes," "Reports," and "Monographs," which report on field visits, observations, theoretical perspectives and the cumulative findings for parts of the project. These were distributed among the principal researchers in each country, to keep them informed about progress in the others. As research progressed, it was found that reports were available in Israel and also in Zambia, Nigeria, and Nepal that were written by agriculturists and planners involved in the various projects in the 1960s. Since none of the settlements still existed in their earlier form, it became evident that important and untold parts of the story concerned the evolution of Israel's *moshavim* and the problems faced by Israeli advisors in the "transfer of technologies"—organizational and agricultural—to less developed areas. This challenge was taken up in the monograph written by Schwartz and Hare (1998; 2000), which focuses on the role of Israeli experts, most of them farmers, in the transfer of technology, including settlement and cooperative farming (*moshav*) experiences. This monograph provided important input to this chapter.

Results

Agricultural settlement projects in Zambia (Kafubu and Kafulafuta)

Background

Israel's agricultural settlement assistance to Zambia started in mid-1966, eighteen months after Independence, and lasted until 1973, when Zambia terminated diplomatic relations with Israel. Thanks to its copper mine revenues, Zambia was relatively wealthy, with a per capita income much above 300 US dollars per annum (Mwanza, 1993), the equivalent of more than 1,800 US dollars in 2008. However, Zambian agriculture was undeveloped, especially in the Copperbelt, where the British had discouraged any activity liable to compete for manpower with the mines (*ibid.*). Also, since not a few white farmers had emigrated from Zambia upon its independence, some commercial farms had been left untended and agricultural production had declined, providing no more than 8% of the GDP in 1965.

Furthermore, it was difficult to bring in food imports. Zambia lacked access to the sea, so that transportation was expensive. The country's only railroad, as well as the main land road, went through Rhodesia. When Rhodesia's white minority declared independence without the UK's consent, international sanctions ensued, including a blockade.

Zambia's political choice to support the sanctions as well as South Africa's black liberation movement had a price tag to it. The country's economy had to bear prohibitive transportation costs, as well as food security problems. Inflation drove prices up by 200% in four years. Under the circumstances, Zambia needed to increase agricultural production, and was especially anxious to maintain a steady supply of white maize, which constitutes, under the form of *nshima* (porridge), the country's staple food. Poultry, dairy and vegetables were also urgently needed.

Beyond the special circumstances of the mid-1960s, Zambia wished to reduce its dependence on copper revenues, which constituted 40% of its GNP, 70% of its foreign currency income and 55% of its government budget (Dumont, 1973). Also, Zambia's urbanization rate (20%) was one of Africa's highest, and its birth rate (3.2% per annum) was explosive. Thus, a declining share of the population was expected to feed the rest.

Prospects to develop Zambian agriculture appeared good. There was an abundant supply of land and enough rain—1,400 mm in the north and 600 mm in the south—to allow for regular cultivation. Also, the country's rising copper revenues, thanks to the Vietnam War, were capable of funding capital investments in agriculture, thus facilitating modernization. Furthermore, Zambia invested massively in education.

On 14 January 1965, President Kaunda launched the "Kafubu Appeal" urging the people of Zambia to enlist in cooperatives. Volunteers would receive credit as well as grants for bush and tree clearing at the rate of 75 Zambian kwacha (ZMK) per ha [37 British pounds (GBP) or 104 US dollars (USD) at the time] and would be entitled to become farmers on the land they had cleared. Thousands of Zambians responded to the appeal. Between the country's accession to independence and the arrival of the Israelis, 400 cooperatives were set up and registered (Dumont, 1973). Their number rose to 635 by December 1969, involving around 60,000 men, women and children (Lombard, 1970).

However, the new cooperatives ran into problems. They had received no guidance and suffered from high input and marketing expenses, while seasonal credit had been provided too late for efficient use, loans were in excess of repayment ability. Moreover, there had been money misuse by officials as well as over-investment in items such as expensive tractors. Moreover, many cooperatives were located on poor soils and physical planning was deficient (Lombard, 1970). Modernizing farming required foreign assistance and thanks to its strategic location near Rhodesia and South Africa and to the demand for copper, Zambia had good access to such assistance. Zambia's growing political role in Africa also enhanced its attractiveness to foreign donors. The country received aid from both East and West. Much of it took the form of agricultural expertise.

Israel's agricultural competence made it an appropriate candidate to provide assistance to agricultural and cooperative extension. Besides, while Zambia's climate is more humid than that of Israel, it does present some similarities with it. Also, Israel had achieved good results with illiterate farmers, and its socialist traditions seemed in tune with Kaunda's aspirations for Zambia's farmers. From the Israeli point of view, partnership with Zambia was also promising, since its copper revenues enabled it to fund ambitious projects extensively, thus improving their likelihood of success.

Characteristics of settlement assistance to Zambia

Israeli settlement assistance in Zambia concentrated on five sites, of which Kafubu and Kafulafuta, located in the western (Copperbelt) province, were the largest. Thus we chose them for our research project. The Copperbelt—a small plateau 160 × 55 km in size and occupying less than 1.5% of Zambia's territory—was its major mining and trading center and contained 20% of the country's population (800,000 people). The region was highly urbanized in 1966 and city life had exposed all groups to a money economy and to British colonial values. Illiteracy was, however, higher than in other provinces. There was no modern farming, and no government experimental farms. Low farmer literacy and the lack of systematic information about the region's agricultural potential made the work of the Israelis harder at both settlement projects.

Kafubu: a false start

The Kafubu project started in 1966 under the aegis of Zambia's Ministry of Cooperatives. The site was located near urban markets, but not on the most suitable area for agriculture. Trees had been felled for use in the nearby copper mines and, as the land was sloped, much of the soil had been eroded by rainfall. The remaining roots were deep and large, making it difficult to plough the soil mechanically. The farmers received stumping subsidies, but the roots were hard to dislodge and many of the farmers, aware that payment was collective, tended to shirk, behaving as "free riders." The project included 23 cooperatives with 10–12 farmers each.

Initial problems were numerous and included the provision of water, communication services and agricultural loans. Extension staff was scarce, and the lack of a farming manager and of a poultry advisor was especially blatant. The tractors provided for bush clearance turned out inappropriate for the area's terrain. Plowing tools and implements broke again and again. There were numerous other "snags" and delays, indicating the inexperience of the Ministry of Cooperatives, as well as its difficulty in achieving cooperation with other Ministries, especially the Ministry of Agriculture. First-year yields were disastrous—not even covering the settlers' expenses. This caused demoralization and many of them abandoned the settlement.

When the Israelis arrived in Kafubu in August 1966, the settlers—most of them unemployed from nearby cities—still lived in tents. Nobody had told them what hardships to expect and their long stay on the margins of urban society had led them to abandon traditional values, setting them to drinking. They had become mistrustful and aggressive to each other, and even more so to persons of authority and to white people.

According to the original plans, eighteen small collectives were meant to practice maize monoculture, while the other five were to grow poultry, producing both eggs and broiler chicken. When the Israelis arrived, they started sowing maize and setting up the infrastructure of the poultry branch (constructing a smooth road to avoid egg breakage during transportation). They also started to form a leadership cadre among the settlers and established a Kafubu Cooperative Union, which united the 23 collective units, to reduce overhead and facilitate the provision of seeds, fertilizer, machines and labor on credit. Next, the Kafubu collectives were regrouped into larger villages, in order to lower the costs of infrastructures such as roads, drinking water and health care. Moreover, the collectives were changed into smallholder cooperatives of the *moshav* type, so as to encourage individual effort.[3] The Zambian government demanded that the settlers grow maize, the country's staple food, and that they sell it below world prices, to feed the cities. Given the artificially low prices and the inadequacy of Copperbelt soils for maize, its cultivation was unprofitable. In order to reduce dependence on maize, the Israelis proposed to introduce new branches, whose produce (milk, vegetables, fruit and other horticultural products) would be marketed in the Copperbelt. The Israeli mission decided to run the project itself, claiming it had to do so because of the lack of discipline and working habits characteristic of Kafubu settlers. Meanwhile, cooperative education was necessary for both the cadre and the rank and file, to enhance "proper cooperative spirit."

It took one year to turn communal organization into a *moshav*-type (family farm cooperative) structure and two more to fully implant the new system. In those three years, the Kafubu settlers depended on their clearing allowance of ZMK 75 (GBP 37 or USD 104) per ha.

The five poultry cooperatives were the most successful. This boosted morale and fostered stability. The members started producing bricks to build their own houses and by the end of the 1968 dry season, 20% of those houses had been built. Success, however, created new problems. It was easy for the settlers to climb on their bicycles and go to sell eggs in the nearby city, thus pocketing the money to hide it from the cooperative and avoid the obligation of repaying their loans. A related problem was cash money. Most settlers had never seen so much of it and wasted it on luxuries and drink. Alarmed by such behavior, the management of Kafubu artificially reduced the amount of cash by repaying debts ahead of time, while investing in production and community development. A third problem was envy. The other Kafubu cooperatives were, after all, part of the same settlement project and Union and felt entitled to some poultry, too. A fourth problem was the neglect of the vegetable garden and other branches. Farmers did not work well.

When a new mission head assumed his post in 1969, the 18 cooperatives received some poultry, and vegetable production started in earnest, thus somewhat reducing the difference between the five poultry and 18 other cooperatives. Maize production also improved. However, Kafubu's initial problems persisted. Shirking and exploitative behavior, such as selling the produce outside cooperative channels to eschew debt repayment, went on. The Kafubu people continued to be characterized as the "bad guys" as opposed to the "good guys" of Kafulafuta.

Kafulafuta: a better beginning

A better start was made at the Kafulafuta site, some 65 km south of the city of Ndola. It was planned entirely by the Israelis and started from the outset as a smallholder (as opposed to collective) cooperative.

The Kafulafuta Union received a land allotment of more than 12,000 ha, much of it covered by bush. The agreement between farmers and government was the same as in Kafubu: for every hectare cleared, the settlers would receive ZMK 75 (GBP 37 or USD 104 at the time). The land was under the authority of a Lamba traditional chief, Senior Chief Mushili, and those allowed to use it owed him deference. Kafulafuta started operating in August 1967. The first crop was successful and the farmers earned money. Morale was good. Economic success was to motivate them to work on, at least as long as the Israelis stayed in Zambia.

The preparations in Kafulafuta were performed at every stage in cooperation with the traditional authorities and local government. All were represented in the Commission dealing with the project on which the head of the Israeli mission also sat. Three hundred Lamba peasants were recruited as settlers. However, bush clearing was even harder than in Kafubu, as trees were taller, and 150 settlers left Kafulafuta in the first year. City people, most of them Tongas, considered more enterprising than the Lamba, were brought in to compensate for the manpower loss. After initial problems to find and train local project cadre, Kafulafuta was run with a modicum of settler participation. It was defined as a "guided self-government cooperative union."

In 1968, 730 tractor-cleared hectares were added to the 400 hand-cleared ones. Kafulafuta practiced animal husbandry, especially pigs. It produced vegetables (mostly greens and carrots) to be sold on the Copperbelt markets and soybeans for feed and oil. Kafulafuta also produced maize, like all other Zambian agricultural cooperatives, and grew poultry, though much less than Kafubu. By 1969, Kafulafuta was also engaged in intensive vegetable culture. Given the distances to "nearby" city markets, it was easier to sell vegetables through cooperative channels. In 1970, a UN expert wrote that "there are now 140 farmers at permanent sites, 110 houses have been built and 25 acres (10 ha) of vegetables are under irrigation; 458 farmers have been planned for at Kafulafuta" (Lombard, 1970).

Kafubu and Kafulafuta, 1970–1973

After the initial period, both projects were highly successful. Their production between October 1970 and October 1971 reached a value of ZMK 189,000 (USD 135,000)[4] in Kafubu and ZMK 100,000 (USD 71,400) in Kafulafuta. Of the aggregate production, 64% was accounted for by poultry produce, 14% by vegetables and 17% by maize. While vegetable production was just beginning, gross income per capita (ZMK 130 or USD 93) was already five times the average rural income per capita, as calculated by Lombard (1970). One year later, in 1971–1972, the two projects doubled their production. The figures testify to their remarkable productivity, as well as to the high incomes of individual farmers, thanks to crop diversification. Both projects repaid in advance part of the loans received from Zambia's Agricultural Finance Company. The success of the Israeli-aided settlement projects convinced President Kaunda that family farm cooperatives were preferable to collectives.

While success overshadowed the differences between the projects, it did not eliminate them. The most obvious was branch structure; Kafubu being based on poultry, while Kafulafuta increasingly relied on irrigated vegetables. A less tangible difference concerned the spirit of cooperation. It remained high at Kafulafuta, while in Kafubu it continued to be marked by "too many chiefs," low work morale, suspiciousness, over-reliance on government, and politicization.

By 1971, Kafubu and Kafulafuta supplied the entire Copperbelt with eggs and vegetables, while the *moshav* had become the model of Zambian collectivism. Despite increasing successes, the relations between the Israeli mission and Zambian bureaucracy deteriorated, especially after 1972, perhaps because of the Israeli mission's tendency to circumvent bureaucratic channels by going straight to the President whenever it disagreed with the bureaucracy. There were also professional disagreements: thus some Zambian officials believed that irrigation and industrial poultry growing were too advanced as technologies, and at least one Israeli expert was of the same opinion (interview with D. Elbaz, who in 1972 headed Israel's mission to Madagascar, 1997).

From the point of view of farmer education, the project was, however, sustainable. Kafubu and Kafulafuta farmers eventually did master modern agricultural technologies and understood the principles of cooperation, even though they lacked the full opportunity to apply them: the Israelis ran the projects until their last days in Zambia. Finally, some Kafubu farmers even became modern entrepreneurs. Most sustainability problems were systemic—caused by the weakness of Zambia's infrastructures, the decline of its economy and instances of predatory bureaucracy.

The post-Israeli period

The departure of the Israelis in 1973 re-emphasized the differences rooted in the formative periods of the two projects. Thus in Kafulafuta, while most of the machines, tractors, agricultural instruments and pumps were repossessed by the bank, the rest of the equipment and the cooperative itself continued to operate after the departure of the Israelis. Gradually, within the next three years, equipment went out of operation, as there was no money for repairs. The Union declined, too. However, even as its contribution to the welfare of its members declined, the latter exhibited "cargo cult behavior" (Worsley, 1957) and stayed on, hoping that the Israelis would return to reward the faithful. After an abortive attempt by the Zambian government to set up a communal cooperative in Kafulafuta in 1978, a multipurpose cooperative was set up with Swedish assistance in 1983. In 1995, it was still operating.

Kafubu had a higher yearly income and some bank loans were repaid ahead of time. When the Israelis departed, cooperative debts had probably been paid up, so that no equipment was repossessed. However, Kafubu farmers displayed an exploitative behavior towards public property. Most of the equipment belonging to the cooperative disappeared, perhaps sold (so it was rumored) by members of the Union leadership for personal gain. The Kafubu Union records were never found and its operation stopped almost completely.

While the collective possessions of Kafubu, such as tractors and agricultural implements disappeared, the large private farms went on for a few years. They included the installations for animal husbandry, such as chicken runs and piggeries. Apparently, big farmers could go on, thanks to the savings accumulated during the Israeli period. However, whenever an animal epidemic occurred, the cost of replacing the animals deeply dented those savings. Thus, big farmers underwent attrition.

In sum, while important differences distinguished Kafubu and Kafulafuta from each other, both declined, losing their ability to sustain the capital-intensive technologies introduced by the Israelis. Decline was not caused by technical or business incompetence. Project farmers had mastered modern technologies, and some were still using them in the early 1990s, twenty years after the Israelis' departure. They also showed considerable business acumen under adverse circumstances. An important factor of the decline of Kafubu and Kafulafuta may have been the lack of management experience among farmers, coupled with the suddenness of the Israelis' departure. Besides, while there were some Zambian managers, it was known that the Israelis were running the projects. This being the case, their sudden departure could not but harm the projects' creditworthiness.

The decline of Zambia's settlements was further speeded up by the worldwide decline of developing economies. Oil prices skyrocketed in 1974, while copper prices sharply declined in 1975, as the Vietnam War was over. This scissors effect ushered in a new era in Zambia with budget deficits, climbing external debt and under-utilization of industrial capacity. Income per capita shrank from USD 588 in 1977 to USD 109 in 1985. The decline of copper mining and shrinking incomes could not help but affect the Copperbelt. Kafubu and Kafulafuta earned their livelihood by providing poultry and vegetables to the cities of the region. As their dwellers' purchasing power was reduced, the two projects incurred further decline in income and production services. Deterioration went on due to shrinking government budgets, which reduced the bureaucracy's ability to keep up infrastructures, such as roads, education and agricultural extension. Finally, the cooperatives were severely afflicted by inflation, which worsened their terms of trade: the prices of modern inputs such as fertilizer and selected seeds grew much faster than those of outputs (crop prices). It also became hard to mobilize investment capital, and disinvestment appeared to be an appropriate strategy. Under such circumstances, progressive farming could not survive.

Lessons from the Zambia experience

The story of Israel's involvement in Zambia starts with difficult beginnings, followed by tremendous success and subsequent decline after the Israelis' departure and the economic downturn of the mid-1970s. How did Israeli assistance manage to overcome initial and later difficulties? What was the contribution of the projects to Copperbelt dwellers, to project farmers and to Zambian agriculture? Why did the projects decline after the departure of the Israelis? Could the projects have been more sustainable? These questions are addressed below:

1. In spite of abundant opportunities for failure, Israeli assistance worked, thanks to several factors:

   - some elements of the *moshav* model were suitable for Zambia, due to the socialist leanings of its elites, and especially of the Zambian president, Kenneth Kaunda.
   - the technical competence and improvisational ability of Israeli field personnel overcame initial hindrances such as lack of food and medical services and lack of water pumps;
   - the initial success of the Kafubu poultry farmers created willingness to give the Israelis time;
   - the successive heads of the Israeli delegations managed to instill *esprit de corps* into their staff, while maintaining, most of the time, acceptable relations with the Zambian Department of Cooperatives and good relations with President Kaunda.

2. The projects contributed to the welfare of Copperbelt dwellers by providing fresh vegetables, poultry and pig meat.
3. The projects directly contributed to the welfare of Kafubu and Kafulafuta farmers by providing them with high incomes, teaching them modern farming (modern animal husbandry, pressure irrigation and use of selected seeds) and boosting their cooperative spirit. However, most farmers lacked the infrastructures and the funds necessary to keep practicing what they had learned. In the long run, modern farming proved unsustainable in Zambia's declining economy.
4. The Israelis' greatest success may well have been the adoption by the Zambian leadership (in 1971) of the Israeli *moshav* (family farm cooperative) model instead of communal cooperatives. That decision happened thanks to the impressive food production of the two projects.

5. The suddenness of the Israeli departure, five years before plan, contributed to the decline of the projects. It is, however, difficult to assess what share of that decline was due to their departure, and what share to the oil crisis and the decline in copper prices.
6. It is likely that simpler technologies (*e.g.* simpler forms of irrigation or less productive, but more robust poultry and cattle species) would have been easier to maintain after the departure of the Israelis.
7. It is doubtful, whether a dynamics of success would have been created without the initial high achievements in the Kafubu poultry branch or irrigated vegetables in Kafulafuta.
8. Without a modern physical infrastructure, without modern garages and veterinary services, without proper extension services and decent government, it is difficult to establish or maintain modern agriculture. Undoubtedly, some Kafubu and Kafulafuta farmers could have been successful in a modern country, but they did not have much of a chance in their own.

Farm settlement in Nigeria (Ohaji)[5]

Background

Initial conditions for Israeli-assisted settlement projects in Nigeria were good. With a per capita income of USD 90 in 1955, the country was relatively prosperous, had an effective administration and an excellent system of higher education. It was hoped that this combination would be enough to keep the economy growing and the federation united. Nigeria prided itself on having a well-developed export agriculture: in colonial days it already was the world's largest exporter of palm kernels and palm oil and a big cocoa exporter. Finally, Nigeria was already well aware of its rich oil reserves, discovered in 1957.

Israeli settlement planning assistance to Nigeria started one year after the latter's accession to independence. Nigeria was Black Africa's largest and most populated country, as well as its likeliest candidate for leadership. The West could hardly afford to let it rely on Soviet assistance. Indeed, soon after Nigeria's independence, Great Britain and the USA offered their assistance, while other western nations followed in their footsteps. Nigeria was also assisted by non-government organizations (NGOs), the largest of which, the Ford Foundation, allocated to Nigeria bigger budgets than to all other African countries put together (Ford Foundation, various documents, 1955–1975).

The two regions selected to receive the benefit of Israel's expertise in agricultural settlement planning were the southeast and the southwest. Both are appropriate for export commodities: the southeast for oil palm and rubber, and the semi-dry southwest for cocoa (Berry, 1982), and field crops, such as wheat and maize. Besides, the southeast and the southwest were both inhabited by relatively well-educated ethnic groups, the Ibo and the Yoruba.

The purposes of Nigeria's resettlement projects were: fostering agricultural development, encouraging profit-oriented production by the family farm sector of food and fiber for local consumption and export, and generating employment for primary school graduates. It was hoped that producing export commodities would contribute to the country's foreign currency income as well as to the farmers' own welfare. To ease the attainment of this goal, the resettlement projects were to be manned by skilled or educated farmers, able to learn quickly how to apply advanced technologies to the production of those commodities. The farmers were to be organized into cooperatives, thus making it easier to use tractors and mechanical implements, and to provide for collective marketing.

In Eastern Nigeria, there was the added purpose of counteracting the spatial disorder of Ibo settlements, which made it hard to endow them with proper infrastructures (roads, electricity, water and drainage). It was hoped that the establishment of several planned settlement projects, each including several villages, would affect the surrounding population and that they, too, would build their villages in an orderly fashion, so as to be able to build modern infrastructures at acceptable costs.[6]

Actually, farmer cooperatives, especially marketing cooperatives, were common in Nigeria, having been massively introduced by the British administration in the years just before independence: there were about 400 marketing cooperatives in 1954; 800 in 1957; and 1,500 in 1959, the year when Nigeria became independent (Osuntogun, 1980). In fact, those cooperatives were used to channel the produce of Nigerian small farmers to the national marketing board and to export markets (Bates, 1981).

The cooperative organization of the Israeli-assisted settlement projects resembled that of other cooperatives in Nigeria. The projects were to achieve a demonstration effect, inducing farmers in nearby villages to use advance methods and grow export crops.

Characteristics of settlement assistance to Nigeria

Farm settlement in Nigeria was essentially a national process, and the Israelis were only involved in its planning stage. In the early 1960s, the government had decided to develop the farm settlements along the lines of the Israeli Lahish settlement cluster. However, it seems that the principles and practices of Israel's *moshavim* were neither understood nor applied. The backgrounds of the settlements in Eastern Nigeria, the principles and norms in rural development, as well as the cooperative structure were all British, and wholly unlike those that could derive from the Israeli settlements. Moreover, Israeli planners were not involved at a level liable to affect settler quality or cooperative practices. Their major contribution was the physical planning of the settlements and settlement clusters.

Land acquisition and allocation

Ohaji in Owerri Province was the largest of seven farm settlement projects[7], comprising a total of 6,085 ha. Its site was one of the best. The holding was almost evenly flat and its arable soils tended to the loose acid sands typical of central Eastern Nigeria. The acquisition of land from different communities was not easy, since they were seldom willing to have their land pooled. There was no agreement as to the terms or modes of compensation. Some local community members were antagonistic to the extent of chasing away government officers who came to survey the site and open it. Ultimately, the government used the police to force entry. Thus the native communities around the farm settlement saw the project as "land snatching" and refused any cooperation. Failing negotiations about compensation caused substantial delays and the Ohaji project started in 1961, some 15 months later than planned.

As compensation to the owners, 40% of the settlers were to be taken from the former land-owning communities. In reality, almost half (46.7%) of the settlers were from this group, while settlers from other parts of the region were no more than 10%.[8]

Each settler received 6 ha, comprising a homestead of 0.2 ha for the living house, a chicken coop and compound gardens, a 5-year rotational farm of 0.2 ha per plot, an oil palm plantation of about 3.2 ha and a rubber plantation of about 1.6 ha. Each village was provided with homesteads for junior staff and self-employed artisans, a training plot for the rubber tapping school, a junior primary school, a community center, a village square and playgrounds, market and village shops, a church and a graveyard, car and motor parks and a communal yam barn.

The settlement nucleus, with offices and living quarters for the settlement officers, was the first to be built. While their houses were constructed, the settlers were quartered under dormitory conditions in the transit camp, receiving training in agriculture and cooperatives. The settlement officers and a group of 120 settlers moved into Ohaji farm settlement in February 1962. The young men worked in four gangs of 30 and attended classes in the evenings on subjects like crop and animal husbandry, land and soil management, simple health science and hygiene, and civics and democratic procedures.

Development of the settlement (implementation phase 1961–1970)

During the first year, roads and staff quarters were built, a rest house, a cooperative shop, a health center and a maternity clinic. In Village 6, 120 settlers' quarters were built. On the settlement nucleus an oil palm nursery and a rubber wood-bud nursery were also developed (the seedlings there would be budded and used to plant the settlers' farms). By 1963, the first batch of settlers moved into their permanent houses.

The second and third batch of settlers were slightly older and, unlike the first batch, they were married, had children, had worked on farms, and showed clear evidence of ability to manage personal resources and stand the rigors of settlement life. During the first year they lived and worked under similar conditions as the first batch, being allowed to bring in wife and children only after moving to permanent houses.

Then, in 1966, the military takeover of the government, the civil disturbances and the Civil War that followed in quick succession, had devastating effects on the scheme. First, the Military Governor of the region ordered a hold on all projects initiated by the past civilian government "until it understood their rationale, cost implications, and method of operation." This delayed the recruitment of the fourth batch of settlers in 1966. It was not difficult to convince the Military of the scheme's value, and it soon pledged its full support to it, making fiscal and logistic provisions for its continuation. However, the civil disturbances and the inability of the Federal Military Government to protect people of Eastern Nigerian origin in other parts of the country led to secession. Eastern Nigeria was proclaimed the Republic of Biafra in 1967, which resulted in war with the Federal Military Government from July 1967 to January 1970. The entire resources of the government were diverted towards prosecuting warfare and trying to survive. The Ohaji farm settlement was now struggling to satisfy expectations of increased food supply.

Performance of settlers from different parts of the region

By the time normal work came to an end, most settlers had established a good part of their projected plantations. There were, however, notable variations in settler output. Settlers from the land-owning communities were less committed to their work than those from other parts of the region. They were more readily diverted by events in their native homes, such as a festival, death of a relative or friend, and general laziness. While settlers from other parts of the country were attracted to the project by access to land, such attraction was less compelling for persons from land-owning communities, who already had a surfeit of it.

Another difficult group was that of settlers from other ethnic communities. Responding to the situation in the country, they began to feel insecure and apprehensive. Thus, they no longer integrated well and their plantation work lagged. Soon most of them abandoned the settlement and went back to their original homes.

The best settlers at Ohaji were those from land-hungry communities around Orlu, Mbaise, Etiti and Mbano. They showed commitment to the project, had the most established plantations, developed their compounds by planting ornamental and fruit trees, and participated actively in the settlement cooperative.

Ohaji Farm Settlement Cooperative Society

All settlers had to belong to the Ohaji Farm Settlement Cooperative Society, the sole cooperative society in Ohaji. This provided the government with a body it could hold responsible for the refund of the loans given to settlers.

Inputs were distributed through the cooperative society. Its officers were to certify the creditworthiness of any settler before he received planting materials, agro-chemicals or even the stipend, because they were provided in advance.

However, most agronomists in charge did not understand cooperatives, and distributed inputs to whoever did well in their plantation work, even if they had failed to pay their cooperative dues or levies. Thus the agronomists undermined the cooperative society and made it harder to recover loans.

Settlers were expected to market their produce through the cooperative society. To this end, the government built and equipped a cooperative shop, an egg store, and a grain-storage silo with grain-drying facilities and otherwise provided the necessary infrastructure. However, many settlers did not use these facilities and sold their produce privately. This was easy since farm produce was in short supply and buyers were combing the village markets for it. The farmers hoped to receive better prices when selling privately (which was often not the case) and to prevent the authorities from finding out about their earning capacity and terminating payment of their stipend.

As settlers were not selling their produce through the cooperative shop, it was difficult to obtain reliable records of their income. However, indirect indicators revealed that those committed to their work were earning an adequate, if not very comfortable, income. Most bachelors in the first batch married; a costly thing in these parts of southeast Nigeria. A good number of settlers bought bicycles, iron beds, improved the furniture in their living rooms, and entertained visitors at the level of intermediate civil servants, without difficulty. They paid their children's school fees with ease and were even able to train one or two additional relatives, whom they supported through the Farm Settlement School. Perhaps the best evidence was in their—and their children's and wives'—well-nourished bodies. This may have contributed to the speed with which the Biafran Armed Forces swooped to conscript the settlers into the army.

The Nigerian/Biafra Civil War and the fate of Ohaji farm settlement

The Civil War had extensive effects on the composition and functioning of the Ohaji farm settlement. Settlers were conscripted into the Army, thus disrupting the plantation work. Refugees, who chose Ohaji as a safe haven, showed little regard for the farms. Palm plantations were neglected, except for occasional visits to harvest the fruits, often carelessly. Moreover, people who could not find accommodation cleared part of the plantations to erect shelter, and palms were cut down to have trunks for roadblocks.

The final push of the Nigerian troops, in December 1969 and January 1970, resulted in massive scattering of persons in the former Biafra. The Ohaji farm settlement was largely vacated. This final push did not last long. Soon, the War came to an end. At the cessation of hostilities, people went back to their former residence and occupation. When the staff and settlers returned to Ohaji farm settlement, they found ravaged homes and plantations. The government of the East Central State, that had inherited the Settlement, was only able (or willing?) to pay the salaries of the staff along with some money to hire labor for government-approved jobs. The settlement authorities, on the other hand, made frantic efforts to rehabilitate settlers and plantations. A few overseas charitable or philanthropic agencies or organizations were approached for aid. Some, notably the Cooperation for American Relief Anywhere (CARE), the British organization OXFAM, and CARITAS, responded with food, clothing materials, as well as some tractors and vehicles. Expectedly, emphasis was placed on food production to supply the war-famished population. Some dressed seeds (paddy rice and maize), cassava sticks and yam sets for planting, were supplied to the settlers through their cooperative society. Sincere efforts were made, but events quickly slipped out of hand as a result of a combination of factors.

First, the settlers were impatient to regain their pre-war opulence. They became agitated when the government rejected their request to reinstate the stipend formerly paid to them. When word of this agitation reached the government, coupled with a noted prejudice against all projects sustained by the erstwhile Biafra, the government decided to discontinue the farm settlement project, turning Ohaji into a big commercial oil palm plantation under the management of a public corporation, ADAPALM Ltd. The first step of ADAPALM Ltd. was to clear the settlers' planting because of uneven development that would have led to uneven maturity and to a difficult harvesting regimen. Thus, Ohaji farm settlement ended without any vestige of the settlers' agricultural work! The only things that survived were the buildings (residential, commercial and industrial), part of the Israeli plan. ADAPALM Ltd. in 1997 was a prosperous corporation with an annual turnover of more than 500 million Nigerian naira (about USD 5,700,000). Most former settlers had gone back to their original homes to live in misery and disappointment. Only six resolved to seek land by themselves, to settle and practice what they were taught in the farm settlement. Without government support, they found life difficult.

It is a matter of conjecture whether the Ohaji farm settlement would have attained the volume of production recorded by ADAPALM Ltd. However, as government was bold enough to take Ohaji over, it is the only project in the former East Central State that is not beleaguered with serious social problems.

Lessons from the Nigeria experience

The government initiated the Ohaji farm settlement and later supervised its destruction. Although there were some deficiencies in both the conception and implementation of the project, the singular act that caused its collapse was the government's determination to scrap it. We may therefore analyze the project as follows:

1. *Conception.* The desire to develop a farm settlement along the line of Israeli *moshavim* was not born out of an understanding of what they really were. Prediction of costs, timing and production seem to have been influenced by hope and enthusiasm more than by adequate knowledge of the factors at play in similar projects. There was also some misconception about the experience and capacity of the project officers.
2. *Implementation.* Too many things were being performed at the same time. People from different backgrounds and circumstances were welded into

one community. This caused difficulties. Plantation development occurred without full appreciation of the problems to be expected. Some agreed to participate so as to gain access to government largesse. There were delays in the supply of inputs, which prevented planned yields from being achieved. Conflicting interests caused avoidable friction, reducing the productivity of both officers and settlers.

3. *Collapse.* The final collapse of Ohaji was caused by factors largely outside the control of project participants. The Civil War caused ravages in the plantations, while the change of government led to policy changes. This led to the demise of the farm settlement. It is, therefore, impossible to use the fate of this, and perhaps any other settlement in East Nigeria, to evaluate the adaptability of the Israeli *moshav* system to other parts of the world.

Israeli resettlement assistance to Nepal: planning non-moshavim (1960–1965)[9]

Background

The Kingdom of Nepal was among the first LDCs to receive resettlement assistance from Israel, the second recipient of such assistance in Asia (after Burma). Nepal was important to the West, even though it was small and utterly poor. Its location between the two giants China and India, and the fears of Chinese communism—especially since Nepal, like India, had a strong Communist Party—made it into a strategic country. So much so, that foreign assistance was said to constitute 60% of the Kingdom's budget (Schwartz, 1996). While Israel had no diplomatic relations with China or India, Nepal was one of its few Asian friends. The USA looked with favor on Israel's assistance to Nepal, hoping that it might help "contain" the Chinese threat and that resettling landless farmers might ease "destabilizing" pressures for radical agricultural reform.

The plans for Israel's settlement projects in Nepal were submitted two years after those for Nigeria. They were authored by the same planner, which indicates that the divergence of their conceptions probably stems from country differences. While Nigeria was rich, uncrowded, and endowed by the British with some good roads, Nepal was poor, overcrowded and lacked such roads. Also, the steep slopes of the Himalayas, the monsoon winds and the 5,000 mm per annum rains, make it harder to build roads or railroad tracks as well as to protect them from rapid soil erosion. Thus, Nepal lacked an essential prerequisite for export agriculture. However, Nepal was short of food and of foreign currency to purchase it on foreign markets. Thus, the main objectives of the Israeli-assisted settlement projects were to increase food production for the country's growing population (then 1.6% per annum), while preserving the *sal* forests, whose hardwood—very valuable as construction material—was Nepal's main source of foreign currency.

Unoccupied farmland was only available in the marshy Tarai region, much of which was covered with virgin forest. Marginal farmers were predominant there, since natural disasters, including monsoon floods, created numerous homeless and landless farmers. Endemic malaria had prevailed, too. As its incidence decreased in the 1950s, thanks to a successful World Health Organization (WHO) program, immigrants from the overcrowded lower mountains rushed into the Tarai, to carve out farms for themselves. They often felled and burned *sal* tree forests, while the authorities stood by, unable to stop them. It was hoped that planned settlement might create some order and limit settlement to marginal forests, preserving the fully productive ones.

Two places were identified for planned settlement in the Tarai: the Nawalpur area in the Nawalparasi district of the Lumbini zone, and the Khajura area in the Banke district of the Behri zone. Additional settlement projects were to be set up by a state-owned, but autonomous, Nepal Resettlement Company established with Israeli assistance on the model of the Settlement Department of the Jewish Agency. In addition to preserving the forests, planned settlement was aimed at increasing food production, reducing population pressure in Nepal's lower hills as well as pressures for agrarian reform, and helping to ensure the country's political stability. It was also expected that Nepal's new settlements might have a demonstration effect, fostering agricultural improvements among the neighboring villages. Finally, it was expected that the Khajura project, set up later on by the Israelis close to Nepal's Indian border, might help protect the Tarai against encroachment by Indian migrants.

Characteristics of settlement assistance to Nepal

The Nawalpur resettlement project

Before the Nawalpur project, the area was under-populated. Households were few and only a small share of the land was cultivated by the Tharu (an indigenous group with natural immunity to malaria). They worked for the landlords, who lived in the hills or in nearby towns. The project was planned to host refugees from Burma and landless or quasi-landless hill people. The original purpose was to resettle 800 families over five years. However, 1,273 farm families were eventually resettled on about 4,300 *bighas* of land (2,900 ha) in the six villages. Thus, the average lot was slightly less than 3.5 *bighas* per family (about 2.3 ha). The resettlement process went on until 1976/77, settling a total of 1,504 families on about 4,733 *bighas* of land (about 3,200 ha).

The settlers came from nine of the fourteen zones and 33 of the 75 administrative districts of Nepal. There were 136 families (9%) from the district itself and 299 families (20%) of Burma repatriates.

The Khajura (Banke) resettlement project

In December 1965, Israel was requested to extend activities to the western Tarai. The expert team selected the vicinity of Nepalganj, a town of some 15,000 inhabitants in the Banke district, as the project area. The transport situation was better than in Nawalpur, as there was a domestic airport nearby, and a sandy track connecting the region to one of the western Tarai's largest towns, at 10 km distance. However, local people had no easy access to these facilities. The soil in the project area is alluvial, but suffers from occasional droughts—even though the annual rainfall is about 1,264 mm.

By January 1967, 40 families had settled. Many more approved settlers went back to their original homes, to fetch their waiting families. The steady influx of newcomers added up to an average of 15 families a day. A total of 265 families received 530 *bighas* of land during the project's first year. Between 1966 and 1972, 1,096 farm families were resettled on about 3,052 *bighas* of land in eleven villages. Resettlement went on in other project area villages until the mid-1970s. Eventually, almost 1,600 families were resettled.

Nawalpur and Khajura compared

Both projects in Nepal were primarily meant for disadvantaged and landless people. Applicants needed to be married and poor and they were required to cultivate the land by themselves. Later on, land was also provided to a few Tharus, the traditional inhabitants of the area, but most of them left their farms feeling discriminated against, in comparison to the Hill people. The selection process became more rigid as applicants increasingly outnumbered available plots. Priority then went to families with sufficient manpower to clear the forest land and to married couples. The settlers were furthermore required to complete clearance within the nine months during which food rations were provided.

In the beginning, both settlers and officials suffered from hardships, since drinking water, transport facilities and health services were almost completely unavailable. The officials also faced some landlords' claims that their land had been taken over by the resettlement project. They were compensated with substitute land. Relations improved later on as the landlords felt that the project had a positive impact. New agricultural technology was introduced, there are five high schools in the area and efforts were made to start a college, whereas old settlements did not even have a primary school.

Nawalpur had some advantages over Khajura. The people were said to be more conscientious, and the cooperative better equipped, e.g., with a mill, and marketing and irrigation facilities.[10] Nawalpur is closer to Kathmandu and physical conditions (soils, rainfall) are better as well. Cotton farming, however, was not as successful in Nawalpur as in Khajura, where it was new to the area when the Israelis introduced it. Indeed from there it spread over the western Tarai, and thousands of farmers used it to improve their lot.

Agricultural production in the resettlement areas was very low in the beginning. Gradually, it increased as the settlers finished clearing the land allotted to them, the use of chemical fertilizers and insecticides grew and irrigation facilities were developed in some of the areas. However, in 1995, land productivity had stagnated for some years as no irrigation was introduced. Large areas still depend on monsoon rainfalls for rice planting.

In the beginning, it was only through cooperatives that settlers could receive their daily necessities and sell surplus produce (if any). As transport facilities improved, the importance of cooperatives declined.

The impact of Israeli assistance and the post-Israeli period

The role of the Israeli experts in the projects was vital. They introduced cotton farming in the Khajura area (where they were involved in technical matters only), while in the Nawalpur project they were involved in management as well. The Israelis introduced modern farming technology and taught farmers about agricultural diversification and crop rotation. They were appreciated as exemplars of hard work and initiative, for their contribution to developing irrigation facilities and for their involvement in managing the agricultural cooperatives and a health clinic.

The thinking of Israeli planners in Nepal was pragmatic, not ideological. Rather than marketing ready-made models, they produced custom-made solutions, taking into account political considerations as well as the local economy and ecology. The settlement projects set up by Israel in Nepal were not *moshavim*, though they did include "patterns similar to the Israeli village of the *moshav* type" (Marton, 1973). In addition to physical planning (of the houses, villages and village clusters), these elements included the equal allocation of land—two plots of 1.5 ha per smallholding family, one near the house and the other at some distance from it. Other elements of the Israeli experience used in Nepal included agricultural technologies, rural centers to serve every few villages (with schools and health centers), and a settlement bureaucracy in charge of establishing new settlements and assisting them in their first steps, the Punarvas Company (similar to the Settlement Department of the Jewish Agency in Israel).

All this, however, did not turn Nawalpur and Khajura into *moshavim*. Their cooperatives were of the Nepali *sajha* variety: unlike the Israeli *moshav* they included several villages. Also unlike the Israeli *moshav*, Nawalpur and Khajura operated as purchasing cooperatives, providing only some inputs and services and processing and marketing only part of the output. It was even proposed that the land be eventually sold to the farmers (while under the *moshav* system, land was to remain nationally owned). Finally, in Nepal cooperative structure was not considered important, and institutionalized mutual assistance was not part of the project structure.

All in all, Israeli experience did not serve as a model but rather as a "repertoire" from which solutions could be drawn, according to the problems one faced. Some solutions, like architecture, were ready-made "packages," while others, such as "farm types" were adjusted to local soil and climate conditions as well as to the limited financial capabilities of Nepal. And in some fields, such as cooperative structure and land ownership, the Israelis adopted previous solutions, locally conceived.

The construction of the East-West Highway in 1972 solved the marketing problem, fostering development. New facilities were set up: schools, post offices, a veterinary center, bank etc. In 1995, some original settlers were still around. Many others have joined them. Large industries were established, driving up land prices and diversifying peoples' livelihood sources. The area is being developed. A large market is just about a fifteen-minute drive away. People are no longer dependent on the cooperative to purchase necessities and market surplus produce. In the 1990s, parts of the Nawalpur area, which are close to Kathmandu, had become popular for residential purposes as some of the rich of the Nepali capital moved "to the country," fleeing pollution.

The Khajura area was also developing fast. By 1995, it had good access roads as well as additional facilities such as schools, health posts, a bank, etc. Here, too, people ceased to be dependent on the cooperative for buying inputs or selling their surplus produce. However, irrigation facilities were still to be developed. The prices of land have also gone up, though much less than in the Nawalpur area. The impact of cotton farming can be seen not only in the resettlement area, but also in some other western Tarai districts. Some people call it "white gold farming."

Facilities, such as the buildings constructed in resettlement days, are not fully used. In Nawalpur, only two small buildings are used, by the farmers' cooperative. In Khajura, the facilities have been transferred to the Cotton Development Board, which uses them. In both areas, the services of the Israeli experts, withdrawn in 1975, had been strongly appreciated. Both projects would not have succeeded without the initial assistance of those experts. Later on, however, the Nepal Resettlement Company did prove able to maintain project performance, at least for a few more years.

Lessons from the Nepal experience

Some useful lessons were learned in the first year of project implementation:

1. Officials realized that roads were a prerequisite of smooth project implementation.
2. It was only through the help of the World Food Program, that most settlers could stay. However, there were cases of misuse of the foodstuff (i.e., farmers who sold canned fish and milk powder in the market to purchase daily necessities like salt and kerosene), and officials claimed that free food distribution damaged work motivation, making close monitoring necessary. Some higher officials also tried to divert some of the foodstuff to other purposes but were prevented from doing so by the experts.

3. Local businessmen were trying to buy the standing crops of the farmers at a lower price. It was then realized that their surplus produce should be marketed only through the cooperatives.

The following can be concluded about the Nepal projects. The two resettlement projects achieved much of their objectives, uplifting the rural poor, maintaining border security mainly in Khajura and increasing production through technical advice, cooperatives and credit facilities. The projects proved sustainable to a degree.

Rather than *moshav* type multipurpose cooperative settlements, Nepal had consumer cooperatives. One Israeli-type cooperative was established at Gaughat Village in the Khajura area. Twelve settlers returned from Burma started a collective farm similar to the *moshav shitufi* (quasi-kibbutz) in Israel, against Israeli embassy advice. However, the Nepali elite feared it would spread "dangerous communist ideologies," fired the official who initiated the project and imprisoned him on trumped up corruption charges (later cleared by the courts, he did not get his job back). The collective was dissolved.

Conclusions

Much of what happened in the three countries studied can be interpreted in terms of strategic goals pursued by the state of Israel in its interactions with the LDC: satisfying the LDC ruling elite and obtaining its support for Israel; avoiding failures, thus upholding Israel's reputation; achieving sustainability; and attaining spectacular, dramatic, agricultural successes. The first two goals implied that maintaining good relations was sometimes done at the expense of realistic planning and administration, especially in the beginning, when Israeli experts sometimes disagreed with local politicians and administrators, but ended up accepting their views against their own better judgment. It also meant that the *moshav* model was implemented only where in line with the interests and ideologies of ruling elites (in Zambia). The last two goals tended to clash, since sustainability in LDCs is more likely to be achieved by avoiding the technologies dependent on developed infrastructures, on which spectacular achievements usually rely.

After overcoming initial difficulties, Israeli-sponsored agricultural settlement projects in Zambia, Nigeria and Nepal achieved high agricultural productivity, while providing improved social services to farmers. However, the projects' contribution to the regions and countries where they were located remained limited. Non-project farmers lacked the access to capital and know-how, which might have allowed them to emulate project farmers. The only exception is cotton growing in Nepal: there, the technology introduced by the Israelis spread beyond the borders of the Khajura project.

Moreover, most agricultural and socio-organizational technologies used in the projects did not prove sustainable. All projects underwent decline: the southeast Nigeria projects after the Biafran war, Kafubu and Kafulafuta (Zambia) after the departure of the Israelis, and even Nawalpur and Khajura (Nepal), as the Israelis gradually reduced their involvement in the projects.

The common limitations evinced by the Israeli-sponsored projects were rooted in the failure of the state in the three LDCs to meet the basic challenges stated by Shils (1963): law and order, effective education, and adequate infrastructures. The limitations were also due to the countries' maintenance of artificially low agricultural prices, undermining farm profitability and to their use of central authority to foster wasteful policies or predatory practices by politicians and administrators. The economic decline experienced by all three countries was worsened by these problems and worsened them in turn, as no budgets were available to maintain education and infrastructures, whereas the struggle for survival fostered lawlessness.

As long as the Israeli experts were deeply involved in the projects in Zambia and Nepal, they were able to attenuate the above problems. They compensated for educational deficiency by managing the projects, providing extension services and mobilizing maintenance services, mechanical or veterinary. The experts solved some infrastructure deficiency by lobbying for road construction, as well as using their influence with government to obtain transportation and communications (trucks, telephones). The same experts were able to forestall some fund misappropriation by officials or agencies. Also, the Israelis' position as outsiders allowed them to be objective, counteracting tribal or familial loyalties, and demanding performance.

Project-relevant differences did not only pertain to the extent of failure of the countries in solving law-enforcement, infrastructure or educational problems. Additional differences were due to geopolitical situations, social structures, cultural backgrounds, natural resources and actual achievements. The latter differences affected development policies, shaped assistance goals, and determined which Israeli experience elements were adopted. Finally, massive historical changes on the world scene and in each country had a profound influence on project fate.

While project goals differed from one country to another, methods always included elements derived from the experience of Israel's rural sector within the 10–15 first statehood years. Some elements were used in all three countries. They included Lahish-type village clusters with rural centers providing municipal services, cooperative organization (though not necessarily of the Israeli type), and mixed farming, including poultry. Other borrowings from the Israeli experience differed from one country to the other. Thus in southeast Nigeria, Israeli participation included the planning elements common to all three countries, in addition to the architectural plans of *moshav* houses; in Nepal, the Israelis also set up a settlement agency modeled after the Jewish Agency's Settlement Department, to plan and implement orderly settlement in the Tarai; and in socialist-oriented and copper-rich Zambia, the Israelis established *moshav*-type cooperative settlements as well as practicing irrigation agriculture.

The piecemeal rather than wholesale adoption of Israeli elements indicates that the Israeli experts tried to fit specific elements to the conditions of every country, as they understood them, rather than mindlessly reproducing the *moshav* model.

Discussion

Scientific relevance

This project provided case studies in Zambia, Nigeria and Nepal, adding new insights into the impact of state support on cooperative behavior and conditions for successful transfer of social and agricultural technologies. The Israeli-sponsored projects in Africa showed the farmers that they could improve productivity and achieve higher incomes. However, this was hardly enough to ensure further development, or even to allow sustainability after the departure of the Israelis, and by 1997, only the memories of better days still lingered.

Seven factors helped account for the different paths taken by the Israeli-planned projects. The first four referred to the situation of the countries while the projects were planned and established. Next come some "Israeli" variables and the formative periods of Israeli assistance. Last, but not least, there were massive developments in the political and economic environment of the three countries. They are briefly summarized below (for additional details, see Schwartz and Hare, 2000).

Wealth, geopolitical situation and access to international assistance

Differences in wealth and access to international assistance affected the three countries' ability to invest in the Israeli-assisted projects and capital-intensive technologies. Income per capita was highest in Zambia (USD 300 per annum). In Nigeria it amounted to USD 90 per annum, while Nepal was among the world's poorest countries with USD 40 per capita in 1955. Access to foreign assistance was good for Zambia (strategic location) and Nigeria (a would-be continental leader). Nepal's strategic position between India and China and the western fear of communism ensured the country considerable foreign assistance. Here the projects were, however, severely short of cash, as authorities feared that "excessive" settler success might awaken "unrealistic" expectations and social turmoil.

Educational level of bureaucrats and farmers

The low education of bureaucrats in Zambia, the thin administrative class of Nepal and the well-trained officialdom of Nigeria were critical in determining the Israeli involvement in running the settlements. In Nigeria, there was no need for Israeli experts, while in Zambia no local expert could be found and the

Israelis had to run the projects themselves. In Khajura, the Israelis assisted in technical matters only, while in Nawalpur the Israelis dealt with administration and organization. These differences also affected the timing of project hand-over to local bureaucrats (from the beginning in Nigeria, late in Nepal and after the departure of the Israelis in Zambia). The education of farmers also varied, affecting the speed at which they learned advanced technologies and the sophistication of their business operations. Ultimately, all farmers, whatever their formal education, mastered the capital-intensive technologies, but few were able to keep using them after the Israelis' departure.

Structure, function, interests and ideology of the elites

There was considerable variation in the ideologies and structures of the LDC elites with whom the Israelis cooperated. The differences shaped planning goals and affected implementation. The Zambian national elites were thin and agreements had to be reached with the highest officials, often President Kaunda himself, who attached great value to cooperatives. The Nigerian elites were more complex, larger and less dependent on expatriates than the Zambian elites. The Nigerian elites were well educated and wealthy, but strongly involved in tribal rivalries, ill prepared for democracy and corrupt. As a caste society and traditional kingdom, Nepal was highly hierarchical. The political elite was made up of a few families of the two highest castes, the priestly caste (Brahmins or Bahuns) and the warrior class (Kshatryas or Chhetris), while the lower castes have very little parliamentary representation, as well as little education and wealth. Under the prevailing system, the Israeli *moshav* did not fit.

Ethnicity and regionalism

All three countries are ethnically and regionally diverse, and in all three, region and ethnicity coincide to some extent. That coincidence affected the Israeli-sponsored settlement projects. The ethnic factor played a crucial role, especially in Nigeria. It led to the war between the Federal Government and the Ibo state of Biafra, which wreaked havoc on the settlement projects. In some projects, the original members came back after the war, but as the parastatal in charge of the state plantations repeatedly tried to repossess the projects, they lost the feeling of owner security. Plantation upkeep declined. Together with the federal neglect of the Ibo states, for example in the matter of road infrastructure, this could not but hinder commercial agriculture.

"Israeli" variables

"Israeli" variables affecting the settlement projects refer to architecture and physical planning, the relations inside Israel's assistance bureaucracies and the involvement of agricultural performance in Israel. With respect to housing and settlement design, Israel used ready-made blueprints, with little efforts to fit them to local conditions. However, no complaints were heard against *moshav*-type houses or the Lahish-type rural service centers of the projects, all of which apparently meant an improvement over previous conditions. In economic planning, differences between the early and late 1960s refer to both produce (more high value-added products) and technology (more advanced technologies such as pressure irrigation and sophisticated poultry sheds). As to the relations inside Israel's assistance bureaucracies, there were problems, especially between field personnel in the countries and desk personnel in Israel. Finally, there was the impact of increased autonomy of the *moshav* organizations due to improved agricultural performance and to political circumstances in Israel, resulting in greater pressure for settlement autonomy in the developing countries in the late 1960s.

The formative periods of Israeli assistance in the three countries

The formative periods of Israeli assistance in the three countries affected subsequent development. The first important issues were the decision to employ Israeli experts and the choice by the national elite of the uses to which they would be put. The next important step was the negotiations about *where,* and with what objectives the Israelis would set up their projects and also *how* they would go about it. The "where" was important, as it affected those who joined the projects and because of the role of regions and their dominant ethnic groups in politics and agriculture. Had the Israelis in Nigeria, for example, set up all settlement projects in Yoruba country in the southwest, the story would have been different, because of distinctive climatic conditions (drier) and cultural values (Yoruba farmers are more hierarchically minded and accepted cooperative discipline more easily than the individualistic Ibos). Moreover, the Biafran war happened in the Ibo southeast. Had another region been chosen for the projects, they might have been more sustainable.

Massive developments in the political and economic environment of the three countries

Important processes in Zambia were the shift from collective to smallholder cooperatives, the decline in copper prices and subsequent economic crisis, and rapid population growth (3.6% per annum).

In Nigeria the foremost development was the Biafran War, which caused destruction, paved the way for military government and resulted in the neglect of infrastructure in the Ibo southeast after the war. Additional developments were the atrophy of much of Nigeria, due to the oil economy, and the strong oil price fluctuations (up in 1974–5, then down, up again in 1980, then down again). These developments may have harmed projects in the southeast even more than elsewhere in the country.

Nepal underwent rapid population growth (about 3.2% per annum in 1977), which multiplied its population from 8 to 19 million in thirty years. Land prices climbed steeply until the mid-1990s and kin often came to live with the farmers, increasing the number of mouths to be fed. The Nepali GNP increase was eaten up by population growth, and the same occurred in many individual farms.

Practical applicability

Could Israeli assistance be used again to go back to the superlative successes of the Israeli-sponsored agricultural settlement of the 1960s and would it be worth doing so? We can speculate on the question. The main point is that everything has changed. Since the 1960s both the Israelis and the citizens of LDCs have changed, and so have LDCs' economies and societies.

In LDCs, the naive enthusiasms of the first years of nationhood have somewhat cooled and some of them have come close to a Hobbesian society making it unlikely that cooperatives or collectives would be very successful. In Israel, the *moshav* of the 1960s exists no more and the vast majority of *moshavim* have lost any cooperative characteristics. It is likely that as a result of these processes in Israel, Israeli experts would demonstrate less missionary zeal to set up cooperative endeavors and also less willingness to work with their own hands than they did in the 1960s.

The field in which present Israeli competence is far above that of the 1960s is, of course, agricultural technology where much has been learned that can improve performance. However, given the large *factor price* differentials between LDCs and developed countries, such as Israel, one should be careful as it might be inappropriate to transfer some novel technologies to LDCs.

Finally, there is the problem of sustainability of western rules of the game in LDC economies. It has not changed from the 1960s. In some countries it may have worsened, which is one reason why so many want to attract foreign businessmen and why they are reluctant to come in.

Recommendations

In the absence of a time machine, proposing what could have been done in 1967 had the future been known, is both unfair and useless, unless such backward-looking proposals are shown to have relevance for the future.

The first proposal of such relevance is to resist temptation and continue abstaining from the expensive holistic community development approach, characteristic of the projects in all three countries. As long as LDCs remain socially, politically or economically unstable, holistic community development is likely to turn into a futile exercise once the experts are gone. Massive events and processes such as the oil crisis, the decline in copper prices or the Biafran war, or the rapid population growth of Zambia and Nepal are not always easy to predict when planning such projects. The same holds true for political changes and fluctuations, characteristic of many LDCs. Indeed, it seems that international assistance agencies did well to abandon the holistic development strategy, pursuing more modest goals. It is probably better to engage in purely technical endeavors, as long as they are compatible with local capabilities and needs (e.g., the development of cotton growing in Nepal).

As far as technology transfer is concerned, those used by the Israelis, especially in Zambia, could have been less ambitious, as was said by some observers in the early 1970s. Perhaps, to paraphrase Dumont (1973), donkeys should have been preferred to tractors. This statement has become even more relevant to technology transfer today than it was thirty years ago. However, one should remember that technology is often important for prestige purposes. Thus, when the Israelis introduced "intermediate technologies," these were often disdained by both farmers and officials.

The economies of the "Third" and "First" World have been drifting apart. The discrepancy of incomes per capita has grown by an order of magnitude. While in the 1950s it was tenfold, it has now become a hundredfold. In the 1950s LDC income per capita varied from USD 40 to USD 500, while developed countries had four-digit incomes per capita dollar incomes. In the 1990s, the incomes per capita in developed countries are five-digit numbers, while those of LDCs have grown less. This creates even wider factor price disparities than before, meaning that the absorption of western technologies, already problematic in the 1950s and 1960s, has become even more so, being liable to lead to stronger distortions in resource allocation or else to fail.

Notes

1. In Nigeria, seven settlements were established in the eastern region, the primary focus of this study, and a comparable number in the western region, one of which was chosen for comparison.

2. The Rehovot Center was liquidated in 1966 and absorbed by the Ministry of Agriculture, under the name of Center for International Development Cooperation, CINADCO. However, its former staff continued to exert influence on agricultural and settlement missions in foreign countries in unofficial and official ways.

3. Apparently, communal agriculture was inappropriate for the Kafubu settlers. In the 1967/68 agricultural season, 250 Kafubu families were unable to plant 250 ha of land in two and a half months when they worked communally. In the following season, the same number of families, working on individual plots, planted around 600 ha in four weeks.

4. The Zambian kwacha went through several exchange rate changes but the best approximation is USD 1.40, the official value in December 1971.

5. This section is based on Nwana (1997) and on the summary chapter he contributed to Schwartz and Hare (2000).

6. Western Nigeria had no such problems. The typical Yoruba villages were spatially well organized, perhaps reflecting the traditions of strong central power and the hierarchical nature of Yoruba society, as opposed to the individualism and absence of central power in traditional Ibo society.

7. The others were Boki in the Ogoja Province, Erei in the Afikpo/Enyang provinces, Igbariam in Onitsha Province, North and South Ulonna in Bende and Okigwe Provinces, and Uzo-Uwani in Nsukka Province. Igbariam and Olunno were also studied, but less extensively than the Ohaji project and will not be discussed here. Information about these projects can be found in Nwana (1997).

8. Other areas of settler recruitment included the administrative division and the administrative province of the settlement land, providing 27% and 25%, respectively, of the total settlers.

9. This section is based on Devkota (1997) and Schwartz and Hare (2000).

10. The irrigation program at Khajura was ambitious. A deep boring system was installed, but did not function for more than a year.

References

Arnon, Y., & Margulies, J. (1984) *The planning of the moshav faces the test of reality: the transformations of the economy in the Lahish Region.* Publications in Regional Development 35. Rehovot: Settlement Study Center (Hebrew).

Bates, R. H. (1981) *Markets and states in tropical Africa: The political bias of agricultural policies.* Berkeley: University of California Press.

Ben Arie, Y. (1970) The Lahish settlement project. *Journal of Economics and Social Geography,* November–December.

Berry, S. (1982) *Fathers working for their sons.* Berkeley: University of California Press.

Devkota, K. P. (1997) *The resettlements and cooperatives in Nepal.* Monograph 4, NIRP Project 92-5.1.

Dumont, R. (1973) *Socialism and development.* New York: Praeger.

FAO (1956) *Study group on problems of individual and group settlements in Israel.* Rome: FAO.

Laufer, L. (1967) *Israel and the developing countries: New approaches to cooperation.* New York: Twentieth Century Fund.

Lombard, S. C. (1970) Farming co-operative developments in Zambia and a note on one communal society. In R. J. Apthorpe (Ed.) *Rural institutions and planned change,* Vol. IV. Geneva: United Nations Research Institute for Social Development.

Marton, S. T. (1973) *Settlement projects in Nepal (Nawalpur and Nepalgunj).* Rehovot: Center for Agricultural International Cooperation.

Mwanza, K. C. (1993) The political economy of food policy and agriculture in Zambia. In R. Mkandawire & K. Matlosa (Eds.) *Food policy and agriculture in Southern Africa.* Harare: Southern Africa Printing and Publishing House.

Nwana, I. E. (1997) Cooperative agricultural settlements in a war ravaged economy. Monograph 3, NIRP Project 92-5.1.

Osuntogun, A. (1980) Cooperative and small farmers in Nigeria. In O. S. Olajuwon, E. J. A. Eweka, & V. E. Bello-Osagie (Eds.) *Nigerian small farmers: Problems and prospects of integrated rural development.* Ibadan: Center for Agricultural and Rural Development (CARD), University of Ibadan.

Shils, E. (1963) On the comparative study of new states. In C. Geertz (Ed.). *Old societies and new states.* Glencoe IL: Free Press.

Schwartz, M. (1996) The beginnings of Israeli resettlement assistance to Nepal: 1960–1965. NIRP Working Paper. Rehovot: Development Study Center.

Schwartz, M., & Hare, A. P. (1998) Agricultural cooperative resettlement: Applying Israeli experience in Zambia, Nigeria and Nepal. Unpublished monograph submitted to NIRP.

Schwartz, M., & Hare, A. P. (2000) *Foreign experts and unsustainable development. Transferring Israeli technology to Zambia, Nigeria and Nepal.* Aldershot: Ashgate.

United Nations (1954) *Monograph on community settlements and report of the survey mission on community development and organization: Israel.* New York: United Nations.

Worsley, P. (1957) *The trumpet shall sound. A study of cargo cults in Melanesia.* London: Mc Gibbon and Kee.

Participating researchers and institutions

Principal researchers:

Prof. A. Paul Hare (project coordinator) and Dr. Moshe Schwartz, The Social Studies Center, J. Blaustein Institute for Desert Research, Ben-Gurion University, Sde Boker Campus, Israel

Prof. I. Eugene Nwana, Biotechnology Research Center, Nnamdi Azikiwe University, Awka, Anambra State, Nigeria

Prof. Oliver S. Saasa, Institute for Economic and Social Research, University of Zambia, Lusaka, Zambia

Mr. Krishna P. Devkota, Agricultural Projects Services Center, Kathmandu, Nepal

Dr. Bart Peperkamp, (formerly): Department of Social Geography, Catholic University, Nijmegen, The Netherlands

Junior researchers:

Mr. Chris Simuyemba. Institute for Economic and Social Research, University of Zambia, Lusaka, Zambia

Ms. Ann Ogbuke. Nnamdi Azikiwe University. Awka, Nigeria

List of members of the Joint Steering Committee

Israel:

Prof. V. Azarya (co-chair). Dept. of Sociology and Social Anthropology, Hebrew University, Jerusalem

Prof. E. Ben-Rafael. Dept. of Sociology and Anthropology, Tel Aviv University, Tel Aviv

Prof. R. E. Isralowitz. Hubert H. Humphrey Institute for Social Ecology, Ben-Gurion University of the Negev, Beer-Sheva

Dr. N. Trostler-Buganim. School of Nutritional Sciences, Faculty of Agriculture, Hebrew University, Rehovot

The Netherlands:

Prof. B. de Gaay Fortman (co-chair). Institute of Social Studies, The Hague

Prof. A. J. Dietz. AGIDS (Amsterdam Research Institute for Global Issues and Development Studies), University of Amsterdam, Amsterdam

Prof. L. W. Nauta. formerly: Dept. of Social Philosophy and Ethics, University of Groningen

Dr. E. B. Zoomers. Center for Latin America Research and Documentation (CEDLA), University of Amsterdam, Amsterdam

Chapter 3

Building Maintenance: Where Did We Go Wrong? Post-occupancy Evaluations of Some Bioclimatic Projects

Isaac A. Meir* and Simon L. Hare†

Introduction

They say that the road to hell is paved with good intentions. If this is true then the story of modern bioclimatic design must definitely be such a road's story. Many are the cases of projects designed so that they will provide optimal indoor conditions with minimum energy input, yet in reality are rarely used properly, and thus do not achieve their goal. Their design and details, aimed at energy conservation and the utilization of alternative energy sources, stop making sense and thus are changed or decommissioned, and systems are dismantled. Such interventions alter the essence of the original design and thus render the building unsuitable for its original purpose regarding energy conservation. This chapter reviews some case studies from Israel in an attempt to assess the reasons that cause such failures, and suggests possible ways to correct the fault. Most of these projects were constructed from the mid-1970s onward.

---

* Desert Architecture and Urban Planning Unit – Dept. for Man in the Desert, Blaustein Institute for Desert Research, Ben-Gurion University of the Negev, Sde Boker Campus, Israel

† The Boston Architectural Center, Boston, MA

Climatic conditions

Despite its relatively small size, Israel has a very diverse climate. Roughly speaking, the country may be divided into four regions: coastal plain, lowlands, highlands and deep valleys (a large part of these valleys being well below mean sea level, with the Dead Sea's coast reaching -400m). However this is a very crude classification, since a significant part of the country is a desert, which affects the climate by drastically reducing relative humidity and exacerbating daily and seasonal temperature swings. Local climatic conditions are also affected by the topographic variability and distance from the Mediterranean coast. Most of the projects described below are located between the lowlands and the mountain range, where climatic conditions may be considered ideal for free running buildings. Table 3.1 presents region-specific climatic data based on monitoring stations (Bitan and Rubin, 1991) or generated by interpolation of existing ones and simulation (METEONORM, 1999).

Building market and limitations

The planning and design of the country was rather centralized until the late 1970s. Most of it was coordinated or even executed by the Ministry of Construction and Housing, which took upon itself the massive construction of public housing projects vital to meet the needs of a large influx of immigrants. Thus housing prototypes of similar or identical type and construction detailing were used in the different parts of the country, disregarding local climatic constraints (Etzion, 1985). Furthermore, the fact that most of the planning and design was done in the Ministry's main offices in the center of the country by architects who happened to have little or no immediate experience with the sparsely populated, arid and hyper-arid south and who perpetuated misconceptions of the desert being a hot area with little or no heating needs. (One of the most interesting exceptions is that of the desert town of Arad, planned and designed by a team that was located on the site throughout the planning period. As a result, the original residential clusters included a number of experimental building types.)

This situation changed in the early 1980s, when construction started becoming privatized. Urban sprawl and a changing economy brought forth the creation of numerous suburban settlements; many of them spreading into the less populated parts of the country, where land values were significantly lower (Bar-Nathan, Beenstock and Haitovsky, 1998; Meir, 1992). This coincided with a diversification in the construction industry, characterized by new materials. The traditional reinforced-concrete and concrete-block construction gave way to aerated concrete blocks, prefabricated lightweight units (Pearlmutter and Meir, 1995), and wood. The insulation standard, devised for residential buildings as the lower insulation threshold, was rarely employed since its implementation was considered by the architects to be unimportant, contractors considered it a bother and of no economic advantage, and clients were unaware of its effect or of the potential energy savings vis-à-vis the initial investment.

Formative changes

The 1970s oil crisis and subsequent changes of design attitudes had a certain influence on design, especially within Israeli academe. Research in energy conserving design and strategies, and the use of alternative energies in buildings was geared up. Three research units were created at the newly formed Institute for Desert Research: i.e., Desert Architecture and Urban Planning (DAUP), Solar Buildings, and Applied Solar Calculations (ASCU), to serve as additional catalysts in the development of scientific and research infrastructure, and to create a region- or climate-specific body of knowledge. Architects who undertook research studies or worked in these units often implemented new strategies and technologies in their projects. In addition, researchers often served as consultants to public and private bodies, and the DAUP unit undertook the designing of different projects in an attempt to create showcase pilot projects. A boost was given by the Ministries of Energy and Housing, which declared two design competitions in the mid-1980s, one for energy-conserving buildings and the other for a solar neighborhood in the town of Mitzpe Ramon, located in the arid highlands of the Negev Desert—an area with hot dry summers and cold winters with occasional snowfalls.

Bioclimatic projects—original design versus user interventions

Several pilot projects evolved out of these processes, some of which are reviewed here. They are presented in order of complexity and size, and are analyzed through a discussion of their current status and evolution over the years, compared with their initial design strategies and intent.

ORT Shoshana Sapir High School, Yeruham (Kones Architects)

Design: This is an addition to an existing school (Figure 3.1). It was designed around an internal courtyard. South facing classrooms were equipped with Trombe-like externally insulated walls at floor level, whereas the northern wing was fitted with similar south facing details on ceiling level. The south wing Trombe-like systems were fitted with low and high openings, which provided a continuous circulation of air between the glazed cavity and indoors. The north wing ceiling systems were equipped with fans. Cross ventilation through the courtyard provided night cooling during the hotter months of the year.

Evolution: A new administration, unaware of the system's function and capabilities, cancelled it once the glazing broke. Without glazing there was no reason for the openings to exist so these were blocked in and the walls were covered with stone cladding as part of a general refurbishment of the whole building. Thus, the southern wing lost a passive heating device and the northern wing was left with no south facing ceiling systems.

Figure 3.1. ORT Shoshana Sapir High School, Yeruham. From top left, counterclockwise: plan; north-south section; south façade with solar heating details; and after refurbishment

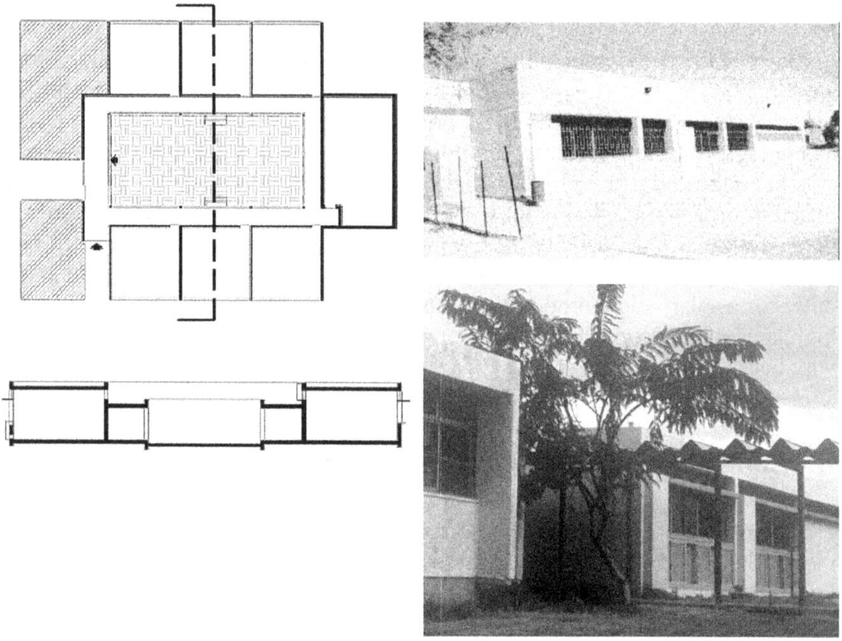

Kindergarten, Netivot (DAUP)

Design: Two identical kindergartens—commissioned by the Ministry of Housing and Construction and intended as a demonstration project—were constructed in the lowlands town of Netivot (Figure 3.2). The climate is temperate with hot humid summers and cold bright winters (see data in Table 3.1). Each symmetrical half of the structure was designed to include a number of low energy strategies. Attached solaria were designed as semi-open spaces for winter use; each one fitted with fans to boost indoor temperatures. The solaria were to be totally open in summer to avoid unnecessary heating of the structure. Lighting of the indoor space emitted from the solarium area through operable louver-fitted French windows was intended to allow for control of summer daylight and winter hot air emission. A wind catcher facing the prevailing evening breeze was fitted with an electric shutter to keep unwanted air from entering in the winter and during hot summer days. Additional features included: south facing clerestory windows with electrically operated external shutters, ceiling fans and selective shading over playgrounds.

Evolution: The project was constructed in a new neighborhood and due to pressure by the residents was put into use by the local authority before the designer actually authorized its completion. Kindergarten attendants were not briefed on the various features of the building, knew nothing about them, and thus did not understand their significance. The protective film covering of the solarium polycarbonate glazing was not removed and the solaria were used for storage. When eventually briefed, the attendants did use the building as intended, but when they were replaced a few years later the different systems were misused again. Using the solaria as closed glazed storage spaces caused overheating in summer and eventually brought about their being refurbished as conventional additions to the building, blocking up the transparent walls. Failure of the wind catcher's electric shutter led to the opening being permanently blocked up.

Figure 3.2. Kindergarten, Netivot. From top left, counterclockwise: plan; south façade; south façade and playground after refurbishment; south façade as constructed (solaria glazing still with protective coating in green)

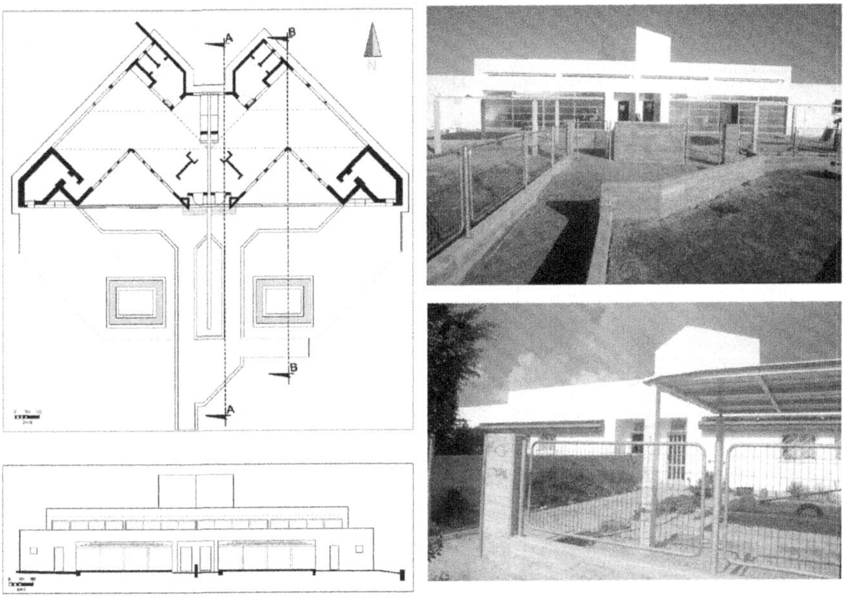

Although the building has almost nothing left of the original systems and details, strangely enough it is still known by the local residents as the "solar kindergarten"—a fact that can only be seen as damaging the image of such buildings since it can only lead to misconceptions regarding essence and form.

Nursery School, Mitzpe Ramon (Tishbi & Rozio, Architects)

Design: The roof of the building was designed as a saw-tooth greenhouse (Figure 3.3). Hot air was drawn from it by a manually operated mechanical system. Auxiliary back-up heating was provided by electric elements inside the fans. During the hot months, night air was used to cool the thermal mass (walls, ceilings and floors), and the roof greenhouses were to be opened. Fans and roof exhaust openings were to be operated through a main switchboard (Roitgur, 1985).

Figure 3.3. Nursery school, Mitzpe Ramon. From top left, counterclockwise: plan; north-south section; roof greenhouse in summer mode (from NE); roof greenhouse in winter mode (from E)

Evolution: The mechanical system was designed by a heating, ventilating and air conditioning (HVAC) company. Its operation demanded continuous involvement of the nursery personnel. Lack of such involvement or mechanical breakdowns (not unusual in a dust-rich environment) caused uncomfortable indoor conditions typical of a region with hot summers and cold winters. The system was operated for a very short period and was abandoned for the easier-to-operate split air conditioning units.

Education Center, Midrasha (Rahamimoff Architects & ASCU)

Design: Built on a cliff overlooking a canyon on the Negev Highlands, this education center was designed as an earth-bermed structure, covered with soil on its north, northeast, and northwest façades (Figure 3.4). The roof was heavily insulated and covered with gravel, except for a series of skylights used as the primary lighting source for the underground spaces. The southeast and southwest façades were fitted with large glazed areas for direct solar gains in winter. Summer strategies included deep pergolas with louvers designed to block all direct solar radiation, as well as a significant part of the diffused and reflected radiation comprising the largest part of global radiation in such an environment. Fans placed in the southern façades were intended to draw in summer night cold air, circulate it through the building and exhaust it through openable clerestory windows (Rahamimoff et al., 1987).

Figure 3.4. Education Center, Sede Boqer. Left - plan; right - south façade with deep selective pergola and fans for night ventilation (below windows)

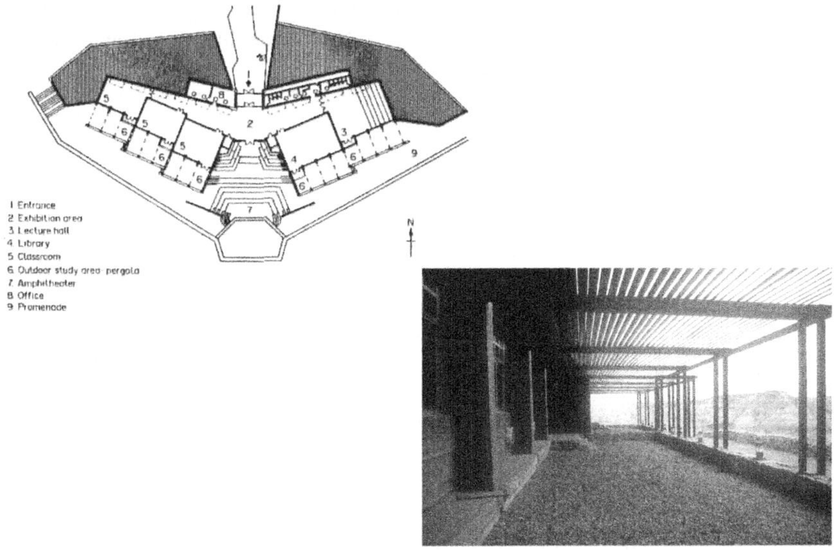

Evolution: Pergolas were removed for maintenance and were replaced backwards, so that the louvers were now facing north, thus blocking the winter sun. Consequently indoor temperatures dropped and auxiliary heaters were placed in the classrooms and the underground spaces. The fans were originally not used. Once the designers explained their importance they were used satisfactorily until the staff changed. Since then they have fallen into disrepair and do not function.

Zuckerman Community Center, Ben-Gurion College Midrasha Campus (DAUP)

Design: A multifunctional building with different size rooms for specific activities; it was originally designed for direct gains in winter and cross ventilation in summer. The brief called for various modes of operation, including full use during hot days, which would demand controlled lighting and possibly comfort ventilation (Figure 3.5). To provide for the varying needs, the southern façade was designed as a double skin. The inner skin included panels with fixed glazing, openable glazed- and opaque-louvers, whereas the outer skin included fixed opaque panels and openable louvers. Both skins were manually operated. Night ventilation was ensured by the openable louvers in the south and a series of fans in the eastern and western façades (Etzion, 1995).

Evolution: Improper operation during the first few months caused damage to various operable parts of the two skins, including the malfunction of the outer skin, originally designed as a series of sliding panels. These were dismantled, the inner skin's openable parts were replaced with fixed glazing and thus the southern façade was left fully glazed and exposed to summer sun. The consequent overheating problems were solved by the introduction of a central air conditioning system, whose ducts were hung within the various spaces, not designed for such additions. This intervention has contributed to the building's deteriorated state.

Figure 3.5. Zuckerman Community Center, Midrasha. From top left, counterclockwise: plan; north-south section; south façade with external operable shutters, summer day mode; original detailing with external operable shutters and internal operable glazing; and south façade after refurbishment

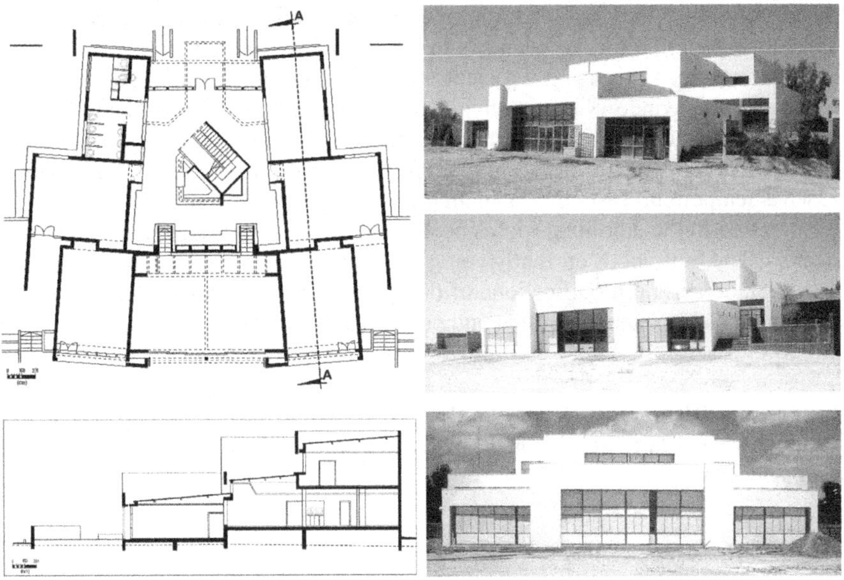

Blaustein International Center for Desert Studies, Midrasha Campus (DAUP)

Design: A multifunctional building including a library, classrooms, lab, lounge, office space and guest rooms, was designed around a central courtyard with glazed roof (Figure 3.6). The three-story complex was built partly underground and partly earth-bermed, the berms covering the western and northern façades up to the height of the second story window sills. The courtyard roof glazing was made of a selective-admission polycarbonate product designed to reflect back to the sky a significant part of the incident direct radiation falling on it at angles close to the normal. Thus the glazed roof functions as a greenhouse in winter and as partial selective shading in summer. Additionally, openable windows positioned at the base and apex of the roof allow for exhausting of air heated under the glazing. A downdraft evaporative cool tower provides almost 1000 kWh of cooling per day. The tower operation was optimized through the use of different water injection systems. In winter, hot air can be sucked from the greenhouse apex and introduced individually into the different guest rooms. This is done by operating small fans in shafts fitted on the southern façade of the rooms. The project was monitored under different operation modes and weather conditions, and modifications and optimization of the cool tower were introduced. These were extensively documented (Etzion, et al., 1997; Pearlmutter, et al., 1996).

Evolution: Having started with a very limited budget, no cleaning system was designed for the roof glazing, which collected dust and eventually reached a much lower reflectivity than the original one. The same limitation caused the design of openable windows to be operated manually instead of mechanically. Thus, it is common for the windows to be left closed long after the beginning of the cooling season, causing deficient ventilation and a rise in indoor humidity. Similarly, the windows often remain open after the beginning of the heating season, thus causing loss of hot air. The budget limitations allowed for only one fan to be purchased for the heating shafts of the rooms. The room fitted with it reaches temperatures of 20°C with an estimated saving of about 90% in electricity invested in the adjoining rooms heated by conventional radiant or convective heaters. The building is currently being redesigned for refurbishment, which will also include certain modifications of original systems to provide for better passive and hybrid system operation and performance.

Figure 3.6. Blaustein International Center, Midrasha. From top left, counterclockwise: plan; north-south section; court with downdraft cool tower and selective glazing roof

Newe Zin Solar Neighborhood, Midrasha Campus (DAUP)

Design: The first—and so far only—solar neighborhood in Israel was designed for a cooperative formed by staff of the Institute for Desert Research, the Ben-Gurion College, the local field school and other institutions located on a remote desert campus (Figure 3.7). The plan provided for solar and wind rights, i.e. unobstructed solar and wind access to each and every unit; weather protection in the open spaces by means of walls built by the house owners between their lot and the public space, and pergolas built by the cooperative on these walls over the pedestrian walkways; and manageable public open spaces with little vegetation—primarily drought- and salinity-resistant plants. Although the bylaws do not dictate the construction of bioclimatic houses, they ban the use of chimneys and strictly define the location of air-conditioning units to prevent noise disturbance. The 79-lot cluster was built in three stages. The bylaws define a building language and geometry, but the individual houses were designed by different architects chosen by the individual owners.

Figure 3.7. Newe Zin Solar Neighborhood, Midrasha. From left, counter-clockwise: plan with actual house footprints; pedestrian path with walls and pergola; solar rights on December 21

Table 3.1. Comparative table of projects: geoclimatic location, floor area, original strategies employed and subsequent changes. Figures under "location" indicate respectively: altitude above mean sea level [m]; distance from the Mediterranean coast [km]; summer air temperature max-min/RH min-max [°C/%]; winter air temperature max-min/RH min-max [°C/%]

project	*location*	*architect* *floor area [m²]*	*strategies*	*design* *construction*	*misuse/intervention*
High school	**Yeruham** 530m/75km 31-17/30-80% 15-3/45-70%	Kones 600	Trombe walls, clerestory windows, direct/indirect gains, cross ventilation	1984 1985	broken glazing-closing of ventilation openings-stone cladding of whole façades
Kindergarten	**Netivot** 150m/19km 31.5-20/45-77% 17.3-8/53-77%	DAUP 278	attached solaria, fans, clerestory windows, operable shutters, wind catcher, selective shading, external insulation, internal thermal mass	1990 1990-1	solaria used for storage-poor ventilation-overheating-glazing blocked up; malfunction of electric shutters-apertures blocked up
Nursery school	**Mitzpe Ramon** 860m/100km 29-18/25-80% 12.5-2/45-80%	Tishbi & Rozio 520	roof greenhouse, active air and water transfer systems	198? 1983-4	systems too complex to operate-occasional breakdowns-systems abandoned-air conditioners installed

Table 3.1. ***Continued***

project	*location*	*architect* *floor area [m²]*	*strategies*	*design* *construction*	*misuse/intervention*
Education center	**Midrasha** 470m/75km 32-17.3/30-80% 16-3.5/48-67%	Rahamimoff & ASCU 1000	earth berming, sun-facing windows, skylights, deep selective pergolas, fans, roof insulation, thermal mass	1982 1983	pergolas dismantled for maintenance-repositioned backwards-auxiliary heating added; fans needed maintenance-decommissioned
Community center	**Midrasha** (see above)	DAUP 736	sun-facing operable windows, operable shutters, cross ventilation, external insulation, internal thermal mass	1986-7 1988-91	shutters misused-broken-dismantled; glazed movable panels fixed-operable louvers replaced with fixed glass
Multifunctional building	**Midrasha** (see above)	DAUP 1800	underground, earth berming, greenhouse, sun-facing windows, skylights, selective glazing, external insulation, internal thermal mass, evaporative and cross ventilation	1987-8 1989-93	Infrequent operation of hot air exhaust windows; unsystematic use of cool tower
Solar neighborhood	**Midrasha** (see above)	DAUP 80,000 (79 units)	solar and wind rights, ban on chimneys, sun and wind protection in public open spaces, manageable green spaces	1985-8 1989-on	poor design of individual houses; additions of air conditioners and oil/gas burners; discontinued construction of public pergolas; redesign of green spaces with water intensive plants

Evolution: The original cooperative was comprised of scientists and staff already living on the campus who were thus acquainted both with the local climate constraints and potential, and with the basic bioclimatic theory and strategies. Several of the first houses were designed by DAUP architects or by architects who were affiliated with the Institute in the past. However, the second and third stage houses were designed for new residents and by non-local designers, many of whom had little or no acquaintance with the possibilities offered by the plan (although free consultancy was provided by DAUP). Thus many of the more recent houses do not comply with the obvious bioclimatic strategies provided for by the plan and its bylaws. Many of these houses have larger fenestration areas facing the specific lots' view, which may be to the north or the west.

A growing number of air conditioning units are being introduced, often after occupation, thus making supervision of their location harder, vis-à-vis the neighbors. This causes a nuisance during summer nights, and, in some cases, promotes the installation of air conditioners in houses that were originally designed for summer night cooling, but whose tenants suffer from the neighboring air conditioner's noise. Improper landscape design causes sun-facing glazing to be shaded by trees in winter, thus affecting the potential for passive heating. At least three of the houses have already installed oil burners, clearly breaching the bylaws.

The external finish originally defined as "smooth stucco in hues of white to beige" has been abused with many of the more recent houses being colored pink, peach, brown and even gray, obviously changing not only the aesthetic and visual character of the cluster, but also the absorptivity of the buildings' envelope. Additionally, as newer members joined the executive institutions of the cooperative, construction of public pergolas intended to provide shading over pedestrian paths was discontinued and public open spaces were redesigned to include large lawns and other water intensive plants.

All of the cases described here have one common malfunction source—a change of user, or user-profile different from that originally intended for the project. It is obvious that public, educational and community buildings which depend on user operated systems fail once the original user-operator is changed. When such a change is not accompanied by a briefing of the new user, regarding the properties of the building and its systems and the potential these have for the creation of more comfortable indoor conditions, the building and its systems become meaningless and obsolete. From this point on, the way is really short to the defacing of the building through refurbishment, decommissioning of the various systems and components, and introduction of conventional conditioning systems. It is also obvious that complicated active systems such as the one in the nursery school described are very problematic, especially when intended for people with little technical orientation. It seems that the best way to cope with such problems is by designing simpler, user-friendly, but also fool-proof buildings, which can still function reasonably even when poorly operated, and which present very few chances of malfunction (such as those inherent in big movable and especially mechanical components).

Discussion and Conclusions

In most of the cases described here, there has been poor communication between the architect and the end users, especially with the passage of time. This may not be essential once bioclimatic principles and strategies become common knowledge, but it seems that such a moment is still quite a long way off. In at least one of the cases—that of the Netivot kindergarten—the building was occupied and operated even before the architect had the chance to ensure proper completion of the building and to properly brief the users.

In the case of the cluster described above, the main fault stems again from lack of education, information and enforcement. Architects, tenants and authorities seem to have a very limited understanding of the need for and the advantages of bioclimatic architecture. Thus, failing to comply with bylaws or the spirit of a plan is not considered a "crime" but rather a "misdemeanor," since these seem to be meaningless—a mere whim of the planner. This also makes the authorities more reluctant to enforce the bylaws, since functionaries, too, do not understand their significance.

In summary, preventing these "crimes and misdemeanors" is primarily a matter of education, aggressive marketing, continuous follow up and diligence in informing the user. It is also a matter of creating the legal framework that will guide architects to comply with bioclimatic strategies, empower authorities to enforce commonly accepted legislation (as opposed to very limited bylaws often perceived as private whims of specific architects), and encourage clients-users to take advantage of these instead of using significantly more expensive and environmentally unfriendly systems.

However, it should be kept in mind that maintenance-intensive systems, such as moveable, openable and operable windows and shutters, present a barrier that, in many cases, is more psychological than real, but nevertheless has negative effects. The concept of "labor intensive" details and systems as opposed to common heating and cooling devices creates an immediate opposition on the part of the non-initiated client or user. The idea that such systems or details are time- and labor-intensive, and demand high maintenance, can be countered in two different and complementary ways. Firstly, it is up to the architect to design details and systems as user-friendly as possible, and to avoid what might be considered by the layman as something that puts extra strain on their efforts and time in the use of the building. Although it is not possible to avoid all movable and operable details, it seems advisable to put as much extra effort as possible in the design of fixed, stationary, totally passive devices—at least until people become more informed. Secondly, it is up to the lawmakers to ensure the economic viability and advantages of passive systems and buildings. Realistic pricing of conventional fuels and electricity, to account for environmental and health burdens imposed by their use, will definitely encourage the use of alternative systems.

Lastly, it seems imperative to establish a framework for systematic and periodic feedback and post-occupancy evaluation (POE). The current lack of such instigates unfavorable changes. It should be in the architect's interest to maintain contact with the client to ensure proper operation and maintenance of the project, but at the moment this is not within any established formal connection between the two. Furthermore, it may be important to promote a more binding relation between architect and future user (or current user that intends to make alterations to the project), though this may be strongly opposed as undemocratic. Nevertheless, it seems important to create the behavioral base for a long-term commitment on both sides—the designer's and the client's.

It is interesting to note that in a number of POE studies undertaken in the region of the Negev desert, processes opposite to those above were recorded and documented. In many of these cases, tenants of conventional housing designed without regard to environmental constraints, made modifications to ground stories and attached open spaces, primarily stemming from the need for better climatic adaptation of buildings to their use (Etzion, et al., 2001). It is possible that a systematic documentation and analysis of different building types and use patterns will yield different results for the different types, thus pointing at different possible strategies.

Acknowledgments

Drawings for this paper were adapted from the originals by K. Nahon. DAUP archives were reviewed with the help of M. Barak. Drawings, slides and information on the ORT Shoshana Sapir High School in Yeruham were made available by Arch. M. Kones. The middle Zuckerman Community Center slide was photographed by Arch. E. Erell. Details on the education center were provided by Arch. A. Rahamimoff. The plan and section of the nursery school were reproduced from Roitgur (1985).

References

Bar-Nathan, M., Beenstock, M., & Haitovsky, Y. (1998) The market for housing in Israel. *Regional Science and Urban Economics*, 28(1): 21–49.

Bitan, A., & Rubin, S. (1991) *Climatic atlas of Israel for physical and environmental planning and design*. Dept. of Geography, Tel-Aviv University; Israel Meteorological Service, Min. of Transport; Min. of Energy and Infrastructure, Div. of R&D, Jerusalem.

Etzion, Y. (1985) Desert architecture—the architecture of the extremes. In Y. Gradus, (Ed.), *Desert development: Man and technology in sparselands* (pp. 81–102). Reidel: Dordrecht.

Etzion, Y. (1995) Experimental projects in desert architecture, Israel. *Journal of Arid Land Studies*, 5S: 81–84.

Etzion, Y. Pearlmutter, D., Erell, E., & Meir, I. A, (1997) Adaptive architecture: Integrating low-energy technologies for climate control in the desert. *Automation in Construction*, 6(5-6): 417–425.

Etzion, Y., Portnov, B. A., Erell, E., Meir, I., & Pearlmutter, D. (2001) An open GIS framework for recording and analysing post-occupancy changes in residential buildings—a climate-related case study. *Building and Environment*, 36(10): 1075–1090.

Meir, I. A. (1992) Urban space evolution in the desert—The case of Beer-Sheva, *Building and Environment*, 27(1): 1–11.

METEONORM 4 (1999) Meteotest: Bern

Pearlmutter, D., & Meir, I. A. (1995) Assessing the climatic implications of lightweight housing in a peripheral arid region. *Building and Environment*, 30(3): 441–451.

Pearlmutter, D., Erell, E., Etzion, Y., Meir, I. A., & Di, H. (1996) Refining the use of evaporation in an experimental down-draft cool tower. *Energy and Buildings*, 23(3): 191–197.

Rahamimoff, A., Rahamimoff, S., Silberstein, A., Faiman, D., Zemel, A., & Govaer, D. (1987) Design considerations for an earth-integrated education center in the Israeli desert. *Tunneling and Underground Space Technology*, 2 (1): 69–71.

Roitgur, A. (Ed.). (1985) *Solar houses in Israel: Passive solar systems for heating residential buildings* (pp.35–38). Jerusalem: Min. of Energy and Infrastructure, Dept. of Energy Conservation.

Chapter 4

Sustainable Architecture

Yair Etzion

Introduction

In developed countries, construction is a significant sector of the economy. Besides providing solutions for housing and other purposes, construction is considered to be one of the leading components of an economy, and it is often regarded as a major indicator of a country's economic health. The construction industry is indeed very sensitive: economic slow-downs are often seen first in this industry, and the first signs of economic recovery are also exhibited by this industry.

Israel's construction industry is typical of most developed countries. In 1996, 7.5% of Israel's labor force was working in the construction industry (www.world66.com/asia/middleeast/israel/economy). In 2005, 5% of the Israeli GDP was the result of this industry's output. This is down from the approximately 30% it held in 1950 (www.mfa.gov.il/MFA/Facts+About+Israel/Economy/ECONOMY-+Sectors+of+the+Economy.htm) It is surprising therefore, that, generally speaking, the building industry is not a pioneer in adopting new technologies that can improve its productivity, its profitability and the quality of its products.

In many modern industries, research and new technologies are the engines for developing new techniques, and industries look to them for ideas and innovations. Regrettably, this is not the case in the building industry, where the percolation of new technologies is slow and far behind technical and scientific developments. One of the leading figures in Israel who was invited to speak at a conference organized by the Center for Desert Architecture and Urban Planning (DAUP), a unit of the Jacob Blaustein Institutes for Desert Research (BIDR), said: "There is high-tech, there is low-tech, there is no-tech and then there is construction." Funny as this statement may be, it pretty much sums up the situation in this industry.

Described below are a few research projects by the DAUP that have technologically matured in the last few years, and the ways they have been implemented—or not—in the building industry in Israel.

Soil Construction

In the early 1980s it seemed very logical to reintroduce soil construction in Israel. Soil construction already had roots in the recent history of the new state: immediately following Israel's establishment in 1948, Israeli scientists were chosen by the UN to lead a research project in soil construction for mass housing (Cytryn, 1957). The newly born state urgently needed to build housing for refugee-immigrants who had fled Europe in the aftermath of the Holocaust: Israel's population had tripled between 1949 and 1952—from about 600,000 to 2,000,000—and Israeli scientists and architects were believed to be able to lead this UN research project.

Indeed, needs in the early 1980s differed from those of the 1940s, but soil construction could still be justified: the energy crisis has just started, but its expected consequences in the future were already clear. At that time, the cement industry alone consumed more than 4% of the country's energy budget, which completely relied (and still relies) on imported fuel. This is a very significant percentage: all the domestic solar water heaters used in Israel in the early 1980s saved a similar proportion of the country's energy consumption. The standard of living had risen in Israel by this time and a single-family dwelling in a remote community had become a significant part of the "Israeli Dream." Many Israelis had traveled abroad, and the houses they began to build all over Israel exhibited a strange mixture of styles and materials, reflecting the various origins of the Israeli population and the architecture of the countries that Israelis had visited. (This mixture of styles and materials was regarded by one professional as "architectural chaos"). The "Santa Fe" building style was easily adopted by Israelis as one of their favorites for private-home construction. This gave rise to the hope that introduction of a new type of building material would be easy and straightforward. Soil blocks, with their lean consumption of embedded energy and their inherent good thermal and acoustic properties, seemed to be a good choice.

The DAUP launched an earth-construction project in the late 1970s, under the supervision of Arie Rahamimoff, architect and then Head of the DAU (Desert Architecture Unit). The project culminated in the building of the "Adobe House" at Midreshet Ben-Gurion (Figure 4.1). This building became a landmark in Israel, and has since been visited by a very large number of students, builders and tourists. Following this project, Rahamimoff left Ben-Gurion University of the Negev and the DAU closed down. It was reopened in 1982 as the DAUP (Desert Architecture and Urban Planning) and was headed by the author of this chapter. One of its first projects was the development of stabilized soil construction technology appropriate for Israel. Previous attempts to stabilize earthen material had failed; the Adobe House was therefore not built of stabilized soil and has required significant annual maintenance ever since.

Figure 4.1. Adobe house (Photo: Wolfgang Motzafi-Haller)

By the 1980s Israel was already a well-developed, Western country. For a soil-construction project to succeed, a pre-condition was that it be based on stabilized soil. Blocks, bricks or walls made of stabilized soil can withstand water damage caused by rain and by other sources. Stabilized soil bricks are usually stronger under compression than are unstabilized blocks. Unstabilized soil construction sustains heavy damage in the presence of water, and less-developed societies that use unstabilized soil for construction invest much of their time and resources maintaining their building stock. Such a routine would not be accepted in Israel.

The easiest way to achieve stabilization was to adopt technology from the United States that involved adding small amounts of bituminous emulsions to the soil mixture. Various manufacturers offer blocks stabilized by this method (i.e., Hans Sampf Co., Inc., Madera, CA). These emulsions impregnate the clay particles that cement the sand and silt particles together in the material, thus preventing water from reaching and replasticizing the clay in the brick and weakening it to the point of collapse. Attempts at Midreshet Ben-Gurion to employ this technology totally failed, and blocks manufactured using these emulsions proved to be even weaker under compression than blocks that had been made using traditional, unstabilized technology. After further investigation the reason for the failure was discovered: soils used in the U.S. for bituminous stabilization are acidic ($pH<7$) and the bituminous emulsions mixed with them are alkaline. The mixture of both results in a very strong, water-resistant block. Israeli Negev soils are alkaline ($pH\sim10$) and mixing them with alkaline emulsions naturally results in total repulsion of the particles rather than adhesion, and a very weak brick, unfit for construction.

A new approach has since evolved: stabilization based on artificial polymers capable of stabilizing blocks made of alkaline soil. Various combinations of concrete and plaster additives (both alkaline) were tested, and a "winning combination" was formulated. A two-year effort culminated in a soil brick that was left submerged in a barrel of water for six month and pulled out undamaged, exhibiting a compressive strength of about 25 kg/cm^2 (Freidin and Etzion, 2004).

Once development of this formulation was complete, attempts were made to commercialize it. This was done on two tracks. The first track was an attempt to convince builders and developers to employ the new technology in their future projects. Interest was expressed by a few builders, but when faced with making the final decision all of them withdrew. The reasons were never explicitly expressed by any of them, but during conversations, a few hints were dropped. The main point raised was that the cost savings offered by the new technology were not sufficiently large for builders to risk using the new and unfamiliar technology. Another point was that the economy—at that time—was a "seller's economy," and builders really had no incentive to lower the cost of buildings. During this period, buildings were sold while still in the design stage, before they were even built, and the lack of competition among builders resulted in no effort to reduce costs. It was clear that all the arguments against implementation of the technology were based on the economy it could—or rather could not—offer. None of the arguments was technical; technical issues were not even raised at this preliminary stage. Similar attempts were made with block manufacturers. Not expecting a high-volume demand for the soil blocks under the prevailing conditions, they also declined to enter into this business opportunity.

The second track was an attempt to convince chemical manufacturers in Israel to manufacture the polymer mixture that had been developed and to export it to regions of the world where unstabilized soil construction was popular. The idea was to distribute the material, either free of charge (funded by international agencies), or to sell it in building-material stores, so builders could obtain small quantities of it to add to the soil mixtures used for construction. This would, of course, upgrade the quality of the structures, decrease risk from water damage or from other disasters, and be of benefit to dwellers of houses that employed this technology. Two Israeli companies expressed interest in the early stages of negotiations, but decided not to take the initiative and withdrew from the negotiations at a later stage.

This brought an end to soil construction research at the DAUP.

Mixed-ash blocks

During the 1980s, the Israel Electric Corporation converted all its power stations to operate on coal, resulting in—aside from an increase in air pollution—generation of an enormous amount of fly ash, the disposal of which became a major ecological problem. Some of the fly ash was purchased by Nesher Israel Cement Enterprises to use as an additive (15%) in the manufacture of Portland cement. The remaining fly ash was simply hauled out to sea and dumped a few kilometers off the Mediterranean coastline of Israel. This caused an even greater ecological problem; and it was an expensive endeavor—US$ 10 per disposed ton of fly ash. (This procedure was discontinued at a later date, following significant opposition by "Greens").

At the same time, experimental production of fuel, based on extraction of oil from oil shale found in large quantities in the Rotem Plain of the Israeli Negev Desert, was underway. A by-product of this process was a large volume of shale ash, which also had to be disposed.

As stated earlier, the cement industry in Israel uses more that 4% of the total energy consumed in the country. This huge consumption rate serves as a permanent incentive for research in this industry toward saving energy.

The DAUP launched a project to address the surplus-ash problem by finding new uses for ash in the building industry. In ancient times, Roman engineers mixed volcanic ash with lime to make concrete long before the development of modern cement. The cementing qualities of ash are therefore well known and expected. Researchers at the DAUP claimed that mixing fly ash with shale ash would yield an even better cementing material.

After a few years of work on this project, an optimal mixture was developed (Freidin and Etzion, 1997).

The researchers were well aware of the potential for harmful radiation from ash, therefore the ash mixture was sent to the Center for Radiation Safety at the Soreq Nuclear Research Center to determine if the radiation levels of the fly ash and shale ash mixture were harmful. Only after determining that the radiation level was within the strictest margins of safety allowed by the German Standard of Building Biology Testing Methods SBM-2003—which at that time was the most demanding standard in the world—did work on the project continue.

A few containers of the already-prepared ash mixture were hauled to a concrete block manufacturing plant to be used in an experimental production run. The manufacturing line normally used to produce concrete was simply rinsed clean with water and then the ash mixture was used to manufacture alternative blocks. It is important to emphasize that no retooling or retraining of workers was necessary, and that the production process was identical to that of conventional concrete block production. The final product was indistinguishable in size, color or weight from the conventional four-cavity blocks that are very popular in Israel. A sample batch of these blocks was sent to the Israel Standards Institute to determine their compressive strength. The results indicated that they exhibited compressive strength higher than required by the Israeli standard for concrete blocks. A test wall was built of these new blocks in the experimental field of the DAUP; 15 years later it still stands, strong and undamaged.

Once proven technologically satisfactory, the marketing effort began. In addition to the ecological benefits, it was estimated that the cost of the mixed-ash block would be cheaper than the conventional block by about 10–15%. It seems, though, that the cost reduction was not meaningful enough to convince block manufacturers to start manufacturing the new blocks. Calculations of the expected savings in the cost of a dwelling unit proved that cost reduction resulting from the use of the new blocks would be just marginal, as the total cost of blocks needed to build such units are marginal at best. In the 1980s, awareness of environmental issues was not sufficiently keen to be a single convincing argument for manufacturers to start manufacturing the new blocks. This being the case, the block manufacturers suspected that neither builders nor consumers would take the risk of using a new material for such small savings, nor view this as a way to contribute to an environment cause.

Still, one manufacturer was ready to take on the challenge. A draft contract between this manufacturer and Ben-Gurion University of the Negev was prepared, and a date for signing the contract was fixed. Not surprisingly, a day or two before this date an article appeared in the leading newspaper in Israel, explaining in detail that building materials containing radioactive fly ash constituted a major health risk … this of course was a deal-breaker; all that remains is to guess who was responsible for planting this article in the newspaper. Just a reminder: Portland cement contains 15% fly ash, and the mixed-ash cement developed by the DAUP was certified to have radiation levels significantly lower than the levels permitted by even the strictest of standards. After the newspaper article, all arguments were ineffectual. No one could blame the would-be manufacturer, who then withdrew his willingness to produce our mixed-ash blocks.

Aerated blocks

Another very popular building block widely used in Israel is made from autoclaved aerated concrete (AAC). This block is popularly called Ytong®, despite the fact that the Israeli subsidiary of Ytong Industries Ltd. is not the sole manufacturer of this block in Israel. This building material complies with Israeli Standard SI 1045 for the insulation of residential buildings set by the Israel Standards Institute. The blocks are light and relatively large (50×30 cm), making them popular with builders.

Production of AAC, though, is extremely energy intensive. Manufacturers contacted by us repeatedly refused to divulge information regarding the energy needed for production, but from observing the production process at one of the manufacturing plants we saw indications that the amount of energy used in the production process is huge. Estimates carried out by the DAUP place the embedded energy in AAC at about 426 kWh/m^3—equal in this respect to that of Portland cement (Huberman, 2007). The process begins with grinding silica sand to a very fine powder. This is then mixed with lime and with aluminum powder, which serves as the expansion agent. Water is added, the ingredients are mixed and the slurry is poured into large molds, where expansion takes place. The aluminum and lime react, freeing bubbles of hydrogen gas that form tiny closed cells in the concrete matrix, and cause the material to expand 2–5 times its original volume. When hardened, the large aerated blocks are sliced in three planes with cutting-wires (reminiscent of a cutting device for hard-boiled eggs) and placed in an autoclave for 12–24 hours for steam-and-pressure curing.

An alternative process was developed by the DAUP, based on a mixture of fly ash and shale ash. Other than a slightly grayer tint caused by the ash mixture, the blocks produced were physically identical to the original aerated blocks manufactured in Israel. The new blocks would pass any standards test for both strength and thermal properties. The uniqueness of the process was that it completely eliminated the need for grinding the silica sand, and—even more importantly—the need to use an autoclave for curing: Blocks could be cured in open air, or at most in a simple low temperature oven. The energy required to manufacture them was a very small fraction of the energy consumed by the conventional process (Figure 4.2).

The first manufacturer—the biggest in Israel—to whom we presented the new blocks, chose to completely ignore them. The second manufacturer had a very interesting argument against utilizing the new technique: as was previously mentioned, the color of the blocks was a very light grey. Conventional AAC is white. The manufacturer's argument: "We manufacture and advertise white blocks." The fact that the block is never visible when the building is complete was not convincing enough. Economic arguments and proposals for separate marketing were not taken seriously, and the project came to a halt.

Figure 4.2. Aerated block manufacturing process (From: http://www.aerblock.com/img/manuf_process_high.jpg)

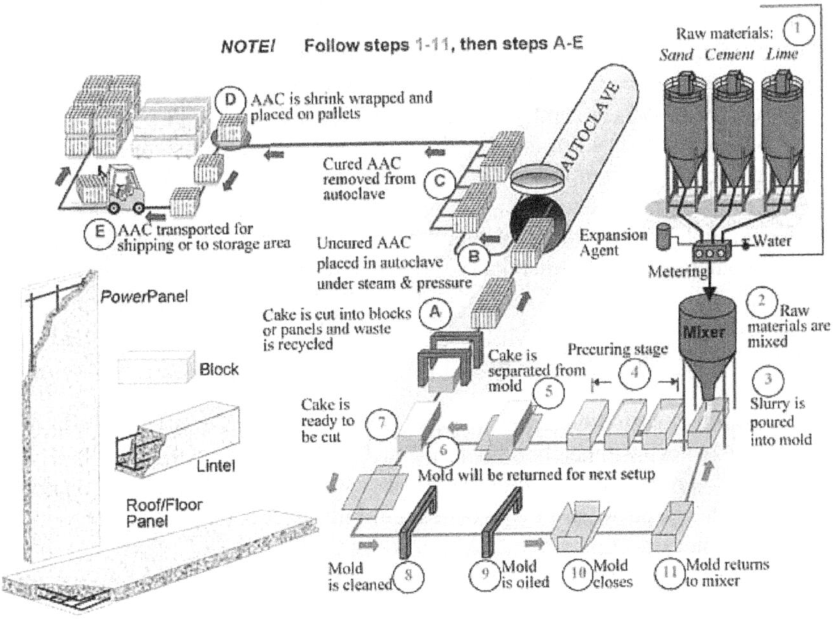

Down-draft evaporative "cool tower" (DECT)

In the mid 1980s the DAUP was commissioned by the BIDR to design a building for its Blaustein International Center for Desert Studies (BIC). The building was designed around a large (~ 25×17 m) glazed internal courtyard, with an average height of 10 m. This courtyard serves as a source of solar energy in winter and as a shaded and cooled space during the summer.

While there was nothing new in employing a glazed courtyard for solar heating, the cooling system of the courtyard was a topic for scientific investigation and innovative design. Near the center of the floor area of the courtyard, a downdraft evaporative "cool tower" (DECT) was installed to take advantage of the very low relative humidity of the desert air during hot summer days (Figure 4.3).

The design of the tower is rather simple: a steel-tube structure about 11 m high supporting a "sleeve" made of corrugated PVC sheets. A wind scoop at the top catches the prevailing northwesterly winds, when available, and a large fan is activated to draw the air in when no wind is available. Water sprayers installed at the top of the tower spray a large volume of mist into the tower's interior. The mist-and-hot-air mixture flow down through the tower, resulting in rapid evaporation of the water and significant cooling of the air—almost to the wet-bulb temperature. Water that has not evaporated is collected and re-pumped to the sprayers at the top of the tower, to be sprayed again. The water volume lost to evaporation is automatically replenished by the system.

On a typical summer day, the DECT is operated for about 10 hours, during which time up to 1.5 m^3 of water is evaporated, the fan and water pump consume 10–12 kWh of electricity and the cooling output is approximately 1,000 kWh. (Embedded energy of evaporated water is not included in this calculation.) On hot days, a volume of about 25,000 m^3 of air is cooled per hour, by up to 14°C; entering the courtyard is a cool and truly pleasant experience (Figure 4.4).

All the criteria set forth for the DECT designers were met: significant cooling with a minimum consumption of energy. Yet, to date, no other DECT have been built in Israel, with the exception of a few small ones (1–2 m high) built on a few residential units at Kibbutz Neot Smadar by dwellers who reinvented the DECT without bothering to learn what the DAUP had already developed so successfully.

"SOLVENT" Glazing System

During the early 1990s, most of the DAUP faculty moved into new private homes in the Neve-Zin neighborhood planned and designed by the DAUP. This is the first bio-climatic neighborhood in Israel (www.bgu.ac.il./CDAUP/). Naturally, the faculty practiced what they preached and built houses that were, among other things, solar houses.

After a period of living in these houses, it became clear that solar houses were certainly an asset, but the benefits came with a cost: In winter, the color of the rooms' contents faded considerably when windows were left wide open to allow the sun in. The level of daylight illumination in these rooms was also much higher than desired and radiative temperatures in sunny spots inside these spaces were much too high. Large south-facing windows also presented a heating problem in summer, requiring that shutters be kept closed during hot periods, rendering many rooms basically windowless.

Figure 4.3. Downdraft evaporative "cool tower" (DECT) (Photo: Yair Etzion)

Figure 4.4. Performance of the downdraft evaporative "cool tower" (DECT) in the Blaustein Institute Center

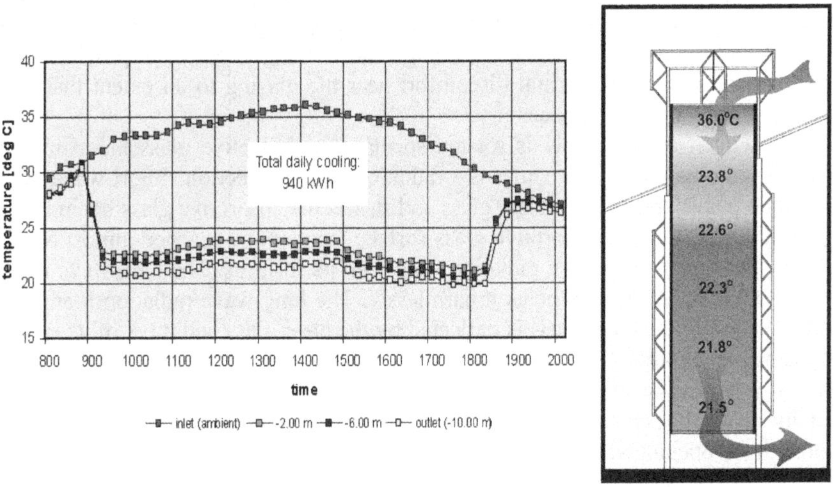

An effort to identify a type of glass that could reduce the magnitude of these problems was fruitless. The demands placed on such glass are inherently contradictory: the glass should transmit as much solar energy as possible during winter to maximize solar heating and block as much solar radiation as possible during summer to minimize solar heating. None of the many new types of glass that change their properties in response to ambient conditions (photo-chromatic glass) or to electric current passed through them (electro-chromatic glass) were able to solve the problem.

A glazing system aimed at overcoming these problems, while preserving the benefits of solar heating by direct gain, was developed by two researchers (Etzion and Erell) at the DAUP. The basic idea was to change the configuration of the glazing system: It should be a "different" window in summer and in winter. The solution was achieved mainly through the conversion of short-wave solar radiation to long-wave radiation and to convective heat.

The system is based on an innovative reversible frame, incorporating a double-glazing assembly: a clear pane of glass to provide a weatherproof seal, similar to a conventional window, and an absorptive pane of glass to provide solar control. The absorptive glass is placed about 4cm from the clear glass, and is open at the bottom and top to let air flow through the cavity created between them.

The double-glazing assembly and the ventilation channel separating the two panes rotate together through 180° to accommodate transformation from winter mode to summer mode (Figure 4.5).

In summer, the absorptive glass is oriented outward. It intercepts incoming short-wave solar radiation and absorbs a major part of it. This raises the temperature of the absorptive glass, causing it to emit long-wave radiation from both its surfaces. Radiation from its outward-facing surface does not affect the building; long-wave radiation from its inward-facing surface strikes the clear glass behind it, which deflects most of it away.

Since the absorptive glass is usually slightly darker, it reduces the light intensity in the space behind it, but it still acts as an open glazed surface allowing an unobstructed view. It lowers the cooling load (saving energy), improves visual comfort and decreases thermal discomfort near the glazing to an extent that may render shading devices unnecessary.

In winter, the assembly is rotated so that the absorptive glass faces inward. Heating is obtained by two means—radiation and convection. Short-wave solar radiation passes through the clear glass and strikes the absorptive glass behind it. In tests at the DAUP, the absorptive glass surface temperature reached almost 60 °C, causing it to emit long-wave radiation from both its surfaces. The long-wave radiation it emits into the room heats it harmlessly. The long-wave radiation it emits in the direction of the clear glass is deflected by the clear glass and most of its energy remains in the room. The warmer absorptive glass also induces air movement in the cavity between the two panes of glass—room air at room temperature enters the cavity at the bottom of the assembly and is exhausted at a higher temperature through the opening at the top of the assembly.

Figure 4.5. "Solvent" reversible-frame window (Photo: Yair Etzion)

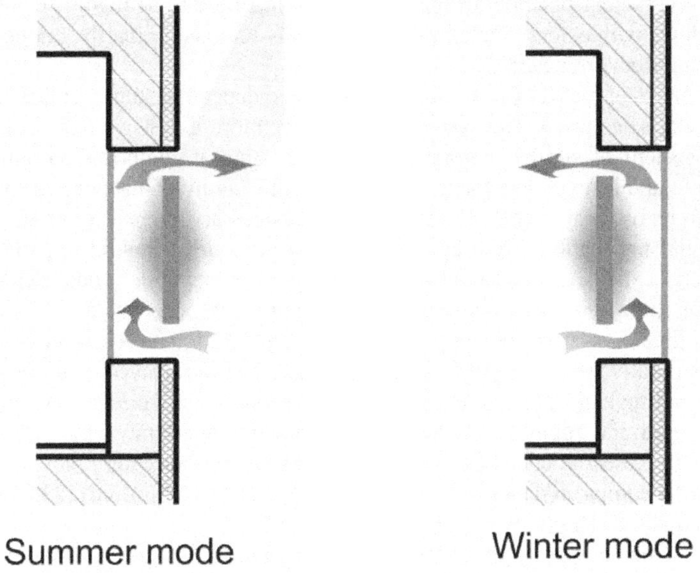

In this configuration, the glazing system provides for solar space heating in winter with a negligible loss of efficiency, compared to direct-gain systems. Visual comfort in the heated space is increased; local overheating caused by exposure to direct solar radiation and damage from solar radiation to furnishings in the interior space are prevented.

The marketing potential of the system was regarded to be high: it is intended for buildings with extensive glazed areas, in sunny regions (primarily, but not only, around the Mediterranean).

Estimates of market size for the system varied depending, among other factors, on the final sales price. The system is inherently more expensive than a standard glazing system of similar thermal performance, since it comprises an additional glass pane and a revolving frame. However, in the sunny climates for which it is designed, its use may render shading devices unnecessary, increasing its attractiveness. Initial installations were expected to be in up-market buildings, particularly offices and expensive residential apartments. Final penetration would depend, of course, on the willingness of customers to pay more for a better product.

The glazing system was tested initially at the DAUP, and when proven feasible, a cooperative research proposal involving Ben-Gurion University of the Negev in Israel and several European institutions in Sweden, Spain, Portugal and Germany was prepared and submitted to the Fifth European Community Framework Programme for Research under the acronym of "SOLVENT" (see the following URL: http://cordis.europa.eu/data/PROJ_FP5/ACTIONeqDndSESSIONeq112362005919ndDOCeq548ndTBLeqEN_PROJ.htm).

Following acceptance of this proposal, the system was tested in various climate regions in Europe and proved very efficient, particularly in warmer regions. In addition to the tangible product, a design tool was prepared to enable designers around the world to determine the proper configuration of glass for their particular location. The project was presented at several scientific conferences and was very favorably received and praised.

Trying to commercialize the system was another story. The first aluminum company—one of the largest in Israel—agreed to manufacture and market products based on the new system, but insisted on submitting a request for a development grant from the Office of the Chief Scientist in the Ministry of Commerce and Industry. A proposal was prepared, initial approval by the Ministry was obtained and funding negotiations commenced. The manufacturer was required to match the grant by investing in the industrial development. At this point, though, it became clear that the funds requested by the manufacturer where far too high, and that the hidden agenda behind the manufacturer's interest in the project mainly involved gaining government funds. When required to match the Ministry's funding, this manufacturer withdrew from the entire deal.

A second manufacturer was then approached. His agreement to develop and market the new system was obtained within twenty-four hours. Yet, three more years passed without one window being produced—not even a prototype. Only now, as this goes to press, and six years from start to finish—three with each manufacturer—has a real aluminum window become available to builders. It has been acronymed, "The Onot (Hebrew for "Seasons") Window" (see: www.bgu.ac.il/CDAUP/onot.wmv).

Design projects

Architecture differs from other "pure" academic fields in that it is also a profession; i.e., architects are professional designers. For architects, "continuing education and development" during their years of professional practice involves reviewing the works of other architects, either by visiting their architectural projects in person, or by examining their works in published professional (as opposed to scientific) literature. This professional literature includes professional articles, and photographs—lots of photographs. The rationale behind engaging the DAUP to design selective architectural projects is two-fold: 1) they serve as scientific experimental tools for testing concepts and techniques developed by the DAUP and 2) they play a major role in the dissemination of know-how related to these concepts and techniques to the community of practicing architects. Two of these projects are described below.

The BIDR Administration Building

The administration building of the BIDR was originally designed by the DAUP in the mid 1980s to serve as an international center for desert studies; as such it contained dormitory rooms, laboratories, teaching facilities, and offices. In 2004 it was converted into the Institutes' main administration building. The main design characteristics of the building were not changed and it still serves as a very successful bioclimatic, or "green," building (Etzion et al., 1997).

The building is arranged around a large glass-enclosed courtyard. The lower two stories of the building are set below ground, to achieve higher thermal capacity and to expose these floors to conditions milder than the ambient desert. The courtyard is cool and pleasant in summer and warm in winter, as are spaces adjacent to it. The large DECT (described earlier) in the center of this courtyard provides enough cool air to maintain thermal comfort during the hottest days. During winter the courtyard serves as a greenhouse. Warm air from the top of this greenhouse is moved into adjacent building spaces to reduce their heating demands (Figure 4.6).

As of this writing, we know of no other building in Israel that has employed any of these techniques.

Figure 4.6. Greenhouse features – enclosed courtyard, Blaustein Institutes of Desert Research administrative building (Photo: Yair Etzion)

Neve Zin

The opportunity to design an entire neighborhood arose when the BIDR faculty organized to build their own homes at Midreshet Ben-Gurion. The site selected for the new neighborhood was an undeveloped parcel of desert within the boundaries of the settlement—part of the Zin Plain. Faculty members intent on build their houses were aware of the work of the DAUP and of the benefits of building "with the environment" rather that against it. This gave rise to the idea of building the first bioclimatic neighborhood in the country—the Neve-Zin neighborhood (Etzion, 1989, 2003).

The new neighborhood comprised 78 lots. A major neighborhood-layout design consideration involved providing solar and ventilation access to each house. This incorporated a grid of relatively wide east-west vehicular roads, whose widths were based on solar geometry, and much narrower shaded north-south pedestrian arteries. This was achieved by clustering four units around a private green area, and arranging these clusters according to strict geometric parameters. The neighborhood plan thus provided shaded pedestrian walkways, separated pedestrian and vehicular roads and included minimal public green areas planted with indigenous greenery. Public green areas in other settlements were seen to suffer from insufficient maintenance, so in Neve-Zin an emphasis was placed on privately owned green landscaping. An attempt was also made to establish a "design language" for individual buildings to adhere to, thus avoiding the "architectural chaos" that so many similar neighborhoods in Israel have succumbed to (Figures 4.7, 4.8, 4.9).

Figure 4.7. Neve-Zin – general layout and clustered grid

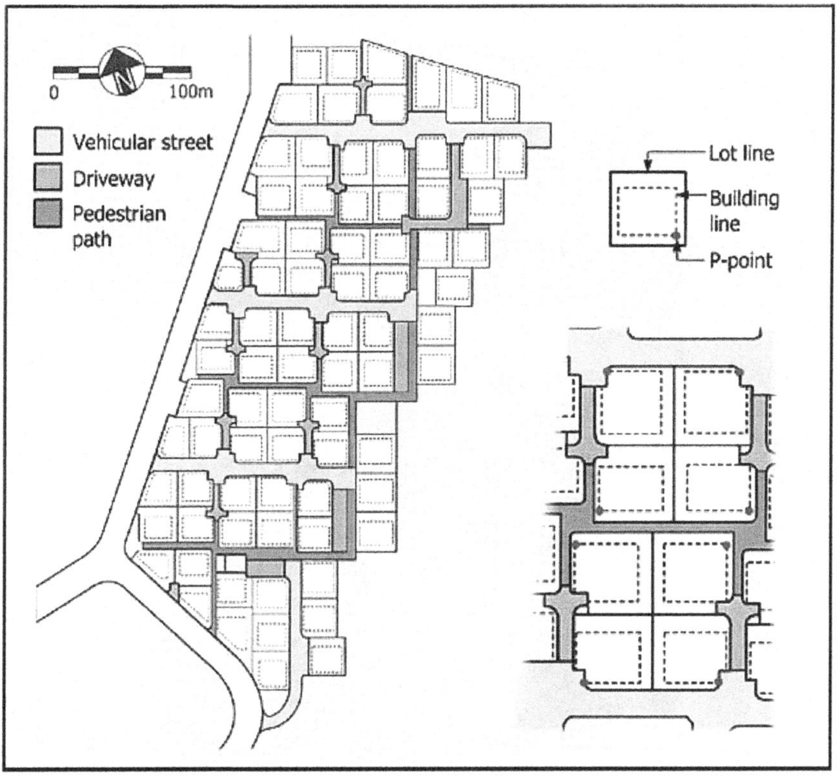

Figure 4.8. Neve-Ziv – East-west vehicular road (Photo: Yair Etzion)

Figure 4.9. Neve-Ziv – Pedestrian alley (Photo: Yair Etzion)

Part of the funding for this experimental neighborhood was provided by the Israel Ministry of Housing. Due to funding arrangements, the neighborhood was built in three phases. The majority of the individuals who built during the first phase were BIDR faculty members—four DAUP members among them—all dedicated to building a bioclimatic neighborhood. The architects commissioned to design the first-phase houses were environmentally oriented designers. The result was that the first-phase neighborhood met the expectations of the original design: a bioclimatic neighborhood with environmentally conscious buildings, and desert-adapted open spaces.

The second and third phases were a completely different story. Once the existence of Neve-Ziv in southern Israel became well known, it attracted people from all over the country. The bioclimatic nature of the neighborhood was not a guiding line (for legal reasons) for the designers of the second and third phase buildings and these were built without even a hint of any consideration for energy-efficient solar-heated or naturally cooled building design. It is important to mention that none of the builders of these homes ever seriously approached the DAUP faculty—the people who originally designed the neighborhood—for advice or guidance regarding bioclimatic home design and construction.

And yet, the Neve-Zin neighborhood is very much in demand; houses are sold at prices much higher than their initial construction costs, and any house offered for sale is sold almost instantly. Neve-Zin is a great success, but the implementation of known technologies for building in the desert was only partially successful.

Another neighborhood—Neve-Boker—was established at Midreshet Ben-Gurion at a later stage. Neither the architects nor the future residences of Neve-Boker have ever consulted with the DAUP faculty regarding the layout of this newer neighborhood or the design of its individual buildings.

The BIDR has added a number of new buildings to Midreshet Ben-Gurion in recent years. Architects from Tel-Aviv, with no prior knowledge of desert architecture, were commissioned by Ben-Gurion University of the Negev to design them. None of them has ever approached the DAUP for technical assistance. Ben-Gurion University of the Negev—which the BIDR serves as a research facility—has also significantly expanded its campus in Beer-Sheva with the help of famous architects from Tel-Aviv, again without involving its own desert architecture research unit.

Here, too, transfer of technology failed to occur.

Conclusion and analysis

Academic research and industrial development push technological know-how in construction far beyond the level at which technologies in this field are implemented. This is the case throughout the world and Israel is no exception. A few examples in the area of sustainability and energy conservation will substantiate this claim. A large variety of thermal-simulation programs exist that have the potential to significantly improve building design, yet very few architects—if any—actually use them. A very large selection of types of "smart glass"—designed to meet many different construction needs—is available, yet their penetration rate into the Israeli building industry has been very slow. Solar-powered air conditioning systems—a few developed in Israel, no less—are available, yet Israeli designers still resort to conventional air conditioning systems and will no doubt continue to do so for a long time to come. Efficient evaporative cooling systems for large buildings are available (e.g., Broad Air Conditioning Ltd.: http://www.broad.com/), yet, to date, none have been installed in any Israeli building. Many systems that support "smart buildings" are available off-the-shelf, but the building industry trails far behind these developments.

The technology is there, yet its utilization is minimal. Were research and development to stop abruptly today, the building industry would still have a long way to go to catch up with the state-of-the-art know-how that currently exists.

Why does the Israeli building industry implement new technologies so slowly?

Accelerating this rate should probably be the main concern of researcher-developers in the field of architecture and construction technology. Researching and developing alone cannot disseminate technologies within the industry. New technology cannot penetrate the market based on its technical virtues alone. Too often, mature technologies simply remain on the drawing board. An industry-wide approach is needed; but alas, the construction industry is very fragmented and complicated. Too many players are involved and cooperation among them is almost impossible to achieve. Government, business, industry, research and consumers must come together to advance the state of this lagging industry. If this does not happen, we will not see a high-tech construction industry.

To date, no serious study in Israel has been undertaken to investigate the very slow rate of transfer of technology from research and development to the construction industry. While new technologies and products of Israel's very advanced high-tech industry—electronics, communications, optics—are sometimes implemented even before fully mature, the exact opposite is true in the building industry.

Attempts by DAUP faculty to obtain funding for studies that would investigate this phenomenon have been doomed to failure. In recent years, several research proposals related to this issue have been submitted to funding agencies and government ministries. None have ever gained funding, perhaps because these institutions and agencies are worried that some of the blame will fall on them.

Since concrete data on this issue do not exist—or are not available—the reasons for this state of affairs still remain to be identified. A few of these reasons have become evident over many years of activity and experience in the architectural research field.

With a few exceptions, construction is generally a very conservative industry, not only in Israel. Most public buildings in Israel are funded by budget-squeezed governmental and municipal agencies or public organizations. Mass housing is built by large construction companies, the goals of which are to maximize profit. Most private residences are built by individuals who, in most cases, invest the better part of their life's earnings in them, along with a substantial mortgage. A standard apartment in Israel would cost the buyer 10 to 15 (or more) years of average earnings to purchase. Building a house requires an even greater investment. The cost factor makes clients—public or private—very cautious. Excessive caution usually leads them to follow the tried and true—and avoid the unknown. Governmental, municipal or public institutions always have an official who is responsible for the construction or purchase of a building. What could possibly motivate such an official to risk trying out new technology? What if it fails? An example: During the 1970s, two residential steel buildings were built utilizing French technology in Givat Olga, a small town between Tel Aviv and Haifa. The buildings were prefabricated and of very poor quality; i.e., poor acoustics, leaky water pipes, poor thermal comfort. It took a few years—during part of which the buildings stood empty—to declare these buildings uninhabitable, after which they were simply torn down. Such occurrences, of course, discourage experimenting with new technologies.

Economic considerations often stand in the way of construction projects that require a higher initial investment. Interested parties become very risk-aversive: They are very careful—even fearful—not to give the designer too much technological freedom. A public official once said to me, candidly, "If you design something unusual and it succeeds you will be praised by many; but if it fails, I will be the one who is held responsible, so why would you expect me to take the risk?"

Architects generally work within the guidelines set by their clients. Excluding the rare "celebrity architect," most architects do repetitive and routine work, lacking in technological innovation; e.g., residences, small public buildings, single-family housing, renovations. Practicing architects normally do not follow technological innovations; even fewer of them study scientific publications in their field—they simply do not need this knowledge for their daily work.

Thus, a lack of "continuous education" is evident. This would not be acceptable in medicine or high-tech electronics, where practitioners must continuously keep up with new developments. As a result, most practicing designers are not even aware of technological developments and are certainly not capable of implementing them in their designs.

In addition, a sociological phenomenon in Israel retards the rate of technology transfer in the building industry. Perhaps it should come as no surprise that in the competitive society Israel has become, individuals and organizations often prefer to invest in conspicuous building features rather than in inconspicuous technological improvements. Thus, for example, more expensive and more fashionable finishing materials that are "in" will have priority over improved glazing systems or better-insulated walls. As stated above, this phenomenon has a sociological or even a psychological explanation rather than technological explanation and thus is beyond the realm of this chapter, despite the major impact it has on technology transfer.

Surprisingly, the Israeli media share some of the "blame" for this situation: TV programs and niche channels dedicated to the "good life" focus on the internal and external "design" of buildings, detailing decorations, expensive building materials and gadgets—and never on their technological qualities. Despite the increasing importance of sustainability in recent years, the media have failed to jump-start any public awareness of the real issues behind sustainability or the building technologies required to achieve it. Very rarely—if ever—will a TV documentary feature technological aspects or innovative improvements in well-publicized buildings designed by famous architects. This sends a very wrong message to the public, which is reflected in the design characteristics most of the public is looking for.

The media—TV, radio and newspapers—frequently discuss sustainability related issues, e.g., recycling of plastic soda bottles, reuse of old tires, conversion of garbage into bio fuels. While all these issues are part of the problem and are certainly important activities, it is not clear why the identity of the main consumers of energy in developed countries—buildings—are not covered by the media. Estimates and data put the share of energy used by buildings in developed counties at 40–50% of energy consumption in these countries, Israel included. The reason may be that investment in buildings and their improvement is a long process, the benefits of which are visible only many years afterwards. The media in our time look for quick and easy fixes; there is no audience for long and exhausting processes. Attempts by my colleagues and me to redirect the attention of the media to this simple fact have met with total failure as we continue to discuss tires and bottles.

Paradoxically, "alternative technologies" do gain some success in Israel, partly thanks to media notoriety. If a designer or builder is ever interviewed by the media, in most cases he or she will be an alternative-technology practitioner. Such technologies are more exotic and imaginative and thus more interesting to the viewers.

There is a small group of designers in Israel who consider themselves "green architects," i.e., architects who design "green buildings" and who offer what they claim to be sustainability-oriented technologies. "Green" pays off nowadays, because "green" is the "in" trend. These sustainability-oriented technologies call for building built with straw bails, used tires, empty glass bottles, unstabilized soil and other exotic inventions. Not only do these techniques have only a marginal effect at best, but they are also impossible to implement on an industrial scale. Still, these techniques have some appeal to a small number of individuals who commission these designers to design for them. Being so unusual, these buildings and their designers attract disproportionate attention from the media and at professional forums. It is important to understand that every unsuccessful building increases the hesitation of organizations and builders to use new and unfamiliar technology, even technologies with solid scientific and technological grounding.

To a certain degree, Israeli schools of architecture—universities, colleges and vocational schools—have gradually distanced themselves from technological issues. Beginning in the late 1960s, a few new schools of architecture opened in Israel, and the Technion - Israel Institute of Technology in Haifa was no longer the only school training architects. When it was the only architecture school in Israel, the training program at the Technion placed a very strong emphasis on engineering, technology and science. Regrettably, in recent years, technology and science have come to be regarded by many architects and architecture students as "too difficult" and "not directly related to design." Architects, alas, are increasingly regarded as artists rather than as technical experts.

The engineering and quantitative aspects of technology are disliked by most architecture students and faculty and are regarded as too difficult to study and practice. Both the students and the faculty of our schools definitely lack the interest, and possibly the ability as well, to deal with quantitative material. For example, training programs dealing with energy, in most cases, lean toward qualitative rather than quantitative aspects. I have witnessed students designing "solar buildings" who don't even know the intensity of the solar radiation striking the absorbing façade.

The competition for students among schools of architecture has led these schools to reduce their emphasis on engineering, technology and science. The easier the program, the higher the number of student applications. This attitude towards technology is reflected later in the designs produced by graduates of these schools. Technological issues are often left to other professionals in the building industry, and new technologies are not sought by the architects.

These are just a few reasons believed to be responsible for the serious impedance of technology transfer from universities and research institutes to the construction industry. These are observations only, based on many years of work in the field of architectural research, and not the results of a serious academic study. That study has yet to take place. Hopefully, when it does, the rate of technology transfer will increase, and technology will have a stronger impact on the industry.

References

Cytryn, S. (1957) *Soil construction: its principles and application for housing.* Jerusalem: The Weizmann Science Press of Israel.

Etzion, Y. (1989) A desert solar neighborhood in Sede-Boker Israel. *Architectural Science Review*, 33:103–109.

Etzion, Y. (2003) Neve-Zin—a bioclimatic neighborhood in the Negev. Conference on energy saving legislations in buildings - EU and Israel, Organized by OPET, Tel Aviv University, September 15–16.

Etzion, Y., Pearlmutter, D., Erell, E., & Meir, I. A. (1997) Adaptive architecture: integrating low energy technologies for climate control in the desert. *Automation in Construction* (special issue: Intelligent buildings), 6:417–425.

Freidin, C., & Etzion, Y. (1997) Building materials based on fly ash and cured at room conditions. Sixth (final) report to the Rashi Foundation, January.

Freidin, C., & Etzion, Y. (2004) Technology for the manufacturing of cementless building components for desert conditions, *Journal of Arid Lands Studies*, 14S, (October): 183–186.

Huberman, N. (2007) Life cycle energy costs of building materials: Alternative for a desert environment. Thesis submitted in partial fulfillment of the requirements for the MA degree, 108 pp.

Chapter 5

Infrared and Solar Energy

A. Paul Hare

This chapter includes excerpts from descriptions of three projects in which Jeffrey M. Gordon of the Department of Solar Energy and Environmental Physics, Jacob Blaustein Institutes for Desert Research was one of the principal scientists:

1. Nonimaging optics (from Gordon and Ng, 2004);
2. Photovoltaic solar collectors (from Gordon, 2006); and
3. Solar surgery (from Feuermann, 2007).

Compact nonimaging infrared systems for decondensing 2000 m^2 of spectator glazing at Singapore Kranji racetrack*

Introduction

The indoor air-conditioned spectator galleries at Singapore Kranji racetrack are often afflicted, at nighttime events, by such extensive condensation on the roughly 2000 m^2 of glazing, through which patrons can watch the races, that

---

* Reprinted with permission from SPIE, with some deletions, from "Compact nonimaging infrared systems for decondensing 2000 m^2 of spectator glazing at the Singapore Kranji racetrack," by Jeffery M. Gordon and Kim Choon Ng. Nonimaging Optics: Maximum Efficiency Light Transfer VII, SPIE Proc. Vol. 5185, pp. 109-117 (2004).

proper viewing is substantially impaired. Our task was to develop a practical, affordable and robust solution. As delineated below, our strategy was to use infrared radiative heating tailored with maximum-performance nonimaging reflectors. Figure 5.1a shows the full facility and Figure 5.1b shows a close-up of the windows (Indeco Consultants, 2000).

Figure 5.1. (a): Full view of the spectator facility. The ground-level area is not enclosed. The upper 3 levels have air-conditioned interiors with spectator glazing 5 to 5.5 m in height. Level 2 is roughly 250 m long, while Levels 3 and 4 are around 150 m in length. (b): Close-up of the spectator glazing at the end of Levels 3 and 4. Level 4 has a forced-convection system that was installed prior to this project, intended to decondense the glazing by blowing hot air from below

(a)

(b)

Figure 5.2 is a cross-section blueprint of the galleries (Indeco Consultants, 2000) and indicates the existence of a track on which a gondola moves so that the windows can be cleaned periodically.

Figure 5.2. Blueprint: cross-section of the spectator galleries

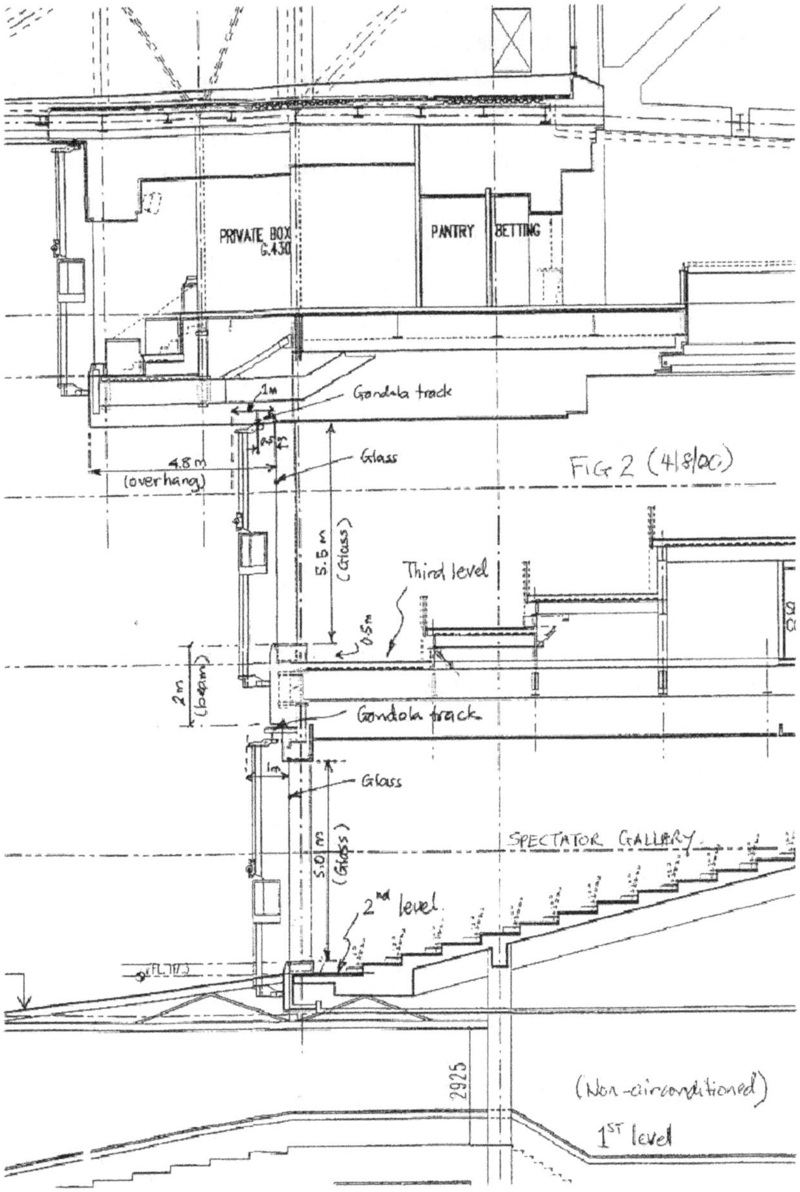

The indoor galleries are tiered, such that spectator views must be accommodated from every height along the glazing (see Figure 5.2). Large side-to-side viewing angles must also be ensured (see Figure 5.1). Respecting unobstructed spectator views of the entire racetrack and not introducing peripheral visual distractions restrict the space available in the vicinity outside the glazings. The vertical placement of any device would need to be high enough to avoid visual distractions, yet below the floor of the level above. Horizontal siting would have to be close enough to the glazing to permit maintenance and access from the existing gondola used for window cleaning.

The condensation on the glazing stems from a combination of the low (21 °C) thermostat setting required for customer comfort and high outdoor wet-bulb temperatures. Significant condensation does not occur during daytime racing events because even the diffuse component of solar radiation heats the outer surface of the glass sufficiently to avert the problem. The introduction of nighttime racing was accompanied by a debilitating condensation problem.

A number of possible solutions had been proposed and rejected, including: (1) double glazing (due to the expense and to the reflections that impede large-angle views); (2) a selective coating (too expensive and not adequately robust for continuous outdoor exposure); (3) wipers (distracting to spectators, mechanically cumbersome, with potentially problematic large moving parts); (4) raising the thermostat for the air-conditioning system (unacceptable to customer comfort); and (5) blowing cold dry air over the windows (additional interior and exterior structure and infrastructure, compounded by a large increment to an already enormous chiller installation).

A hot-air blower system had been installed for Level 4 (the top level) prior to our involvement (see Figure 5.1b). The discontent of the facility's engineering staff with this scheme derived from: (1) the failure of the system to prevent condensation during frequent nighttime cross winds; and (2) the excessive installed electrical capacity required. Our task was restricted to new retrofits for the 2000 m^2 of glazing in Levels 2 and 3.

Infrared Radiative Heating

We suggested infrared radiative heating as a possible remedy for the 2000 m^2 of windows in Levels 2 and 3. Psychometric and thermal balance calculations indicated that the glazing exterior necessitated only 5 °C of heating to eliminate nighttime condensation. With reasonably uniform irradiation of the glazing (e.g., a flux map with an average to minimum ratio AVG/MIN of about 2), the required electrical power—per unit area of glazing—would be no more than 1/3 that of the hot-air blower system. There would be no moving parts, and maintenance would be limited to occasional cleaning of the infrared heating modules, which could be performed from the gondola already installed for window cleaning.

Flux uniformity on the windows is crucial, because the *entire* glazing area must be decondensed. This is a consequence of the arrangement in which spectators watch racetrack events from every height along the glazing and along its full lateral extent. Hence the requisite electrical power is dictated by the region of weakest flux on the glazing. For example, simply introducing off-the-shelf heating elements, and containing them within the allowed space outside the windows (no more than around 1.5 m above the windows and extending no more than about 2.5 m horizontally from the windows) projects an AVG/MIN ratio in excess of 10. The problem is exacerbated by a substantial fraction of the emitted radiation missing the target unless reflectors are introduced. Furthermore, markedly inhomogeneous flux on the glazing would translate into local hot spots that could cause discomfort to spectators standing close to them.

We approximated the rate at which thermal power should be absorbed for decondensation as 0.75 kW per horizontal linear meter of window, based on the following parameters: (1) 5 °C of required window heating; (2) a heat loss coefficient from the window to the environment of about 20 $W/(m^2{\cdot}K)$; (3) a window height of 5.5 m (Level 3); and (4) approximately 75% of the electrical power input to the infrared heaters being utilizable for window heating, i.e., not being lost by natural convection, heat losses through the back insulation, transmission through the window, absorption in the reflectors or reflections off the glazing. The required input power to the heating element must be further multiplied by the AVG/MIN ratio of the flux map on the window (a function of reflector design).

Standard 245 mm × 60 mm insulated ceramic rectangular heating elements were purchased for our prototypes, installed 4 to a linear meter in a horizontal row. If, for example, an AVG/MIN ratio of 2 can be realized, then each element should be operated at close to 0.4 kW, which corresponds to a heater temperature near 850 K.

A related issue is then the correct brightness range for the radiating elements. High source temperatures allow small, easily accommodated heating elements; but if the source temperature is too high then non-negligible transmission through the glazing results and spectators experience deleterious direct heating. Fortunately, the commercial heating elements noted above can be driven up to approximately 1000 K.

If reflectors can be designed that satisfy the efficiency and flux uniformity stipulations, then the infrared heater solution offers the advantages of: (1) no moving parts; (2) adopting off-the-shelf inexpensive components; (3) modularity for both the heating elements and the reflectors; and (4) easy controllability of system operation via temperature sensors and feedback control.

Maximum-Performance Nonimaging Reflectors

Nonimaging reflectors offer the possibility of simultaneously satisfying: (1) maximum radiative efficiency (a minimal fraction of emitted radiation missing the glazing); (2) an acceptably uniform flux distribution on the glazing; and (3) compact devices that can fit in the allowed space just above the glazing, which means projecting over a highly asymmetric field of view. The reflectors must face downward due to frequent rain and toward minimizing dirt accumulation. Hence there is no possibility of irradiating the windows from below.

This is a nominally two-dimensional (2D) far-field design problem (although we will have to ensure that adequate flux reaches the end windows—a point addressed at the end of this section). Maximum radiative efficiency augurs in favor of reflectors with sharp angular cutoffs. Hence asymmetric compound parabolic reflectors should be excellent candidates (Welford and Winston, 1989); Gordon and Kashin, 1994). Their only shortfall is the flux uniformity on the target.

Far-field flux uniformity from compound parabolic luminaires can be improved considerably, at no sacrifice in sharp angular cutoff, by judiciously introducing gaps between the source and the reflectors. So we considered a range of system designs for which the variables are: (1) the tilt of the luminaire, (2) the placement of the luminaire and hence the extreme rays for which it is designed, and (3) the gap sizes. The source width was 60 mm (for the standard heating element noted above), and manufacturing constraints prescribed a luminaire depth not exceeding about 40 cm.

The AVG/MIN ratios are 1.7 and 2 for Levels 2 and 3 respectively (the glazing of Level 3 is 11% taller than that of Level 2, hence requiring a broader asymmetric field of view that is more difficult to accommodate).

The *horizontal flux* map would be nearly perfectly uniform (in both cases) except for small end effects. But the nominal end losses can be compensated fully. Thanks to the simple electrical controls, one can roughly double the power input to the last 2 m of luminaries at the two ends, and thereby satisfy the flux uniformity constraint. The incremental total power is less than 1%.

Test of Concept

Based on our designs, prototypes covering a horizontal extent of 6 m were fabricated and installed for testing (Figure 5.3).

Prototype viability was demonstrated for a worst-case scenario: a hot and very humid night, with the air-conditioning thermostat lowered to 19 °C (2 °C below the usual setting). There was near-total condensation on the windows—completely obscuring spectator view (Figure 5.4a). The infrared luminaries were lit at their design input power, and within seconds there was complete decondensation (Figure 5.4b).

Figure 5.3. During installation of the 6-m long infrared luminaire prototypes. (a) Front view from the spectator gallery, with 1.5 m of luminaries installed. The entire exit aperture of the luminaries appearing lighter than the surroundings stems from the full optical "unfolding" of the source (light-colored when unfired) by the compound parabolic reflectors. (b) Top view to the level below, showing the luminaire encasements

(a)

(b)

Figure 5.4. Testing the decondensing ability of the installed prototypes. (a) Before, with severe condensation on the window exterior. (b) After, with complete decondensation achieved

(a)

(b)

Summary

Infrared radiative heating was identified as the most promising method, among a number of varied possibilities, for the practical, robust and affordable decondensation of 2000 m^2 of spectator glazings at the Singapore Kranji racetrack. The challenge was tailoring reflectors to off-the-shelf heating elements such that a high degree of flux uniformity would be generated on the windows at maximum radiative efficiency. The large highly asymmetric field of view for the luminaries, as well as severe restrictions on the allowed space for luminaire installation, only added to the exigency.

The importance of efficiency and flux homogeneity derive from the need to limit the required installed electrical capacity and to avoid hot spots on the windows that would disturb spectators. Furthermore, only relatively low-brightness infrared elements were tenable toward avoiding excessive transmission of infrared radiation through the glazing.

We developed asymmetric compound parabolic luminaire designs with gaps between source and reflectors (the gaps being introduced mainly to improve flux uniformity on the windows, but also to avoid thermal bridges and to protect the reflectors). These devices provide: (1) sharp angular cutoffs for the far-field target, (2) maximum radiative efficiency, and (3) flux maps with satisfactorily low AVG/MIN ratios. A 6 m-long prototype was fabricated, installed and, when tested under worst-case conditions, performed successfully.

How innovative optics are serving the photovoltaic revolution[†]

The confluence of recent advances in photovoltaic technologies and advanced optical design has resulted in unprecedented capabilities for ultra-efficient, yet affordable solar electricity generation at high concentration.

A quiet revolution in solar electricity generation is underway, thanks to progress in the parallel tracks of advanced photovoltaic (PV) materials and innovative optics. Commercial multi-junction PV technologies have already demonstrated solar cell efficiencies of about 40%. These efficiencies can only be realized at flux levels of hundreds to thousands of suns. At such elevated concentration values, the cost of these new ultra-efficient cells becomes attractively

---

† Reprinted with permission from SPIE, with some deletions, from: "How innovative optics are serving the photovoltaic revolution" by Jeffery Gordon. SPIE Newsroom (The International Society for Optical Engineering, 10.1117/2.1200602.0132. (2006).

small, even though they are far more expensive on an area basis than conventional silicon and thin-film PVs. The burden to provide a practical and affordable system then shifts to the optical design.

High-flux PV systems (all of which require dual-axis solar tracking) have rapidly evolved to modular one-concentrator/one-cell configurations. Both the concentrator and cell are miniaturized, with cell areas not exceeding 1 cm^2 and concentrator unit areas up to 400 cm^2. Large systems are assembled from numerous identical modular units. With inexpensive lens-based optics, net (delivered) flux concentration in commercial PV installations has been limited to about 100–400 suns (one sun referring to maximum clear-sky solar normal-beam irradiation, about 1 mW/mm^2). Lens geometric and chromatic aberrations have combined with incompact designs (depth-to-diameter ratios of around 1.0) to limit the achievable flux and the practicality of many such systems. These factors preclude development of optical systems that simultaneously satisfy high collection efficiency, high flux, and compactness.

Thermal engineering is also a factor, because material integrity and cell efficiency typically suffer with excessive heating. Large concentrators, which by their very nature ordain sizable PV targets, mandate problematic forced-circulation water cooling, as well as noticeable parasitics. In addition, since voltage requirements dictate connecting many cells in series within a single module, the inhomogeneous flux map typical of large concentrators can result in substantial power dissipation as the module operates at the current of the most weakly irradiated cell.

We recently developed tailored optics that are aplanatic and planar, with efficient high-flux concentration (Gordon and Feuermann, 2005; Winston and Gordon, 2005) (see Figures 5.5 and 5.6). Two mirror contours are tailored to achieve the complete elimination of spherical and comatic aberration. The aim, however, is unrelated to image formation and instead focuses on efficient maximum flux transfer. The achromatic dual-mirror aplanat can also achieve the fundamental limit for concentrator compactness: an aspect ratio of 1/4.

If the concentrator is filled with transparent dielectric of refractive index n, (e.g., glass with n ~1.5), then with proper optics, either the concentration can be increased by n^2 (provided the cell is optically coupled to the concentrator), or overall optical tolerance can be relaxed by a factor of n, thereby reducing system cost. Because the entrance aperture is the only refracting interface and is normal to the solar beam, chromatic aberration is negligible. The dielectric slab concentrator is essentially achromatic.

Figures 5.5–5.7 illustrate the evolution of our optical designs for two generations of these aplanats to experimental and commercial realization by the H_2Go/SolFocus Corporation of Saratoga, CA, with commercially available Spectrolab solar cells. Generation 1 was tailored to a square 100-mm^2 triple-junction cell. The air-filled concentrator incorporates a tapered glass rod that contributes toward a liberal optical tolerance and flux homogenization, as well as allowing the cell to be conveniently sited near the vertex of the primary mirror. A protective glazing tops the unit and a thin metal sheet encloses the back of the module, serving as a passive heat sink that limits cell temperature to no more than around 20 °C above ambient. Net delivered flux values of 500 suns were measured in the field. The geometric concentration is 650×.

The success of Generation 1 prompted a thin, all-glass second generation, predicated on a 1.0 mm^2 cell of higher efficiency, thanks to lowered internal resistance. Amenable to existing precision mass-production techniques from the semiconductor industry, prototype fabrication and testing are currently underway: first manufactured for a net flux of 500 suns, then scaled to 1000–2000 suns, subject to economic optimization.

Figure 5.5. (a) Cross-section of the optical design (dual-mirror aplanat) of Generation 1 H_2Go/SolFocus PV concentrators. It consists of two tailored mirror contours, a protective entry glazing, and a tapered glass rod, at the narrow end of which the cell is optically bonded. This example achieves the fundamental 1/4 aspect ratio (compactness) limit. (b) Engineering exploded view, including the metal-sheet heat sink. (c) Photograph of the 100-mm^2 triple-junction cell around which the concentrator was designed for a net flux of 500 suns (650×)

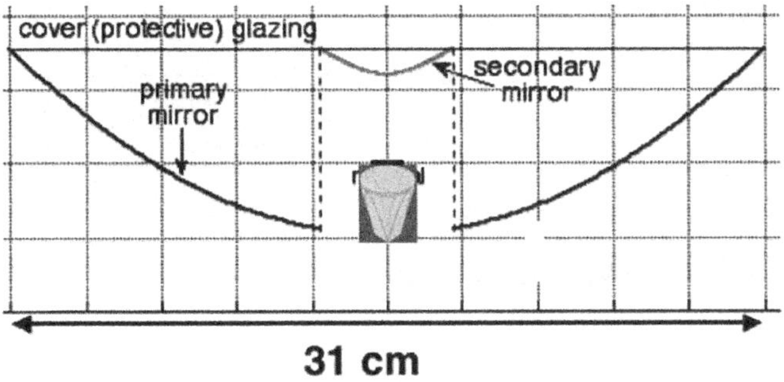

(a)

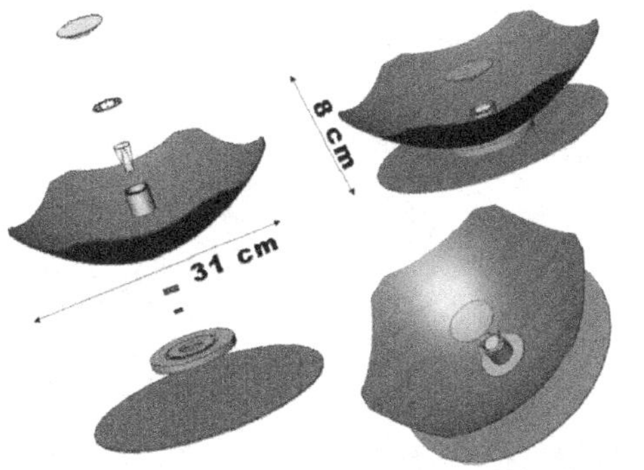

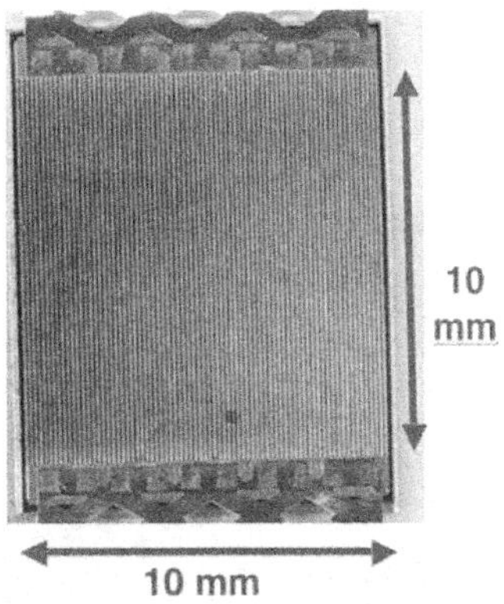

(b) (c)

Figure 5.6. (a) Cross-section and raytrace of the glass-filled achromatic aplanat of Generation 2 H_2Go/SolFocus PV concentrators (blackened lines indicate the unmirrored glass surfaces). (b) Schematic view of the concentrator unit. (c) Photograph of the top view of a sample module. (d) Photograph of the 1.0-mm^2 triple-junction cell for which the concentrator is designed

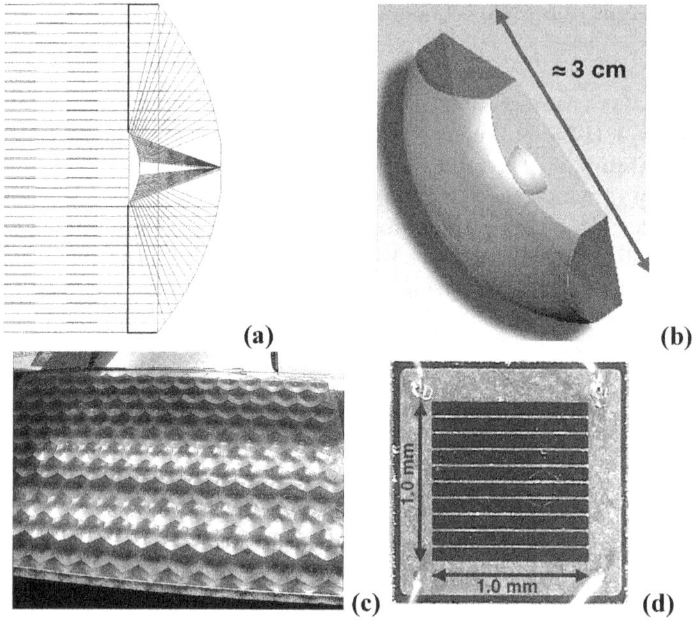

Figure 5.7. Prototype modules of Generation 1 (right, on a two-axis solar tracker) and Generation 2 (left)

Solar Surgery‡

The project involved using solar energy in place of laser surgery. We needed to design optics that would take radiation from one place to another with minimal loss of radiation. As part of our search for this newly developed theory we went to Soroka hospital to see the surgeon who uses lasers to operate on people. Jeff Gordon made the contact. The surgeon invited us to see what he was doing so we could judge whether there was anything we could do to help improve the efficiency of the system. At that time we were not thinking of solar applications. We attended two operations in the operating theater. We discussed with the surgeon his needs and how his equipment could be improved. He explained all the benefits of laser surgery: less bleeding, less infection, and reduced recovery time. There are many benefits when lasers are used for surgery as opposed to using a knife. You can cut people with a knife. You can take out tumors. With lasers you can get into the body through fibers. It is minimally invasive, because you just need to cut a small hole. Of course you have to have some imaging system to locate and observe the surgery. Then you can kill tumor cells by heating them up. Then they die off.

The main reason this is not being done very frequently in Israeli hospitals is because there are only a limited number of lasers. They are so incredibly expensive that hospitals avoid buying these instruments. They cost about $100,000 to $150,000, including special infrastructure. You also need a technician. The lasers pull a lot of power. The regular 220-volt outlet is not sufficient. You need a cooling system. So the main reason for not using lasers is because they are too expensive.

That was the initial visit. Then, coming back home, Gordon noted that we had been working with solar energy at high concentrations. The power level that the surgeon needs is several watts coupled into a small diameter fiber. This could be done with the sun. But maybe the sun is impractical, because there are clouds during the day, and there is night. Why not use a light source. So we investigated the properties of existing light sources. At the time, the brightness or power density that you can achieve was not sufficient for the power level required for laser surgery. In principle, lamps can go to a very high power level but not to sufficiently high power *density*. The power density of the light sources was an order of magnitude below what the surgeons wanted, at least at that time. But the sun could do ten times better. That was about 1997. Light sources were

‡ Based on an interview with Daniel Feuermann, a mechanical engineer and a member of a small team of scientists who develop optics for special applications (see: Feuermann, D., 2007).

not strong enough. Not in the sense of enough watts but they did not have the intensity, the power density at the surface of the light. So this brought us to the idea of using the sun. A quick calculation showed that we could do it very cheaply. We could concentrate sunlight into optical fibers. Then the question was, is it worth it, given that there are clouds and nighttime? When this was discussed with the surgeons in the hospital, they said that if we could produce a much cheaper replacement for lasers, it would be worthwhile. If a solar system would only operate 180 days a year, since this is not emergency surgery, but rather scheduled surgery, then they could schedule surgery on days when the sun is shining. They would have a window of about 6 to 8 hours a day when they could perform solar surgery. They could tell the patients they would be called in for surgery the next day it is sunny. The nice thing about this is that this type of surgery is like an outpatient procedure. They can send the patients home the same day. They don't have to keep them in the hospital. However, when they cut the patients open, they must keep them in the hospital for several days. This is how the story developed.

We first did a theoretical paper on the subject and published it. CNN and *Nature* did a report on our publication. Once it appeared in the high-exposure media others thought it was a great idea. Someone representing a fund said that they would provide the money and suggested that we should produce a prototype. We did. That is how we became involved in doing solar surgery on rats. The rats were sacrificed at the end of each procedure to investigate the tissue to determine if the procedure was effective. This involved pathology. There had been surgery on the livers of the animals. The animals lived after the operation, but only for a few days before it was necessary to see what had happened.

Our current project is on a side issue. When we began, electric lamps were not strong enough. However, with the development of the PC projector, there was an incentive for manufacturers to produce smaller and smaller and brighter and brighter lamps. The smaller and more compact the system, the more precise the picture. Ideally, you would want to have a point source. You want to make it as small as possible and still have high power. In the last few years, lamps have been developed that approach the power density we need to do the surgery. So now we have another project. We are now developing the optics to use light from such a light source and concentrate it into a fiber and then all this should fit into a little box that we can give to a surgeon. There will be a little fan in there for cooling (like a PC projector). This will have only a low cost, about the same as a PC projector, about $2,000. All the elements are in a PC projector. But now the light would be focused onto a fiber tip. So, although the solar surgery got so much attention, it will be a passing episode.

Suppose the electricity went out? When we were experimenting with solar, doing surgery on chicken livers, the electricity did go out on occasion. However, we had photovoltaic cells supplying power for the machinery that tracked the sun with the solar dish (collector), so the operation was not interrupted. With the hospital it is a different story. There are other things going on at the same time, such as monitors on the patient, that depend upon electricity, and for that reason the hospital has large emergency generators. So electricity outage should not be a problem.

There were all kinds of side issues. We realized that there was delayed tissue death. Under high radiation a certain fraction of the tissue would die immediately. But if you let the animal live for a few days, you find that the volume of killed tissue becomes larger. This was something that not many people had observed. This has given rise to another publication that had nothing to do with solar energy.

Also we needed to decide which method to use to determine which tissue is truly dead. The usual way is for the pathologist to use stains. This does not detect true tissue death. There was a discrepancy with what we actually saw because there is a change in the color of the tissue. It becomes whitish. However, under the microscope, from the point of view of pathology, the tissue did not look dead. That had to do with the wrong method for checking the tissue.

In the beginning, there was a tense situation because the pathologist said she knew what she was doing. But we said that cannot be so. It is as if we measured something with a ruler and we say it is ten millimeters and you say it is only six. If the tissue is changing color, something must be happening. She then found another method that was more suitable and which confirmed our observation.

The same people are still interested in the lamp surgery development. However, first we have to develop the lamp and the optics and have that measured to make sure the technology works. Once that is done, we will build a prototype as we did for solar surgery. Then the surgeons will come in and do experiments on tissue and on animals.

Selection of fibers? Before the first project we had not worked with fibers. If you just look as the manufacturer's data, they are not quite what you get.

On being creative? The basic framework to be creative and come up with ideas is pressure. We have a problem and we cannot go on without finding a solution. Or we are trying to do something and it doesn't work. For example, we try to put light through a fiber and we find that there is much less coming out than we expected. So what is going on? You dig into that and you find new things that occur. One nice example is what happens when measuring photovoltaic cells at extremely high flux; at flux levels for which there are no previous measurements. We find that there is absolutely something strange going on. We see when we take measurements that the curves look very wrong. It is not due to something wrong with our experiment; there is just something strange. We have our technician take the whole thing apart and put it together again and it still does the same thing. There were four people involved. We were all wondering what was going on to account for this strange behavior. I was already on my way to America so I didn't even have time to join in the discussion. Gordon was about to leave for sabbatical. Eugene Katz insisted that this was not an experimental error. Two days later when Gordon was already in the States he sent an e-mail and said, "I've got it!" Gordon remembered something he had read several years ago and said that this must be the reason. Sure enough, that was the right guess. That was indeed the problem. His idea was readily accepted. Gordon understood solid-state physics. There are papers that describe this particular part of physics that interfered with our experiment, but not in conjunction with photovoltaic cells. No one had ever tested photovoltaic cells at 10,000 suns (10,000 times the usual intensity of the sun at midday). It is not a new idea but a new piece of physics. A paper about this has been accepted for publication with a remark: "excellent science." In the literature we found reports of measurements on photovoltaic cells that show there is a contradiction in what the authors say. They did not understand the data and they described it in a strange way. They did not understand what was going on. Now this discovery can explain what they saw.

Out of 100 ideas that come up, 99 are junked. The criteria for making a decision are very simple. Either the physics does not allow it or it is an idea but it is not practical. Or it is an idea but the economics are going to kill it. We are not working in basic physics but applied research. So these are the questions: Can it be done? and, Is it worth it?

Do you recall a time when you contributed an idea to this project? Sometimes you are having a discussion and a few weeks later people do not remember who said what. Some people think they said it and others think they said it. I have my own version of what I contributed. For example, when we discussed using a fiber—this is more of a technical nature, I am an engineer—we all knew what to do: you take a parabola, you get a focus and then you have the fiber placed at the point of the focus. You have to hold the fiber very accurately in place, so I added a little mirror. Now, the fiber could come from below—from the back. With the mirror you loose about 4 or 5%. The extra loss is compensated by the ease in cleaning the glazing without having a fiber sticking out.

We discussed the possibility of using this parabola with fiber for power production. There could be panels with lots and lots of little (parabolic) dishes in them. Then fibers would come out from each individual dish. If you bundle these fibers together, you have a string of fibers. Again, the problem was what does one do with the fibers that are sticking out. Now you can have a flat piece of glass without all the fibers sticking out.

References

Feuermann, D. (2007) Solar surgery. In A. P. Hare & G. M. Kressel (Eds.) (2008). *The desert experience in Israel*. Lanham, MD: University Press of America.

Gordon, J. (2006) How innovative optics are serving the photovoltaic revolution. *SPIE Newsroom*. 10.1117/2.1200602.0132.

Gordon, J. M., & Feuermann, D. (2005) Optical performance at the thermodynamic limit with tailored imaging designs. *Applied Optics*, 44:2327–2331.

Gordon, J. M., & Kashin, P. (1994) Achieving uniform efficient illumination with multiple asymmetric compound parabolic luminaries. *Optical Engineering*, 33:267–272.

Gordon, J. M., Kashin, P., & Rabl, A. (1992) Nonimaging reflectors for efficient uniform illumination. *Applied Optics*, 31:6027–6035.

Gordon, J. M., & Ng, K. C. (2004) Compact nonimaging infrared systems for decondensing 2000 m^2 of spectator glazing at Singapore Kranji racetrack. Nonimaging Optics: Maximum Efficiency Light Transfer VII. *Proceedings of SPIE*, 5185:109–117.

Indeco Consultants (2000) *Star track: Designing and building the Singapore racecourse*. Singapore: Epigram Books.
Welford, W. T., & Winston, R. (1989) *High collection nonimaging optics*. San Diego: Academic Press.
Winston, R., & Gordon, J. M. (2005) Planar concentrators near the étendue limit. *Optics Letters*, 30:2617–2619.

Chapter 6

Remote Sensing

Arnon Karnieli

This chapter contains three articles published previously by Arnon Karnieli with others, which have been adapted for republication in this book. The first two relate to Mongolia, the third to Kazakhstan. Each article demonstrates the importance of remote sensing and satellite imagery to assessing the status of dryland vegetation in these countries.

Do Vegetation Indices Reliably Assess Vegetation Degradation? *[1]

Introduction

Grazing domestic animals on native vegetation in rangelands is a major form of land-use practiced all over the world. According to the FAO, permanent grasslands extend over about 26% of the land surface of the Earth. Therefore, range management and monitoring, especially over vast and remote areas, based on traditional field survey and measurement, might be problematic, since they

* Presented by A. Karnieli, Y. Bayarjargal and M. Bayasgalan at the International Conference on Remote Sensing and Geoinformation Processing in the Assessment and Monitoring of Land Degradation and Desertification, September 7–9, 2005, Trier Germany.

are expensive, manpower demanding and time-consuming. Alternatively, satellite remote sensing has been intensively used for a large number of vegetation applications in range regions, due to the large surface area covered and the frequency of observation. Recently, the Enhanced Vegetation Index (EVI)—with improved sensitivity in high biomass regions and improved vegetation monitoring, while correcting for canopy background signals and reducing atmospheric influences—was developed to optimize the vegetation signal (Huete, 1988; Liu and Huete, 1995). The EVI is based on the normalized difference vegetation index (NDVI), the soil-adjusted vegetation index (SAVI) and the atmospherically resistant vegetation index (ARVI); it uses functionalities from each of these to overcome soil and atmospheric interferences. The EVI is formulated as:

$$\mathrm{EVI} = \mathrm{G}\frac{\rho_{NIR} - \rho_{red}}{\rho_{NIR} + \mathrm{C1}\rho_{red}\mathrm{C2}\rho_{blue} + \mathrm{L}} \quad (1)$$

where ρ are atmospherically-corrected or partially corrected (Rayleigh and ozone-absorption) surface reflectances in the respective spectral bands, L is the canopy background adjustment term, and C1and C2 are the coefficients of the aerosol resistance term, which uses the blue band to correct for aerosol influences in the red band. The coefficients adopted in the EVI algorithm are: L = 1, C1 = 6, C2 = 7.5 and G (gain factor) = 2.5.

Overgrazing is considered to be the key cause of rangeland degradation (Thomas and Middleton, 1994) while rangeland degradation is almost entirely a matter of vegetation degradation (Dregne and Chou, 1992). The latter is directly related to reduction in biomass and/or decrease in species diversity (Eswaran, Beinroth and Virmani, 2000). However these tendencies can be more complicated, since vegetation degradation might be measured qualitatively rather than quantitatively. For instance, invasion by or increase in undesirable brush species may actually increase biomass production on degraded rangelands, resulting in loss of palatable pasture grasses as they are replaced by unpalatable species (Dregne and Chou, 1992; Brown and Archer, 1999). From the remote-sensing point of view, implementing the above-mentioned vegetation applications that are related to quantitative variables is a common task and widely used in rangelands. Bastin, Pickup and Pearce (1995) examined the potential of spaceborne systems for rangeland-degradation mapping around Australian watering points and noted that distinguishing among different plants or changes in species composition is not possible. However, taking advantage of hyperspectral image spectroscopy technology, several studies in recent years have aimed to map the distribution of some biological invaders (Lass et al., 2002; Underwood, Ustin and DiPietro, 2003; Lass et al., 2005) and evaluate changes in canopy chemistry and other canopy characteristics caused by invasion (Asner and Vitousek, 2005).

In Mongolia, from historic times, animal husbandry has been the mainstay of the economy; 99% of Mongolian territory has been used as natural pastures. During the last 70 years, the population density in Mongolian drylands has more than tripled and total domestic livestock herds (sheep, goats, cattle, horses and camels) have increased over 2.3 times in size, reaching 30 million animals. Consequently, irrational anthropogenic activities, such as overgrazing, have accelerated, causing vegetation degradation to become the main type of rangeland degradation (Adyasuren, 1998; Batjargal, 1999). A few case studies—at plot-scale level—drew attention to the severe decrease in vegetation cover caused by overgrazing near settlement water sources. Yonghong and Jargalsaishan (1993) noted that plant community abundance (composition and richness) decreased as grazing pressure increased and that native vegetation was replaced by exotic species in the northeast pastureland of the country. They found that succession series along the grazing gradient were *Stipa grandis* + *Leymus chinensis* in the lightly grazed sites, *Stipa kreylovii* + *Artemisia frigida* + low grasses in the moderately grazed sites, and *Carex duriuscula* + *Artemisia scorparia* + annuals in the heavily grazed sites. Based on ground observations over two years, Fernandez-Gimenez and Allen-Diaz (1999) and Fernandez-Gimenez (2000) found that vegetation patterns (e.g., species composition, biomass) changed along grazing-gradients from the watering points in response to increased grazing pressure in the Mountain-Steppe and Steppe zones of Mongolia, while no consistent changes due to grazing were observed in the Desert-Steppe. Also, it was noted that vegetation changes over degraded and eroded areas were significant and that unpalatable plants or weeds fully occupied these areas. However, no shrub encroachment was associated with degradation of Mongolian grassland (Fernandez-Gimenez and Allen-Diaz, 1999; Fernandez-Gimenez, 2000; Tserenbaljid, 2002).

Advantage was taken of a unique phenomenon in Mongolia. A railway, constructed in the 1960s and >1000 km in length, crosses the country from the northern border with Russia to the southern border with China (Figure 6.1). It has been protected since then by fences along its entire length to prevent animals from crossing the railway. Thus, no grazing has been allowed inside the fences, while intensive grazing characterizes the surrounding area. Where the railway passes through the steppe biome, it facilitates investigation of anthropogenic-induced rangeland degradation. The distance between fences can be as wide as 4 km where the train tracks form winding curves, thus enabling remote sensing research to be carried out with high resolution imagery (Figure 6.2). This study attempted to explore the ability of remote sensing techniques to assess vegetation degradation in rangeland regions of Mongolia.

Study Area

Mongolia has a continental climate, characterized by cold and dry winters and warm and wet (rainy) summers. The current research is focused on the Mongolian steppe biome (excluding the desert steppe) (Figure 6.1). The area occupies about 4.5×10^8 ha. Mean annual precipitation ranges between 150 and 300 mm and the mean annual temperature ranges between -3 and +3 C°. The aridity index (ratio of precipitation to potential evapotranspiration) ranges between 0.2 and 0.5, indicating a semi-arid environment. The southern part of the region is characterized by flat plains and rolling hills covered in feather grass and shrubs. Typical species of grass include *Stipa krylovii* and *Agropyron cristatum*; unpalatable shrubs such as *Caragana* spp. and *Artemisia* spp. are abundant. Mountains in the north are characterized by coniferous forests on the northern slopes,

Figure 6.1. Geo-botanical map of Mongolia, showing the railway that crosses the country from the northern border with Russia to the southern border with China, along which the study sites were located

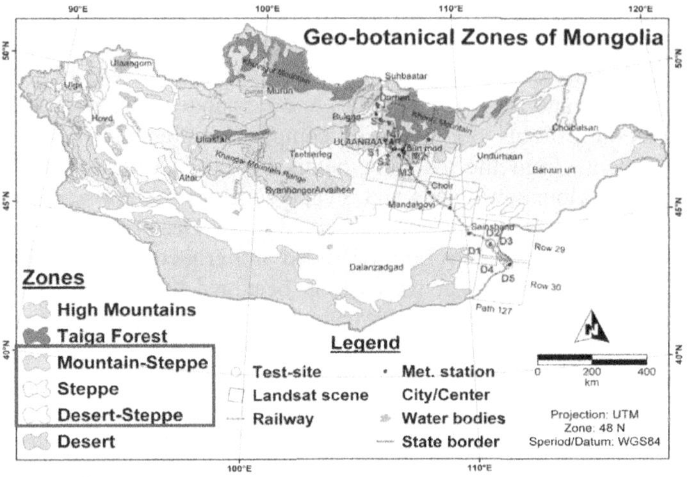

Figure 6.2. Example of a Landsat ETM+ image (RGB=4,3,2) showing a study site (M1). The area between the fence and the railway is the ungrazed side while intensive grazing characterized the surrounding area

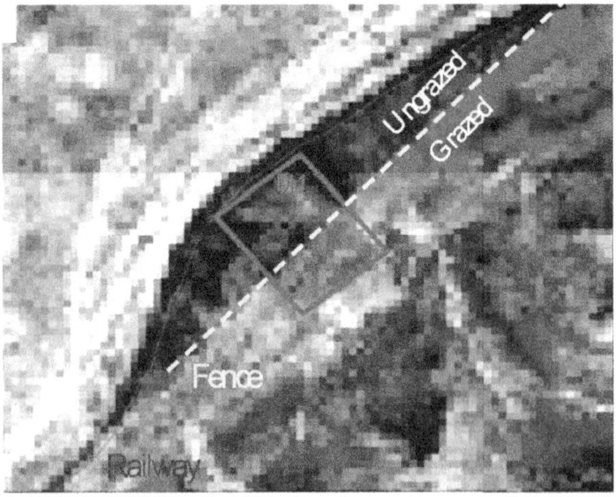

while the southern slopes are covered by open steppe vegetation. The vegetation is therefore a combination of Siberian taiga forest and Mongolian steppe flora, including such species as pine (*Pinus sylvestris*), aspen (*Populus tremula*) and edelweiss (*Leontopodium ochroleucum*).

Methodology

The research was conducted at six study-sites selected along the railway (Figure 6.1). Three of the sites were located in the Mountain-Steppe zone (sites M1, M2 and M3) and three were in the Steppe zone (S1, S2 and S3). Each site consisted of pairs of study polygons—ungrazed (inside the fenced-off area) and heavily grazed (outside the fence). All the sites were large enough for spatial resolution on Landsat images; e.g., 30 × 30 m, and all were characterized by flat topography.

Four Landsat-7 Enhanced Thematic Mapper Plus (ETM+) images, acquired in the early 2000s, were used. To reduce image-to-image variations related to sun angle, differences in atmospheric condition and vegetation phenology, all images were selected during the vegetation growing season, when the opportunity to discriminate between vegetation and soil cover was optimal. In addition, cloud cover was minimal in all images. Digital number values were converted to radiance and ground-leaving reflectances were created from the radiances with the 6S algorithm. Later, the four images were merged to create a continuous scene. The EVI (Equation 1) was computed from the reflectance values. This index was selected to reduce uncertainty in soil background and atmospheric effect along the entire study area. Approximately the same number of pixels was sampled in the grazed and the adjacent ungrazed polygons.

On-site verification activities were conducted during two field-campaigns in the summers of 2002 and 2003 within the framework of the Joint Russian-Mongolian Complex Biological Expedition, with the participation of the University of Moscow, the Remote Sensing Laboratory at the Blaustein Institute for Desert Research (Ben-Gurion University of the Negev), the Institute of Botany (the Academy of Science of Mongolia) and the National Remote Sensing Center (Ministry for Nature and Environment of Mongolia). Biophysical variables among dominant and co-dominant plant species were sampled and measured, including plant density (number of plants per unit area), composition, dry biomass, and percent of vegetation cover. The study polygons were precisely located by a Global Positioning System (GPS).

In conjunction with the above, spectral reflectance measurements were implemented with the FieldSpec-HandHeld Spectroradiometer, manufactured by Analytical Spectral Device (ASD, 2000), at wavelengths of 325–1075 nm, with a spectral resolution of 2 nm. A High Intensity Contact Probe device with a fiber optic cable was attached to the spectroradiometer. This device has an independent light source (about 2 × solar intensity) that makes it feasible to take measurements under all weather conditions. The contact probe was attached to clipped plants and soil samples. Measurements of a white reference panel

(Spectralon plate, Labsphere Inc.) were taken immediately before each spectral measurement.

Results and Discussion

It was hypothesized that intensive grazing would reduce plant density, biomass and cover. It was further assumed that the EVI would have lower values outside the fences due to vegetation degradation. As expected, between 23 and 34 plants per unit area were recorded inside the fence while 12 to 25 plants per unit were recorded outside. Also as expected, higher plant density characterized each ungrazed polygon, relative to its adjacent grazed polygon. Similarly, the average inside-the-fence dry biomass was twice that of the outside-the-fence dry biomass. This ratio is higher in the Mountain Steppe polygons than in the Steppe polygons. The same trend can also be observed with plant cover. Greater cover was observed in ungrazed polygons than in ungrazed polygons. However, a revised trend was revealed by analysis of the spaceborne data. Landsat-ETM+-derived EVI showed significantly higher values outside the enclosures than inside. This phenomenon occurred at each of the study sites. These unlikely results, i.e., negative correlation between the biophysical variables and the vegetation index, should lead to further discussion about the composition, phenology and palatability characteristics of these plants.

Mountain-Steppe Zone

Different perennial grasses were dominant in the ungrazed areas while forbs dominated the grazed areas with little contribution from grasses. The species composition in ungrazed areas consisted of *Stipa krylovii*, *Halerpestes salsuginosa*, *Leymus chinensis*, *Agropyron cristatum*, *Poa attenuata*, *Galium verum* and *Agrostis mongolica*. These perennial native grasses have good palatability values for animals during the summer, especially *S. krylovii* and *A. cristatum*, which are highly nutritious and very digestible plants for all livestock throughout the year (Jigjidsuren and Johnson, 2003). They bloom in early August and their seeds mature in September. Also, communities such as *P. attenuata*, *L. chinensis* and *A. mongolica* are highly palatable for all livestock, especially for small animals (i.e., sheep and goats) throughout the entire season; in addition the other native *Poaceae* grasses, these are highly-valued and the dominant plants of Mongolia's pastureland (Tserenbaljid, 2002). During the blooming period in August, the main perennial grasses (e.g., *S. krylovii*, *L. chinensis* etc.) have bright-gray and brown-gray flowers with very straight spikes that are 30–70 cm tall and 1–1.5 cm wide at the top.

Since these needle grasses grow relatively uniformly and cover 20–30% of the fenced-off areas in each study-site, the surface looks relatively brighter to the human eye (Figure 6.3A). In the false color composite of the Landsat image the ungrazed area looks dark and no indication of photosynthetic activity is observed (Figure 6.2). In the grazed areas, *Artemisia frigida, Artemisia dracunculus, Potentilla acaulis, Glaux maritime, Bupleurum scorzonerifolium, Allium bidentatum, Agrostis mongolica, Caragana pygmaea, Leymus chinensis, Koeleria cristata, Gallium verum, Potentilla acaulis* and *Iris bungei* are dominant. Livestock, especially sheep, can hardly graze these perennial forbs and sub-shrubs in summer while goats moderately grazed them in autumn. Because of their indigestibility value during the mid summer, the unpalatable perennial forbs (e.g., *A. frigida* and *A. dracunculus*), the weedy annuals (e.g., *P. acaulis*) and the sub-shrubs (e.g., *C. pygmaea*) in areas dominated by these plants appear relatively green (Figure 6.3A). Nevertheless, these dense bunch-forming semi-shrubs are very nutritious for livestock in early summer and late autumn when toxic values may be low (Mandakh Bayart, personal communication). Gunin, et al. (1999) noted that several species, such as *Artemisia* (*A. scoparia, A. frigida*) and *Echinopsilon divaricatum* are indicators of rangeland degradation and of human-induced desertification processes.

Figure 6.3. General view of the research sites: (A) Mountain Steppe site (M1): association in the ungrazed area – *Halerpestes salsuginosa* + *Agrostis mongolica*; grazed area – *Glaux maritima* + *Agrostis mongolica.* Note the darker tones in the grazed area are due to the wide spread of *Iris bungei.* (B) Steppe site (S1): association in the ungrazed area – *Stipa krylovii* + *Buplerum scorzonerifolium* + *Cleistogenes squarrosa*; grazed area – *Carex duriuscula* + *Artemisia adamsii.* Note the brighter tones in the fenced-off area due to the *S. krylovii*

Steppe Zone

Among the three study-sites in the Steppe zone, communities of *Stipa krylovii, Allium senescens, Agropiron cristatum, Festuca sibirica, Stipa grandis, Cleistogenes squarrosa* and *Buplerum scorzonerifolium* were dominant in the ungrazed areas. All these perennial grasses have very high nutritional values, so they are invaluable forage plants (Jigjidsuren and Johnson, 2003). In contrast, *Carex duriuscula, Artemisia adamsii, Artemisia frigida* and *Potentilla acaulis* were dominant in the grazed areas. As noted by Fernandez-Gimenez and Allen-Diaz (1999) and by Fernandez-Gimenez (2000), these species dominate in response to different levels of degradation in the Steppe zone of Mongolia. Visually, ungrazed areas in the Steppe appear to be brighter than the grazed areas due to the prevalence of *S. krylovii* (Figure 6.3B).

Based on the abovementioned hypotheses, a one-tail t-test was performed for each of the variables. A significant difference can be seen between the grazed and ungrazed data series, and, indeed, grazing had the expected significant effect on the biophysical variables, but not on the EVI.

Most of the dominant species in the fenced-off area were good palatable plants, i.e., *S. krylovii*, *B. scorzonerifolium*, *A. senescens* and *A. mongolica*; these have lower reflectance levels in the NIR range of the electromagnetic spectrum. Contrary to this, unpalatable species such as *G. maritime*, *L. chinensis*, *P. acaulis* and *I. bungei*, which occupied the grazed areas, have higher reflectance levels in the NIR range. Figure 6.4 demonstrates these differences with two representative plants. *S. krylovii* is a good, palatable grass, representative of the protected area. By mid-summer, this grass has turned yellow, its cells have lost their water and its refractive index has decreased; hence, its reflectance in the NIR range has decreased. *I. bungei* is representative of the grazed area. This plant is a succulent; therefore, it is characterized by a high refractive index that produces high reflectance values.

Figure 6.4. Dominant species in the Mountain Steppe zone. (A) *Iris bungei*, representative of grazed areas, is a succulent plant characterized by a high refractive index that produces high NIR-reflectance values; (B) *Stipa krylovii*, representative of protected areas; is a good, palatable grass. During mid-summer it turns yellow, its cells lose water and its refractive index decreases; hence its NIR-reflectance decreases

Conclusions

On-site observations along the Mongolian railway confirmed a previous range-condition model of vegetation dynamics (e.g. Dyksterhuis, 1949). The model predicted that as herbivore numbers increased, plant biomass and cover would decline and species composition would shift from dominance by perennial grasses and forbs ('climax' species) towards dominance by unpalatable forbs and weedy annuals. When grazing is decreased or removed, biomass and cover are predicted to increase and species composition should shift back towards late-successional stages. Although the common remote-sensing-based vegetation-indices models assumed higher index values as biomass and cover increased, the current observations showed the opposite. The reason is the difference in leaf structure and phenological stage between the palatable species inside the fenced-off area and the unpalatable species outside the fences. The palatable species were mainly grasses that turned yellow during mid-summer, while the unpalatable species were succulent plants characterized by a high refractive index that produced high reflectance values.

Comments on the use of the Vegetation Health Index over Mongolia[†2]

Introduction

The Vegetation Health index (VHI) is based on a combination of products extracted from vegetation signals, namely the Normalized Difference Vegetation Index (NDVI) and from the brightness temperatures, both derived from the NOAA Advanced Very High Resolution Radiometer (AVHRR) sensor. VH users rely on a strong inverse correlation between NDVI and land surface temperature, since increasing land temperatures are assumed to act negatively on vegetation vigor and consequently to cause stress. This article explores this hypothesis with data from Mongolia incorporating information from six different ecosystems.

---

† From: "Comments on the use of the Vegetation Health Index over Mongolia," by A. Karnieli, M. Bayasgalan, Y. Bayarjargal, N. Agam, S. Khudulmur, and C. J. Tucker (2006). International Journal of Remote Sensing. 27(10): 2017–2024. DOI: 10.1080/01431160500121727

Evolution of the Vegetation Health index

Spaceborne data have been widely used for estimating herbaceous biomass accumulation in grasslands and steppes. The first satellite application for assessing biomass was in northern Senegal in 1981–1983 (Tucker et al., 1983; Tucker et al., 1985). Subsequently, other investigators have expanded this work throughout the Sahelian zone and elsewhere, and reported similar results. Among the various available sensors, the Advanced Very High Resolution Radiometer (AVHRR), onboard National Oceanic and Atmospheric Administration (NOAA) satellites, has been the most suitable and has been used for this purpose. This instrument provides two major land products: the Normalized Difference Vegetation Index (NDVI) and the Land Surface Temperature (LST). The NDVI is the most frequently used vegetation index, and is based on the ratio between the maximum absorption of radiation in the red (R) spectral band and the maximum reflection of radiation in the near-infrared (NIR) spectral band (Tucker, 1979). Since the Earth's surface temperature influences vegetation growth (Running et al., 1995; White, Thornton and Running, 1997; Tucker et al., 2001; Badeck et al., 2004), LST values have been used as criteria, in addition to the NDVI, for evaluating the status and development of vegetation.

Using these AVHRR-derived products, various researchers have developed algorithms for time-series analysis, or for relating a specific period of interest to a long-term statistic, e.g. the NDVI Anomaly Index (Liu and Negron-Juarez, 2001) and the Standardized Vegetation Index (Anyamba, Tucker and Eastman, 2001; Peters et al., 2002). Following this approach, Kogan (1995) has suggested the Vegetation Condition Index (VCI):

$$\mathrm{VCI} = \frac{(\mathrm{NDVI} - \mathrm{NDVI}_{min})}{(\mathrm{NDVI}_{max} - \mathrm{NDVI}_{min})} \tag{2}$$

This equation relates the NDVI of the composite period of interest (which can be a week, dekad, month, or year) to the long-term minimum NDVI (NDVI_{min}), normalized by the range of NDVI values calculated from the long-term record of the same composite period. The VCI values range from 0 to 1, the low values representing stressed vegetation conditions, middle values representing fair conditions and high values representing optimal or above-normal conditions.

On the presumption that the LST provides additional information about vegetation condition, Kogan (1995) adapted the VCI normalization approach to LST and developed the Temperature Condition Index (TCI):

$$\mathrm{TCI} = \frac{(\mathrm{BT}_{max} - \mathrm{BT})}{\mathrm{BT}_{max} - \mathrm{BT}_{min}} \tag{3}$$

where BT represents the brightness temperature derived from the AVHRR band 4. Note that, in order to apply the TCI for determining temperature-related vegetation stress, it was formulated in reverse ratio to the VCI, based on the hypothesis that the higher the temperature, the worse the conditions for vegetation. Consequently, low TCI values (close to 0) indicate harsh weather conditions (due to high temperatures), relative to the composite period, middle values reflect fair conditions and high values (close to 1) reflect mostly favorable conditions.

Several authors have used the combined responses of reflected (e.g. NDVI, VCI) and thermal (e.g. LST, brightness temperature) products of the NOAA-AVHRR to provide a more ecological and physical interpretation of remotely sensed data for examining vegetation conditions (e.g. Gutman, 1990; McVicar and Jupp, 1998; Karnieli and Dall'Olmo, 2003). This innovative approach assumes a strongly negative correlation between NDVI and LST, due to an increase in evaporation, along with a decrease in soil moisture, caused by higher temperatures, resulting in a decline in the vegetation cover (Nemani and Running, 1989; Lambin and Ehrlich, 1996). For example, McVicar and Bierwirth (2001) use the ratio of LST and NDVI (LST/NDVI) to provide a rapid means of assessing drought conditions.

Following the above-mentioned hypothesis, Kogan (1995) proposed another index, the Vegetation Health Index (VHI), which is an additive combination of VCI and TCI:

$$\mathrm{VHI} = \alpha \mathrm{VCI} + (1-\alpha)\mathrm{TCI} \tag{4}$$

where α is the relative contribution of VCI and TCI in the VHI. In most published analyses, α has been assigned a value of 0.5, assuming an even contribution from both elements in the combined index, due to the lack of more accurate information (Kogan, 2000). The VHI has been applied for different applications, such as drought detection, drought severity and duration, early drought warning (Seiler, Kogan and Sullivan, 1998), crop yield and production during the growing season (Unganai and Kogan, 1998), vegetation density and biomass estimation (Gitelson et al., 1998), assessment of irrigated areas (Boken et al., 2004) and estimation of excessive wetness (in contrast to drought) (Unganai and Kogan, 1998). These applications have been demonstrated in various scales: global (Kogan, 1997, 2000), regional (Liu and Kogan, 1996) and national (Seiler, Kogan and Sullivan, 1998)—in many parts of the world.

Recently, Kogan et al. (2004) dealt with applying the VHI for drought detection and derivation of pastoral biomass in Mongolia. The VCI, TCI and VHI were computed from the long-term NOAA Global Vegetation Index (GVI) dataset for the period 1985–2000, in 16 km × 616 km resolution (Kidwell, 1997). Due to a lack of more accurate information on the influence of VCI and TCI on the VHI in Mongolia, the α-coefficient of the VHI equation was fixed at 0.5. Spatial results of the VH, for the three relevant years, from the Gobi desert in the south (41°N), to north of the Lake Baikal in Siberia (56°N), are presented (Kogan et al., 2004).

The prime objective of the current article is to investigate the VHI-based hypothesis that increasing temperatures act negatively on vegetation vigor and consequently cause stress. The territory of Mongolia (about 1.5 million km^2) can serve as a good example for such research, as this country is located in the cold desert belt of central-east Asia. Mongolia is characterized by mostly natural vegetation, without anthropogenic influences, such as urban heat islands, industry, agricultural crops, etc. The north–south transect across the country is relatively short (approximately 1000 km), but covers six different ecosystems, namely: Taiga, High Mountains, Forest Steppe, Steppe, Desert Steppe and Desert, from the north southwards (Figure 6.5). Mean annual temperature increases gradually from 27°C in the north to 7°C in the south, while mean annual precipitation ranges from more than 350 mm in the north to less than 75 mm in the south.

Dataset and methodology

The Pathfinder AVHRR Land (PAL) NDVI and brightness temperatures, in bands 4 and 5, were used in this study. Data are composed of monthly maximum values, with an 8-km spatial resolution, in geographical (lat/long) projection, spanning a period from 1981 to 1999. The PAL dataset was generated from the NOAA satellites 7, 9, 11 and 14 (Agbu and James, 1994) and was obtained from the Goddard Space Flight Center (GSFC) Distributed Active Archive Center (DAAC).

Figure 6.5. Ecosystem map of Mongolia

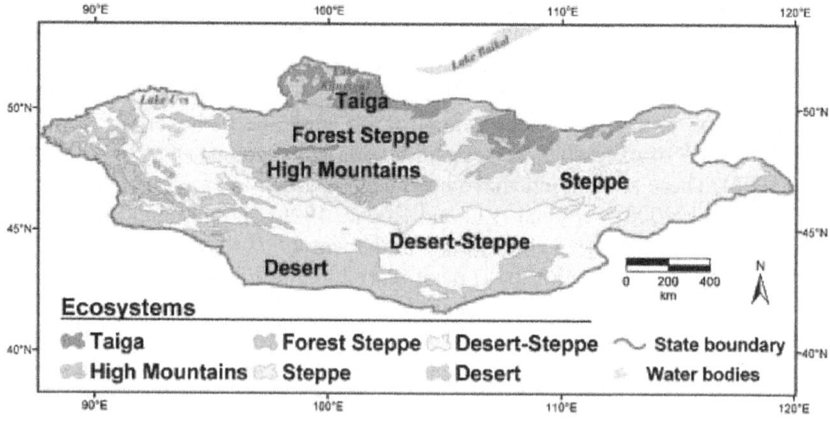

NDVI values were extracted directly from the PAL archive. LST values were computed from the brightness temperatures in the thermal bands, by a split-window algorithm (Price, 1984) of the form:

$$LST = BT_4 + A(BT_4 - BT_5) + B(\varepsilon) \tag{5}$$

where BT_4 and BT_5 are brightness temperatures in bands 4 and 5, respectively, A (52.63) is a coefficient related to atmospheric transmittance, being dependent on the atmosphere type, and $B(\varepsilon)$51.27 is the emissivity effect, which depends on both the channel surface emissivities (ε_4 and ε_5) and atmosphere type. Price (1984) assumed that the emissivity of most of the land surface and vegetation cover is equal to 0.96, so this value was used in the current research.

Analysis, results and discussion

Scatterplots of the NDVI vs. the LST values are presented in Figure 6.6. Linear regression analysis of the entire dataset reveals a significant ($F<0.001$) inverse relationship between NDVI and LST. This trend is well documented on regional and continental scales (e.g. Nemani et al., 1993). However, the regression results of the six separated distinct clusters, representing the six different ecosystems, reveal a different situation. Individual regression analysis results of these clusters reveal negative relationships between NDVI and LST for the southern ecosystems (Desert, Desert Steppe and Steppe); a flat relationship for the Forest Steppe ecosystem; and positive relationships for the northern ecosystems (High Mountains and Taiga). Note that the regressions of the Taiga, Steppe and Desert Steppe were found to be statistically significant ($F<0.05$). Gradual transition from the most negative (Desert), to the most positive (Taiga), relationships can be observed.

Figure 6.6. Scatterplot of LST against the NDVI. Note an overall significant negative relation between the two variables. When examining each ecosystem separately, the northern ecosystems are characterized by a positive trend

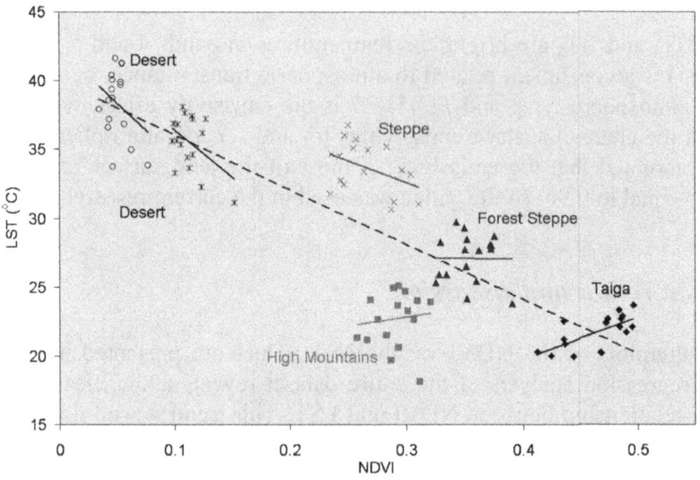

A time series of NDVI and LST values for the Desert Steppe ecosystem, as representative of out-of-phase relationships, is presented in Figure 6.7a. A mirror reflection of the two trends can be seen. By contrast, the two variables progress almost in-phase along the study period in the Taiga ecosystem (Figure 6.7b).

These results are consistent with previous observations showing a substantial change in the correlation slopes between NDVI and LST (e.g. Lambin and Ehrlich, 1996; Tateishi and Ebata, 2004). Low latitude regions of the Northern Hemisphere are characterized by negative correlations, as water is the main limiting factor for vegetation growth. On the other hand, mid and high latitude regions, where energy is the major limiting factor for vegetation development, are characterized by a positive correlation, implying that rising temperatures favorably influence vegetation activity. Temperature is the main driver of many biological processes, namely chemical (enzyme-catalyzed) reactions, which usually increase plant maturation (Badeck et al., 2004). Mongolia, located in the cold desert belt of central-east Asia, exhibits along a relatively short distance, both precipitation-dependent ecosystems in the south and temperature-dependent ecosystems in the north. Therefore, the VHI cannot be applied to the entire territory.

Figure 6.7. Time series of NDVI and LST values for the (*a*) Desert Steppe ecosystem and (*b*) Taiga ecosystem. Note out-of-phase and in-place relationships in (*a*) and (*b*), respectively

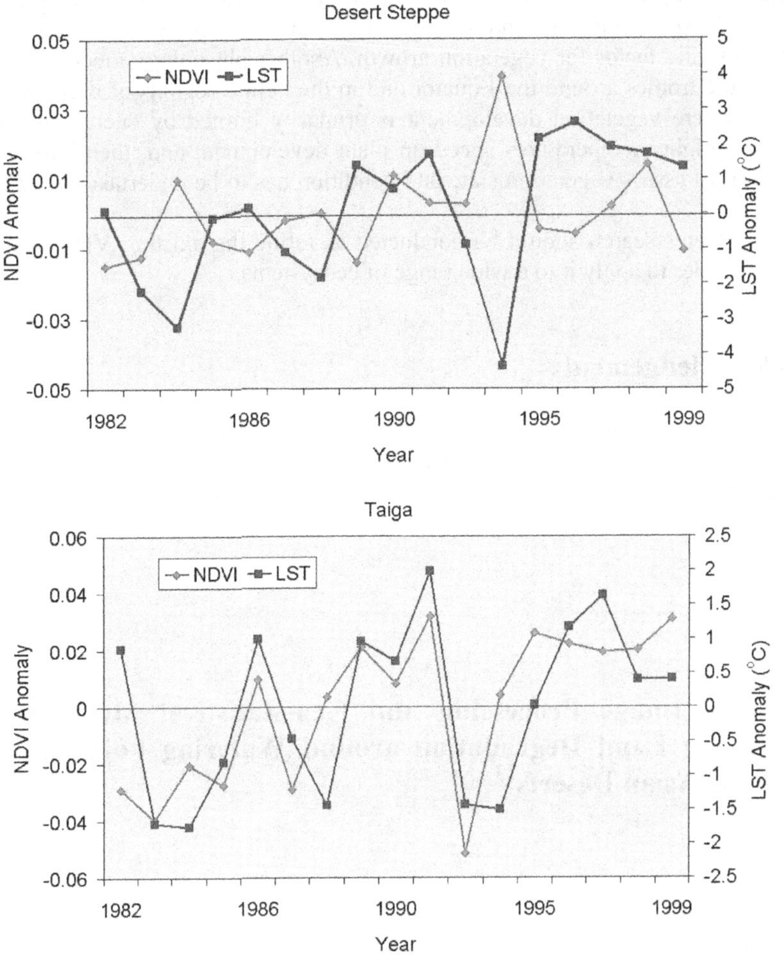

Conclusions

This study attempted to investigate the VHI-based hypothesis that increasing temperatures act negatively on vegetation vigor and consequently cause stress. It is shown that analysis of spaceborne-derived vegetation indices, such as NDVI/LST and VHI, which are based on the NDVI and LST, requires a good understanding of the relationships between these variables in different ecosystems. Since the role of the α-coefficient in the VHI is to determine the individual contributions of the NDVI and the LST to the vegetation condition, it is generally expected that a correlates to the slope of the regression of these two factors. As was demonstrated, however, not only does the magnitude of the slope vary across different ecosystems, its direction can be reversed. Consequently, in its present form, the VHI can be successfully applied only in the low latitudes, mainly in arid, semi-arid and sub-humid climatic regions, where water is the main limiting factor for vegetation growth. Another physiology mechanism exists in the tropics around the Equator and in the humid regions of the high latitudes, where vegetation development is primarily limited by energy. In these regions, higher temperatures speed up plant development and, therefore, using the VHI to assess vegetation state and condition has to be undertaken with caution.

Further research should be conducted to refine the existing VHI formulation, in order to apply it to a wide range of ecosystems.

Acknowledgements

The authors wish to express appreciation to the United States Agency for International Development for funding this project (Grant No. TAMOU-00-C20010).

Satellite Image Processing and Geo-statistical Methods for Assessing Land Degradation around Watering Points in the Central Asian Deserts[‡3]

Introduction

According to the United Nations Environment Program (UNEP) the term *overgrazing* refers to a practice of allowing a much larger number of animals to graze at a location than it can actually support. As a result, overgrazing by different types of livestock is perhaps the most significant anthropogenic activity that degrades rangelands and causes desertification in terms of plant density, plant chemical content, community structure, and soil erosion (Manzano and

‡ Presented by A. Karnieli, U. Gilad, M. Ponzet and T. Svoray at the 1st International Conference on Remote Sensing and Geoinformation Processing in the Assessment and Monitoring of Land Degradation and Desertification, Trier, Germany, 7–9 September, 2005.

Navar, 2000). In arid and semi-arid environments, land (soil and vegetation) degradation is particularly related to the area surrounding point-sources of water, either natural or artificial, i.e., wells or boreholes (Lange, 1969). Pickup and Chewings (1994) were the first to define the term '*grazing gradient*' as "spatial patterns in soil or vegetation characteristics resulting from grazing activities and which are symptomatic of land degradation."

Domestic animals (sheep, goats, cattle, camels, yaks and horses) prefer to graze in the vicinity of a watering point. When food is depleted in this area they move away from the source of water but return regularly for drinking. Consequently, a higher number of individuals is frequently concentrated around the watering points; density then decreases gradually with increasing distance from water (Pickup, Chewings and Nelson, 1993; Friedel, 1997). Livestock grazing-distance limits depends on the water demands of the different animals, the season, the weather conditions and the quality of the forage. Typical distances are 4–6 km, but they can increase to 10 or even 20 km under extreme conditions (Hodder and Low, 1978).

Since the radial pattern around watering points is well observed from space, most of the recent studies are based on the interpretation and modeling of remotely sensed data, which can be analyzed in a semi-automated and repeatable way over vast and remote areas. Various remote-sensing based models have been developed to estimate the spatial distribution of the different variables around the watering points; these are: the PD54 (Pickup, Chewing and Nelson, 1993), the Normalized Difference Vegetation Index (NDVI) calculated from the Advanced Very High Resolution Radiometer (AVHRR) (Hanan et al., 1991) or the probabilistic linear spectral mixture model, called AutoMCU, based on Monte Carlo analysis (Harris and Asner, 2003).

The above-described ground and spaceborne observations have demonstrated not only that the grazing gradients are characterized by concentric circles around the watering points, but also that the spatial nature of the measured biotic, abiotic and environmental variables are distributed in a common fashion. Each variable (e.g., track density; vegetation cover; grass and annual plant production; bush encouragement; soil pH; organic content; phosphate, nitrate and soil nutrient concentrations—particularly potassium and phosphorus) has a low (or high) value near the center that changes exponentially as the distance increases. Moreover, most of these observations show that the rate of improvement (or decline) of each variable does not change after several kilometers (usually five, as mentioned before) from the watering point.

This article presents another approach for assessing and mapping the grazing impacts around watering points. The primary goal of the study was to characterize spatial and temporal land degradation/rehabilitation in drylands of Central Asia, in term of vegetation and soil patterns, at different time periods, and with respect to the socio-economic changes before and after the collapse of the Soviet Union. More specific objectives of the study were: (1) to develop a geostatistical model, based on the kriging technique and using high-resolution satellite image processing to assess spatial and temporal land-cover patterns in three different key time periods (mid-late 1970s, late 1980s and 2000); (2) to conduct a change-detection analysis, based on geo-statistical products, in order to assess the direction and intensity of changes between the study periods; and (3) to link the previous findings to the socio-economic situations before and after the collapse of the Soviet Union that influenced the *grazing gradients* and hence the land-use/land-cover state of the study sites.

Methodology

Study site

The study site, about 160,600 km^2 in size, is located in the Ust-Urt desert plateau between the Caspian and Aral Seas in Central Asia and occupies the southern part of Kazakhstan, the northern part of the Karakalpak Republic, and Turkmenistan (Figure 6.8). It rises to between 150 and 300 m AMSL. Its semi-nomadic population raises sheep, goats and camels. It consists primarily of stony desert with grey-brown soils. The area has a semiarid continental climate with hot summers and cold windy winters. Vegetation consists of *Artemisia terrae-albae* and *Anabasis salsa*. The North Caspian basin is a still relatively unexplored petroleum-rich region with large oil and gas reserves.

Figure 6.8. Location of the study site in Ust-Urt plateau, Kazakhstan. The inset shows the area in Landsat-TM image (RGB=4,3,2)

Image processing

Three Landsat images of the study sites were used, acquired by different sensors (MSS, TM and ETM+) and three different time periods 1975, 1987 and 2000. From the original images, subsets of the study fields were made, bounding an area of 1,184 km^2.

Reflectances were used to calculate several vegetation indices for each image subset. Performance analysis—in terms of standard deviation and stretch of index values within the dynamic range—was applied to all indices. These analyses revealed that the Tasseled-Cap-derived Brightness Index (BI) produced the best contrast and, consequently, was selected for further analysis. Besides the statistically significant results, and that this index was originally designated to examine soil properties, its advantage is in its ability to compare among different sensors having different spectral bands, as it reduces their different spectral bands to one normalized layer of BI values. Also, it was found that the BI is uniquely independent of the spatial dimension, thus all image frequency histograms were spectrally matched.

The resolution of the images was reduced by a factor of 3 and 6, for the MSS and TM/ETM+ images, respectively. The resulting pixel size of 171 m is processed easily, as the sub-scene size was reduced from approximately 1.5 million pixels to only 40,000 pixels.

Geostatistical analysis

Geostatistical technique, namely *kriging*, was used to model and map the spatial variations of the soil brightness values in the study site. Spatially dependent variation may be treated statistically and described through a number of parameters derived from a *semi-variogram* that is the function relating the semi-variance to the directional distance between two samples. The semi-variance is defined as half the mean squared difference between two samples, a given direction and distance apart (Equation (6)). The direction and distance are defined by the vector h that is commonly referred as the *lag*.

$$\gamma(h) = \frac{1}{2N(h)} \sum_{i=1}^{N(h)} \left(z_{xi} - z_{xi+h}\right)^2 \qquad (6)$$

where $\gamma(h)$ is the semi-variance at lag h, $N(h)$ is the number of sample-pairs a distance h away, and z_i is the value of the regionalized variable at location i. In addition to the lag, the variogram is characterized by three other parameters—the nugget, range, and sill. The nugget is variability at zero distance and represents sampling and analytical errors. The range of influence designates the extent of distances, say a, beyond which autocorrelation between sampling sites is negligible. The sill represents the variability of spatially independent samples.

An empirical semi-variogram can be calculated from the given set of observations and than fit to several common theoretical models. Once the theoretical semi-variogram has been chosen, several criteria can be applied to determine the correctness of the model and to adjust its parameters.

The rationale for choosing the kriging technique lay in the similarity in spatial structure of most of the abovementioned variables—gradually increasing (or decreasing) as a function of the increasing distance from the watering point until reaching the limit of no grazing effects—and in the typical shape of the variogram. The kriging technique has recently become very common for analyzing spaceborne data.

Image differencing change detection analysis

A post-processing change detection method—namely BI Differencing—was implemented to compute the degree and direction of the changes in each site and between the imaging periods. The general function of the change can be considered as:

$$\text{change} = \begin{cases} 0 \text{ if } \left|BI_t - BI_{t+1}\right| \le T \\ 1 \text{ if } \left|BI_t - BI_{t+1}\right| > T \end{cases} \tag{7}$$

where t and t+1 represent the two time periods and T determines the threshold value. A common way to assess changes is based on determination of a threshold in terms of standard deviation (SD) levels below and above the mean of the difference between the BI values (Δ<BI>) of the images under study. In this manner, one can distinguish between changed and unchanged pixels as well as between negative and positive changes. In the current study, one SD from the Δ<BI> was defined as the threshold, and steps of one SD beyond this threshold determined the magnitude and direction of the change.

Results and Discussion

Figure 6.9 presents the false color composite image subsets for the Ust-Urt site in 1975, 1987 and 2000, as acquired from the Landsat MSS, TM and ETM+ sensors, respectively. Watering points can be recognized as light spots spreading over most of the MSS and TM images, but not the ETM+ one. In the latter image, many of the watering points have disappeared; however, a wide bright area exists in the center of the image (Figure 6.9C). Figure 6.10 represents the respective BI products as calculated by Equation (6) with the appropriate BI coefficients. The same features can be seen in these products, but here the brightness levels were equally stretched for the different sensors and ranged between 0.65 and 0.96. High values correspond to bare soil while low values indicate vegetation.

Figure 6. 9. Subset images used for the kriging analysis: (A) Landsat-MSS, 1975; (B) Landsat-TM, 1987; (C) Landsat-ETM+, 2000

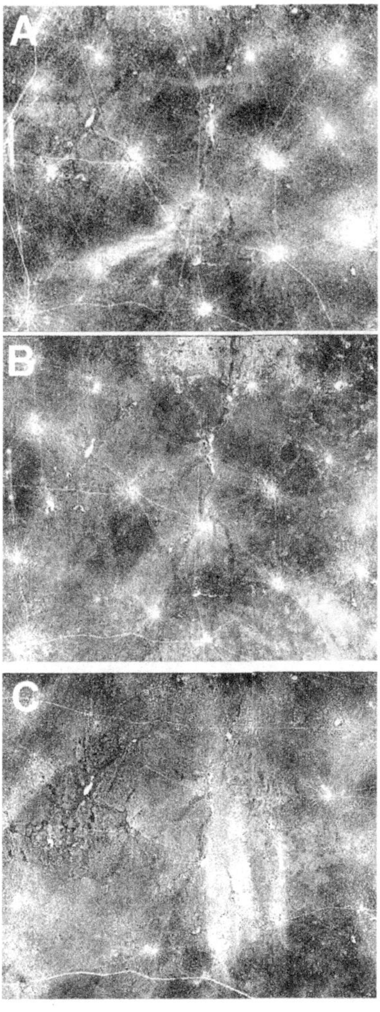

Figure 6.10. Brightness Index products for the matching images shown in Figure 6.9

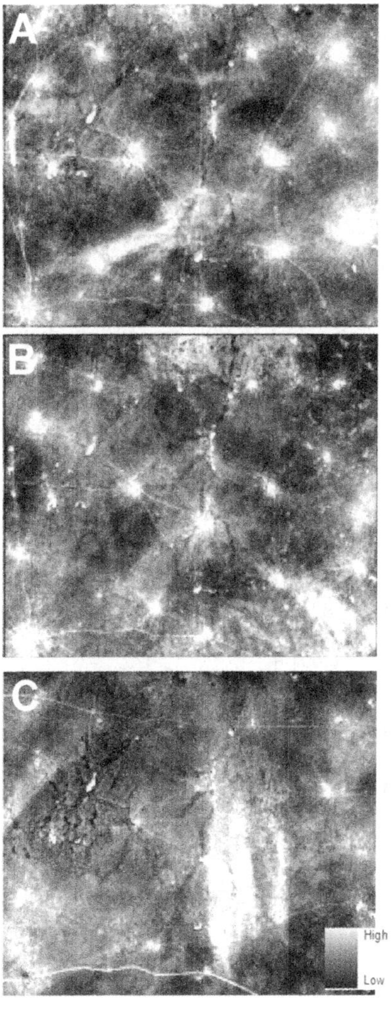

The three BI images were used for the kriging analysis. The first step in this course of action was to establish the empirical semi-variogram based on approximately 40,000 pixels in each image, for the three periods. Subsequently, several theoretical models were examined and the exponential model was selected, due to the best cross-validation results. Thus, all empirical models were fitted with an equation of the form:

$$\gamma(h) = C_0 + C_1\left(1 - \exp\left(-\frac{|h|}{a}\right)\right) \tag{8}$$

where: a is the range, h is the lag, C_0 is the nugget and C_0+C_1 equals the sill. The fitted exponential models are illustrated in Figure 6.11. All variograms were processed with 16 lags of 1,000 m each. Visually, the 1975 and 1987 variograms look quite similar, having a typical variogram shape. Note that when the shape of the variogram is more rounded it can be associated with more symmetrical features in the image. Contrarily, the 2000 variogram exceeds the same sill level only beyond 16,000 km, and its shape is more linear than round.

Figure 6.11. Theoretical variograms based on exponential models for the images shown in Figure 6.10

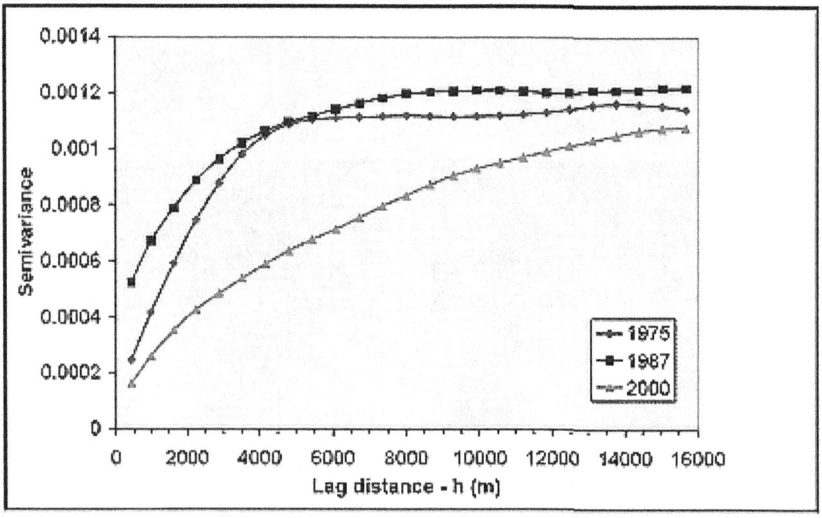

In the next step, kriging interpolation maps were established based on the exponential models. Since the grazing impact was assumed to be an isotropic feature and the direction has no influence on the spatial variation, the maps were derived from the omni-directional variograms. Figure 6.12 depicts the final products for the distribution of the BI values for the three periods. In the 1975 and 1987 maps, one can observe the belts around the watering points, indicating progressive land degradation radiating from the wells, i.e., the grazing gradient. The dark areas in the images corresponding to a BI of 0.83–0.94 are related to zones where the grazing impact is the predominant feature that has a strong effect on the spatial variation. These areas can be considered as the center of the grazing impact, denoted as the 'sacrifice zone' (SZ) by (Perkins and Thomas, 1993). The surrounding lighter-shaded belts are referred to as a 'mixed zone' where grazing impact and natural variability overlay each other or create a stable balance. This zone can be compared to an edge zone of the grazing impact and highlights rather the principle migration routes of livestock. The dark areas corresponding to a BI of 0.69–0.75 are considered areas where natural variability overbalances the grazing impact, denoted as 'grazing reserve' (GR) by Perkins and Thomas (1993). The radial pattern, related to the grazing gradient, does not appear in the 2000 map. Instead, the dominant feature in the middle of the scene is shaded in darker tones relating to strong grazing impact.

As opposed to this favorable land cover change, a degradation process characterizes considerable portions of the area (30.7%) during the second period (2000 vs. 1987); while rehabilitation occurred only in minor portions (12.5%).

Two different processes govern the land-use and land-cover changes in the Ust-Urt Plateau. On the one hand, desertification processes have developed due to recent exploration and exploitation of the gas and oil reserves in the region (Ulmishek, 2001). Consequently, large areas have undergone intensive technological desertification, meaning that large amounts of heavy-duty equipment, large-scale plants and vehicular use have damage the soil surface. On the other hand, following independence of the former Soviet states in 1991, and the imposition of difficult economic conditions with transition reforms, several major socio-economic changes occurred. Strong centralized government subsidy programs were terminated, including the practice of guaranteed supplemental forage in cold winters and drought years. Farmers could no longer feed their livestock during the harsh winters, water wells were demolished, pumps were stolen or broken and means of transportation to convey the animals to the markets in the central cities were no longer available (Antonchikov et al., 2002). Consequently, drastic decline in livestock populations were observed after 1991 that resulted in much less grazing pressure and hence recovery of the natural vegetation and rehabilitation of the land.

Image differencing change detection analysis, based on the BI values, was performed on the two pairs of kriging maps—1987 vs. 1975 and 2000 vs. 1987. Results are illustrated in Figure 6.13. Figures 6.13A and 6.13C are the change maps while Figures 6.13B and 6.13D are the respective frequency histograms of the change categories. The SD lines are also presented. The difference map computed from 1987 and 1975 shows that, despite the area that was considered as 'no change' (64.0%), more area underwent a process of rehabilitation than degradation (22.1% vs. 13.8%).

Figure 6.12. Kriging maps of Brightness Index values, established upon variograms presented in Figure 6.11, for the images shown in Figure 6.10

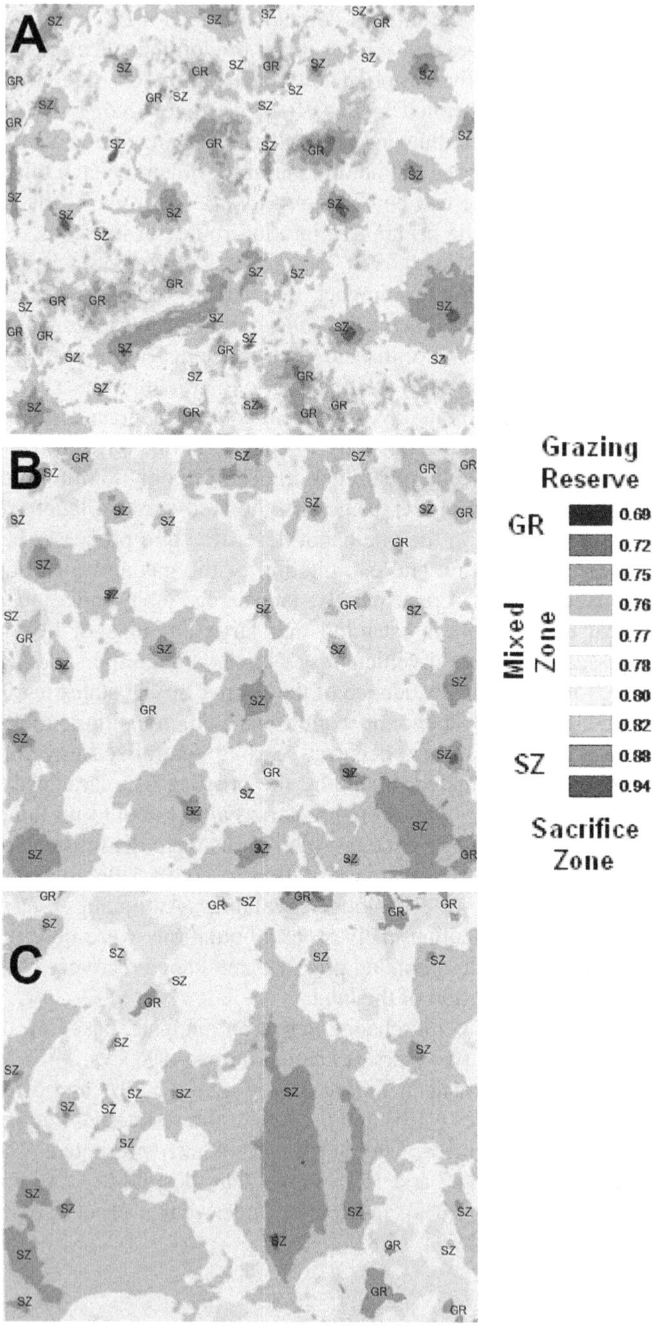

Figure 6.13. Change detection products, based on the Brightness Index Differencing algorithm. (A) Difference image between 1987 and 1975; (B) Frequency histogram for the degree of change between 1987 and 1975; (C) Difference image between 2000 and 1987; (D) Frequency histogram for the degree of change between 2000 and 1987

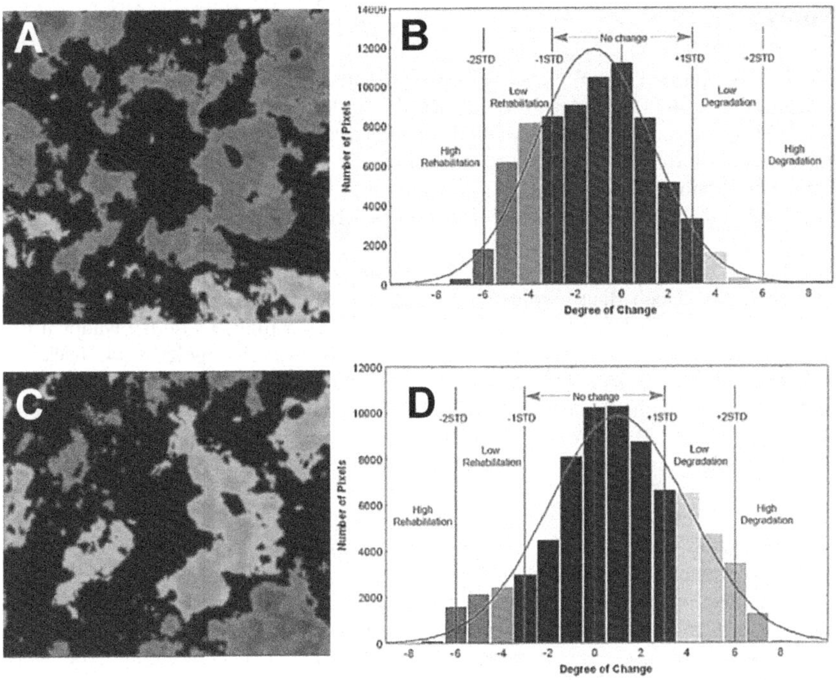

Conclusions

The Tasseled Cap-derived Brightness Index (BI) was selected to describe the spatial surface patterns since it produced the best contrast. This index was originally designated to examine soil properties and its advantage is in the ability to compare between the different sensors having different spectral bands, as it reduces their different spectral bands to one normalized layer of BI values. It was also found that the BI is uniquely independent from the spatial dimension and thus all image frequency histograms were spectrally matched. Geostatistical analysis, based on the Kriging interpolation technique was found to be a suitable method for assessing the spatial patterns of land-use land cover, especially around watering points in arid and semi-arid regions. The reason is the similarity between the shape of the variogram and the directional change of many biotic, abiotic and environmental variables along the grazing gradient. Temporal changes were effectively conducted by the index differencing technique. In the study site, a certain part of the area was degraded due to recent exploration and exploitation of the gas and oil reserves, while another part underwent rehabilitation processes due to a dramatic reduction in the grazing pressure after the collapse of the Soviet Union.

Notes

1. A. Karnieli and Y. Bayarjargal are affiliated with the Remote Sensing Laboratory, Jacob Blaustein Institutes for Desert Research, Ben-Gurion University of the Negev, Israel; M. Bayasgalan is affiliated with the National Remote Sensing Center, Ministry of Nature and Environment, Mongolia. This paper was presented at the International Conference on Remote Sensing and Geoinformation Processing in the Assessment and Monitoring of Land Degradation and Desertification, September 7–9, 2005, Trier (Germany)

2 A. Karnieli, Y. Bayarjargal and N. Agam are affiliated with the Remote Sensing Laboratory, Jacob Blaustein Institute for Desert Research, Ben-Gurion University of the Negev, Israel; M. Bayasgalan and S. Khudulmur are affiliated with the National Remote Sensing Center, Ministry of Nature and Environment, Mongolia; C. J. Tucker is affiliated with the NASA Goddard Space Flight Center, USA. This paper was previously published in the International Journal of Remote Sensing Vol. 27, No. 10, 20 May 2006, 2017–2024 (ISSN 0143-1161 print/ISSN 1366-5901 online # 2006 Taylor & Francis. http://www. tandf.co.uk/journals. DOI: 10.1080/01431160500121727)

3. A. Karnieli and U. Gilad are affiliated with the Remote Sensing Laboratory, Jacob Blaustein Institutes for Desert Research, Ben-Gurion University of the Negev, Israel. M. Ponzet is affiliated with the Department for Land Cultivation and Environmental Protection, Rostock University, Germany. T. Svoray is affiliated with the Department of Geography and Environmental Development, Ben-Gurion University of the Negev, Israel. This paper was presented at the International Conference on Remote Sensing and Geoinformation Processing in the Assessment and Monitoring of Land Degradation and Desertification, September 7th to 9th, 2005, Trier (Germany)

References

Adyasuren, T. S. (1998) State of drought and desertification in Mongolia. *Environment and development issues in Mongolia*. Ulaanbaatar, Mongolia.

Agbu, P. A., & James, M. E. (1994) *The NOAA/NASA Pathfinder AVHRR Land Data Set User's Manual*. Goddard Space Flight Center, Greenbelt: Goddard Distributed Active Archive Center, NASA.

Antonchikov, A. N., Bakinova, T. I., Dushkov, V. Y., et al. (2002) *Desertification and ecological problems of pasture stockbreeding in the steppe regions of Southern Russia*. IUCN. Moscow: World Conservation Union.

Anyamba, A., Tucker, C. J., & Eastman, J. R. (2001) NDVI anomaly patterns over Africa during the 1997/98 ENSO warm event. *International Journal of Remote Sensing*, 22:1847–1859.

ASD (2000) *Hand held spectrometer user's guide*. Field Spec UV/VNIR. Boulder, Colorado: Analytical Spectral Devices.

Asner, G. P., & Vitousek, P. M. (2005) Remote analysis of biological invasion and biogeochemical change. *Proceedings of the National Academy of Sciences of the United States of America*, 102(12): 4383–4386.

Badeck, F. W., Bondeau, A., Bottcher, K., Doktor, D., Lucht, W., Schaber, J., & Stich, S. (2004) Responses of spring phenology to climate change. *New Phytologist*, 162:295–309.

Bastin, G. N., Pickup, G., & Pearce, G. (1995) Utility of AVHRR data for land degradation assessment - a case-study. *International Journal of Remote Sensing*, 16(4): 651–672.

Batjargal, Z. (1999) Desertification in Mongolia. In *RALA Report 200*. (pp. 107–113). Icelandic Agricultural Research Service. Reykjavik, Iceland.

Boken, V. K., Hoogenboom, G., Kogan, F. N., Hook, J. E., Thomas, D. L., & Harrison, K. A. (2004) Potential of using NOAA-AVHRR data for estimating irrigated area to help solve an inter-state water dispute. *International Journal of Remote Sensing*, 25:2277–2286.

Brown, J. R., & Archer, S. (1999) Shrub invasion of grassland: Recruitment is continuous and not regulated by herbaceous biomass or density. *Ecology*, 80(7): 2385–2396.

Dregne, H. E., & Chou, N-T. (1992) Global desertification dimensions and costs. In *Degradation and restoration of arid lands*. Lubbock: Texas Tech. University.

Dyksterhuis, E. J. (1949) Condition and management of range land based on quantitative ecology. *Journal of Range Management*, 2:104–115.

Eswaran, H., Beinroth, F. H., & Virmani, S. M. (2000) Resource management domains: a biophysical unit for assessing and monitoring land quality. *Agricultural Ecosystems and Environment*, 81(2):155–162.

Fernandez-Gimenez, M. E. (2000) The role of Mongolian nomadic pastoralists' ecological knowledge in rangeland management. *Ecological Applications*, 10(5): 1318–1326.

Fernandez-Gimenez, M. E., & Allen-Diaz, B. (1999) Testing a non-equilibrium model of rangeland vegetation dynamics in Mongolia. *Journal of Applied Ecology*, 36(6): 871–885.

Friedel, M. H. (1997) Discontinuous change in arid woodland and grassland vegetation along gradients of cattle grazing in central Australia. *Journal of Arid Environments* 37:145–164.

Gitelson, A. A., Kogan, F., Zakarin, E., Spivak, L., & Lebed, L. (1998) Using AVHRR data for quantitative estimation of vegetation conditions: Calibration and validation. *Advances in Space Research*, 22:673–676.

Gunin, P., Vostokova, E. A., Dorofeyuk, N. I., Tarasov, P. E., & Black, C. C. (1999) Assessing present day plant cover dynamics. In P. Gunin, E. A. Vostokova, N. I. Dorofeyuk, P. E. Tarasov, & C. C. Black (Eds.). *Vegetation dynamics of Mongolia* (pp. 131–163). Dordrecht, The Netherlands: Kluwer Academic Publishers.

Gutman, G. (1990) Towards monitoring droughts from space. *Journal of Climate*, 3:282–295.

Hanan, N. P., Prevost, Y., Diouf, A., et al. (1991) Assessment of desertification around deep wells in the Sahel using satellite imagery. *Journal of Applied Ecology*, *28*:173–186.

Harris, A. T., & Asner, G. P. (2003) Grazing gradient detection with airborne imaging spectroscopy on a semi-arid rangeland. *Journal of Arid Environments*, 55:391–404.

Hodder, R. M., & Low, W. A. (1978) Grazing distribution of free-ranging cattle at three sites in the Alice spring district, Central Australia. *Australian Rangeland Journal*, 1:95–105.

Huete, A. (1988) A soil-adjusted vegetation index (SAVI). *Remote Sensing of Environment*, 25:295–309.

Jigjidsuren, S., & Johnson, D. A. (2003) *Forage plants in Mongolia.* Ulaanbaatar, Mongolia: Admon Press.

Karnieli, A., Bayarjargal, Y., & Bayasgalan, M. (2005) Do vegetation indices reliably assess vegetation degradation? *Proceedings of the 1st International conference on remote sensing and geoinformation processing in the assessment and monitoring of land degradation and desertification*, Trier, Germany, 7–9 September, 2005.

Karnieli, A., Bayasgalan, M., Bayarjargal, Y., Agam, N., Khudulmur, S., & Tucker, C. J. (2006) Comments on the use of the Vegetation Health Index over Mongolia. *International Journal of Remote Sensing.* 27(10): 2017–2024. DOI: 10.1080/01431160500121727

Karnieli, A., & Dall'Olmo, G. (2003) Remote sensing monitoring of desertification, phenology, and droughts. *Management of Environmental Quality: An International Journal*, 14:22–38.

Karnieli, A., Gilead, U., Ponzet, M., & Svoray, T. (2005) Satellite Image Processing and Geo-statistical Methods for Assessing Land Degradation around Watering Points in the Central Asian Deserts. *Proceedings of the 1st International conference on remote sensing and geoinformation processing in the assessment and monitoring of land degradation and desertification*, Trier, Germany, 7–9 September, 2005.

Kidwell, K. B. (ed.). (1997) Global vegetation index user's guide. *NOAA Technical Report,* Camp Springs, Maryland: NOAA/NESDIS.

Kogan, F. N. (1995) Application of vegetation index and brightness temperature for drought detection. *Advances in Space Research*, 11:91–100.

Kogan, F. N. (1997) Global drought watch from space. *Bulletin of the American Meteorological Society*, 78:621–636.

Kogan, F. N. (2000) Satellite-observed sensitivity of world land ecosystems to El Nino/La Nina. *Remote Sensing of Environment*, 74:445–462.

Kogan, F., Stark, R., Gitelson, A., Jargalsaikhan, L., Dugrajav, C., & Tsooj, S. (2004) Derivation of pasture biomass in Mongolia from AVHRR-based vegetation health indices. *International Journal of Remote Sensing*, 25:2889–2896.

Lambin, E. F., & Ehrlich, D. (1996) The surface temperature–vegetation index space for land cover and land-cover change analysis. *International Journal of Remote Sensing*, 17:463–478.

Lange, R. T. (1969) The piosphere: sheep track and dung patterns. *Journal of Range Management*, 22:396–400.

Lass, L. W., Thill, D. C., Shafii, B., & Prather, T. S. (2002) Detecting spotted knapweed (*Centaurea maculosa*) with hyperspectral remote sensing technology. *Weed Technology*, 16:426–432.

Lass, L. W., Prather, T. S., Glenn, N. F., Weber, K. T., Mundt, J. T., & Pettingill, J. (2005) A review of remote sensing of invasive weeds and example of the early detection of spotted knapweed (*Centaurea maculosa*) and babysbreath (*Gypsophila paniculata*) with a hyperspectral sensor. *Weed Science*, 53(2): 242–251.

Liu, H. Q., & Huete, A. R. (1995) A feedback based modification of the NDVI to minimize canopy background and atmospheric noise. *IEEE Transactions on Geoscience and Remote Sensing*, 33:457–465.

Liu, W. T., & Kogan, F. N. (1996) Monitoring regional drought using vegetation condition index. *International Journal of Remote Sensing*, 17:2761–2782.

Liu, W. T., & Negron-Juarez, R. I. (2001) ENSO drought onset prediction in northeast Brazil using NDVI. *International Journal of Remote Sensing*, 22:3483–3501.

Manzano, M. G., & Navar, J. (2000) Process of desertification by goats overgrazing in the Tamaulipan thornscrub in north-eastern Mexico. *Journal of Arid Environments*, 44:1–17.

McVicar, T. R., & Jupp, D. L. B. (1998) The current and potential operational uses of remote sensing to aid decisions on drought exceptional circumstances in Australia: a review. *Agricultural Systems*, 57:399–468.

McVicar, T. R., & Bierwirth, P. N. (2001) Rapidly assessing the 1997 drought in Papua New Guinea using composite AVHRR imagery. *International Journal of Remote Sensing*, 22:2109–2128.

Nemani, R. R., & Running, S. W. (1989) Estimation of regional surface resistance to evapotranspiration from NDVI and thermal-IR AVHRR data. *Journal of Applied Meteorology*, 28:276–284.

Nemani, R., Pierce, L., Running, S., & Goward, S. (1993) Developing satellite-derived estimates of surface moisture status. *Journal of Applied Meteorology*, 32:548–557.

Perkins, J. S., & Thomas, D. S. G. (1993) Spreading deserts or spatially confined environmental impacts? Land degradation and cattle ranching in the Kalahari desert of Botswana. *Land Degradation and Rehabilitation*, 4:179–194.

Peters, A. J., Walter-Shea, E. A., Ji, L., Vina, A., Hayes, M., & Svoboda, M. D. (2002) Drought monitoring with NDVI-based standardized vegetation index. *Photogrammetric Engineering and Remote Sensing*, 68:71–75.

Pickup, G., & Chewings, V. H. 1994: A grazing gradient approach to land degradation assessment in arid areas from remotely sensed data. *International Journal of Remote Sensing*, 15:597–617.

Pickup, G., Chewings, V. H., & Nelson, D. J. (1993) Estimating changes in vegetation cover over time in arid rangelands using Landsat MSS data. *Remote Sensing of Environment*, 43:243–263.

Price, J. C. (1984) Land surface temperature measurements from the split window channels of the NOAA 7 Advanced Very High Resolution Radiometer. *Journal of Geophysical Research*, 89:7231–7237.

Running, S. W., Loveland, T. R., Pierce, L. L., Nemani, R. R., & Hunt, E. R. (1995) A remote sensing based vegetation classification logic for global land cover analysis. *Remote Sensing of Environment*, 51:39–48.

Seiler, R. A., Kogan, F., & Sullivan, J. (1998) AVHRR-Based vegetation and temperature condition indices for drought detection in Argentina. *Advances in Space Research*, 21:481–484.

Tateishi, R., & Ebata, M. (2004) Analysis of phenological change during 1982–2000 Advanced Very High Resolution Radiometer (AVHRR) data. *International Journal of Remote Sensing*, 25:2287–2300.

Thomas, D. S. G., & Middleton, N. J. (1994) *Desertification: Exploding the myth.* Chichester: John Wiley & Sons.

Tserenbaljid, G. (2002) *Colour atlas of anthropophilus plants of Mongolia.* Ulaanbaatar, Mongolia: Admon Press.

Tucker, C. J. (1979) Red and photographic infrared linear combinations for monitoring vegetation. *Remote Sensing of Environment,* 8:127–150.

Tucker, C. J., Vanpraet, C. L., Boerwinkel, E., & Gaston, A. (1983) Satellite remote sensing of total dry matter production in the Senegalese Sahel: 1980–1984. *Remote Sensing of Environment,* 13:461–474.

Tucker, C. J., Vanpraet, C. L., Sharman, M. J., & Van Ittersum, G. (1985) Satellite remote sensing of total herbaceous biomass production in the Senegalese Sahel: 1980–1984. *Remote Sensing of Environment,* 17:233–249.

Tucker, C. J., Slayback, D. A., Pinzon, J. E., Los, S. O., Myeni, R. B., & Taylor, M. G. (2001) Higher northern latitude normalized difference vegetation index and growing season trends from 1982 to 1999. *International Journal of Biometeorology,* 45:184–190.

Ulmishek, G. F. (2001) Petroleum Geology and Resources of the North Caspian Basin, Kazakhstan and Russia. *U.S. Geological Survey Bulletin 2201-B.* USGS, Denver.

Underwood, E., Ustin, S. L., & DiPietro, D. (2003) Mapping nonnative plants using hyperspectral imagery. *Remote Sensing of Environment,* 86:150–161.

Unganai, L. S., & Kogan, F. N. (1998) Drought monitoring and corn yield estimation in Southern Africa from AVHRR data. *Remote Sensing of Environment,* 63:219–232.

White, M. A., Thornton, P. E., & Running, S. W. (1997) A continental phenology model for monitoring vegetation responses to interannual climatic variability. *Global Biogeochemical Cycles,* 11:217–234.

Yonghong, L., & Jargalsaishan, L. (1993) Grazing dynamics of *Stipa grandis* steppe in the northeast Mongolia plateau. *Proceedings of the International Symposium on Grassland Resources.* Huhehot, Inner-Mongolia, China.

Chapter 7

Gully Incision: Natural Desertification in Arid Environments, an Example from the Negev Highlands of Israel*

Yoav Avni

Introduction

Soil erosion, caused mainly by gully incision eroding loess and alluvial sediments, is a well known feature in arid and semiarid regions around the world, i.e., southwest North America, Australia, Africa, Mediterranean Europe, China and the Middle East (Bull, 1997; Fanning, 1999; Oostwoud Wijdenes et al., 1999; Vandekerckhove et al., 2000; Vandekerckhove et al., 2001; Vandekerckhove et al., 2002; Shi, Dian and You, 2002; Poesen et al., 2002; Poesen et al., 2003; Billi and Dramis, 2003; Vandekerckhove, Poesen and Govers, 2003; Avni, 2005). This phenomenon is critical in the Negev Highlands desert of southern Israel (Figure 7.1). The region, a predominantly rocky terrain consisting mainly of marine carbonates (limestone and chalk) of Late Cretaceous to Tertiary age, reaches an elevation of 600–1000 m above sea level (Figure 7. 1) and is dissected by numerous valleys 30–200 m wide.

---

* Based on "Gully incision as a key factor in desertification in an arid environment, the Negev highlands, Israel," by Y. Avni. Originally published in CATENA 63: 185-220 (2005), copyright Elsevier B. V., and later modified and updated in 2007 with new data.

Figure 7.1. A. Location and rainfall distribution map of the Negev Highlands (after Evenari et al, 1982). B. Topographic map of the Negev Highlands and the study sites (after Atlas of Israel, 1970)

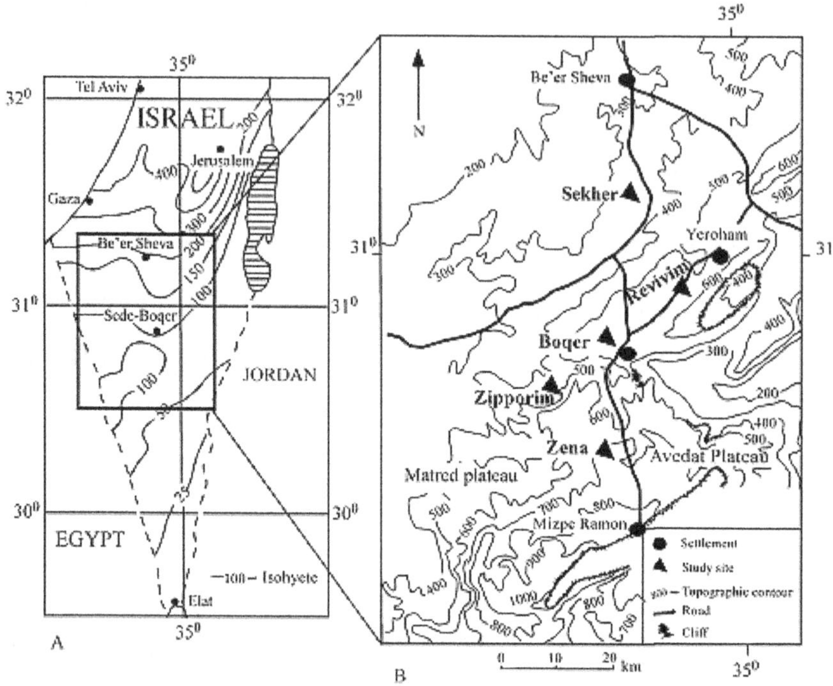

In this region, the land that sustains most of the natural biomass and that has agricultural and grazing value is limited to the valley floors (Seligman, Tadmor, and Raz, 1962; Figure 7.2). The special importance of the valley bottom in this region is based on the following:

1. Average annual precipitation in the Negev Highlands ranges from 80 mm in the south to 120 mm in the north of the region (Figure 7.1). However, the flooded areas in the valleys receive the equivalent of 300–500 mm (Evenari, Shanan, and Tadmor, 1982; Yair, 1983). This value varies from one basin to another, depending on the rainfall/runoff coefficient in each basin. In several studies within the region (Evenari et al., 1982; Yair, 1983; Yair and Kossovsky, 2002), it was estimated as 15–20% of the annual rainfall. This extra water was the key factor in the establishment of agricultural activities in the desert, which were based on runoff-harvesting techniques (Evenari et al., 1982), and contributes to the enrichment of natural vegetation along the valleys, which serve as the best grazing land in the region (Seligman et al., 1962).

2. The alluvial soil deposited along the valleys originated mainly from redeposited loessial sediments forming alluvial aridisol with silt-loam and sandy-loam texture. The soil is relatively deep (2–4 m) and contains a high percentage of fine particles (60–80% silt and clay) and a relatively high concentration of nutrients (Bruins, 1986; Ward, Feldman, and Avni, 2001, Ninari and Berliner, 2002).
3. Soil salinity is relatively low (0.20–0.70 dS m^{-1}) as a result of leaching by floodwater percolating into the alluvial section (Yair, 1983; 1987; Bruins, 1986; Ward et al., 2001).
4. The topography along the valleys is relatively flat and the longitude gradient of 1–3% makes this land easy to cultivate.

Figure 7.2. General view of rich vegetation concentrated along the valley floor of Nahal (wadi) Eilot in the Negev Highlands west of Mizpe Ramon. Note the sharp decrease in vegetative cover on the side of the valley

For these reasons, the valleys in the Negev Highlands have been used for grazing land for the past 5000 years (Bruins, 1986; Avni, 1996). During periods with strong central governments, i.e., the Iron Age (3000–2700 BP) and the Roman, Byzantine and Early Islamic periods (1800–1100 BP), permanent settlements were established in the desert region and the settlers subsisted on a combination of herding and farming (Rosen and Avni, 1993; Avni, 1996). The settlers terraced and leveled the valley floors, which decreased flood velocity and facilitated optimal distribution of water for irrigating the fields (Evenari et al., 1982).

At present, some of these ancient fields are cultivated by nomadic Bedouin of the Negev, who grow mainly wheat and barley (Figure 7.3). Fields that have not been cultivated recently have a relatively rich vegetation of desert shrubs (e.g., *Atriplax halimus, Artemisia herba-alba* Asso, *Retama raetam)* accompanied by a large number of perennial species of herbaceous wadi vegetation and geophytes (Seligman et al., 1962). This rich plant assemblage is grazed by the Bedouin herds (Figure 7.4) or by the small population of gazelles and ibexes that still exist in the area. Similar conditions are present in several regions of the Middle East, such as eastern Syria, southern and eastern Jordan, the mountainous areas of the northern and central Sinai Peninsula, and northwest Egypt.

Gully incision is a well-known phenomenon developing in loessial soils and alluvial sediments in the entire desert region of southern Israel, and in other desert regions of the Middle East. The gullies form vertical walls 1.5–4.5 m high (Figures 7.5 and 7.6) and dissect the fine-grained alluvial cover of the valleys. The gullies concentrate runoff into well-defined channels that incise the alluvial fill and restrict the floodwater available for vegetation on the rest of the valley floor. Coarse gravel covers the bottoms of the gullies. No recovery effects of the gully channels appear in the nearby region downstream and the lower segments of the valleys develop wide channels covered by gravel (Figure 7.7). During floods (Figures 7.8 and 7.9), the gully heads migrate upstream, causing intensive erosion of the valley soils and damage to the agricultural fields, which reduces the vegetative biomass available for pastoral herds and other grazing animals (Figure 7.5). This phenomenon causes ongoing degradation of soil and biomass.

Similar processes of gully incision and land degradation are active in semi-arid and arid regions around the world, e.g., the American southwest (Bull, 1997), southeast Australia (Fanning, 1999), Africa (Botha, Wintle and Vogel, 1994; Botha, 1996; Eriksson, 1998; Billi and Dramis, 2003), Mediterranean Europe and the Middle East (Oostwoud Wijdenes et al., 1999; Vandekerckhove et al., 2000, 2001, 2003; Poesen et al., 2002, 2003) and central Asia, including major parts of northern China (Shi et al., 2002).

Soil degradation, gullying and desertification

The term "desertification" is used worldwide to describe the result of complicated processes that cause degradation of soil, biomass and biodiversity. In this chapter the term "desertification" is defined according to the United Nations Conference on Environment and Development (UNCED, 1992) as "Land degradation in arid, semi-arid and dry sub-humid areas resulting from various factors, including climatic variations and human activities." An example of "climatic variation" is the severe multi-annual drought in the Sahel region of the 1970s and 1980s that caused the collapse of the natural environment. In addition, it is widely accepted that human pressure on marginal lands in semi-arid environments from overgrazing, unsustainable agricultural activities and environmental damage brought about by modern infrastructures are the causes for desertification in these regions (Mainguet, 1991; Bruins and Berliner, 1998; Bruins and Lithwick, 1998). Two documented examples of the results of human intervention in a semi-arid environment are the massive land degradation and gully incision caused during the last 140 years by European settlers in Australia (Fanning, 1999) and the dramatic shrinking of the Aral Sea during the last decade (Mainguet and Le'tolle, 1998). However, in many regions, studies of geological records show several natural Quaternary cycles of shifts from aggradation and deposition to land degradation, erosion and gully incision—long before the anthropogenic impact started (e.g., Botha et al., 1994; Botha, 1996; Bull, 1997; Eriksson, 1998; Avni and Porat, 2002; Clarke et al., 2003; Avni et al., 2006). These studies indicate that there are natural long-term geologic-geomorphologic processes within drainage basins that lead toward land degradation and soil erosion. These processes are well documented in the Negev Highlands of southern Israel, where gully incision and headcut retreat is one of the most important geomorphologic agents causing soil erosion and land degradation and decreasing agricultural and range potential, leading to natural desertification (Avni, 2002; Avni and Porat, 2002; Avni, 2005; Avni et al., 2006).

The aim of this chapter is: (1) to describe the ongoing gully-head retreat in the Negev Highlands at present; (2) to discuss the implications of gully retreat for the ongoing environmental change operating in the region that is leading toward increasing natural desertification and (3) to present a conceptual model for the geological and geomorphological control of the long-term natural desertification in the Negev Highlands region as an example for other arid and semi arid regions around the globe.

Geological and historical background

The Negev Highlands (Figure 7.1) is a rocky terrain located at an elevation of 600–1000 m, consisting of marine sediments—mainly limestone, dolomite, chalk and chert layers of Upper Cretaceous to Tertiary age (Arkin and Braun, 1965; Braun, 1967; Zilberman, 1981, 1991; Avni, 1991). Several parallel anticlines and synclines dominate the structure of the northern part of the region. These structures consist of hard carbonates (e.g., limestone, dolomite) of the Cenomanian–Turonian Judea Group exposed in the anticlines, and soft carbonates (e.g., chalk, marl) and chert of the Senonian–Paleocene Mount Scopus Group exposed in the synclines (Arkin and Braun, 1965; Zilberman, 1991). To the south, a large uplifted plateau (the Avedat and Matred plateau, see Figure 7.1) developed, consisting of Tertiary marine limestone and chalk sediments of the Avedat Group (Braun, 1967). The area is dissected by numerous streams, draining northwestwardly to the Mediterranean and northeastwardly to the Dead Sea basin. The valleys are incised to a depth up to several tens of meters below the hills, and the valley bottoms are 30–200 m wide. In most cases, the valley bottoms are filled with late Pleistocene sediments, mainly fluvial loess and gravel, deposited in the region during the last glacial interval, some 73–14 ka BP (Avni, 1991; Zilberman, 1992; Avni and Porat, 2002; Avni et al., 2006). These originally deposited sediments were eroded and redeposited along the valleys in several cycles evolving during the end of the Pleistocene and into the Holocene.

Figure 7.3. Bedouin wheat fields in the upper valley head of Nahal Besor, near Nahal Zipporim

Figure 7.4. Bedouin herd grazing on rich vegetation developed within the upper segment of a valley in the Negev Highlands, near Sede Boqer

The present landscape of the Negev Highlands developed during the Pliocene and the early Pleistocene as a geomorphological rebound to the main morphotectonic phase, which uplifted the region and generated the evolution of its present drainage network (Avni, 1991; Zilberman, 1991). The valleys developed during several phases of erosion, alternating with phases of accumulation of alluvial sediments within the valleys. During the late Pleistocene, desert loess sediments, transported by the wind from the northern margins of the Sahara desert (Offer, Zangvil and Azmon, 1992), accumulated in the Negev, covering both hill slopes and valleys (Bowman et al., 1986; Bruins and Yaalon, 1992; Zilberman, 1992). Most of these sediments were brought to the region in several phases during the last glacial cycle—oxygen isotope stages (OIS) 4 and 3—some 70–26 ka BP (Zilberman, 1992; Avni and Porat, 2002; Avni et al., 2006). After their initial accumulation, these sediments were redeposited along the drainage system, mixing with alluvial clastic sediments of local origin and accumulating to a sectional thickness of 5–10 m (Avni, 1991; Zilberman, 1992). This accumulation blocked the drainage network and decreased runoff (Yair, 1987). However, since OIS 2, and mainly under the current arid Holocene climate, intensified erosive processes in the region have been observed removing part of the late Pleistocene sediments. This shift in geomorphologic processes is related to the major post-glacial climatic change, whereby the amount of dust flux has decreased and the amount of precipitation has decreased but the intensity of rainfall events has increased, producing the well-known phenomenon of desert "flashfloods." Thus, sediments that had accumulated by the late Pleistocene under a pluvial climate are now being removed during the present Holocene climate.

Human settlement in the Negev Highlands continued from the Early and Middle Paleolithic period to medieval times. The main periods of habitation attested to in the archaeological record are the Epipaleolithic (approximately 15,000–10,000 BP), the Early Bronze age (5200–4800 BP), the Intermediate Bronze age (4200–4000 BP), the Iron age (3000–2700 BP) and the Byzantine and Early Islamic periods (1700–1100 BP) (Evenari et al., 1982; Bruins, 1986; Lender, 1990; Haiman, 1992; Avni, 1996; Avni, forthcoming). Throughout history, the Negev Highlands have been a peripheral region, the pattern and intensity of their settlement closely linked to the fortunes of the more sedentary north. During the Byzantine and Early Islamic periods, settlements reached their zenith in terms of number of sites and density of population (Rosen and Avni, 1993; Patrich 1995; Avni, 1996; Avni, forthcoming). The agricultural population terraced the wadis and leveled the valley floors to decrease flood velocity and to enable optimal distribution of water for irrigating the fields. They also built walls on the hill slopes to collect runoff and direct it to the fields. These ancient agricultural fields cover large areas of the Negev Highlands and appear in several forms: from minor terraces and dams, built along riverbeds of secondary order, to large agricultural farms, which were constructed along the main valleys of the region. Most of the farming installations were abandoned during the latter part of the Early Islamic period (1400–1100 BP). However, for the past several hundred years, pastoral Bedouin herders, who cultivate wheat and barley during rainy years, have utilized some of the remaining ancient farming systems.

Figure 7.5. Nahal Zipporim gully head, general view. Note the sharp drop in vegetative cover below the headcut (vehicle as scale)

Figure 7.6. Nahal Zipporim central headcut: coarse conglomerate exposed at the base of the alluvial loessial section (people as scale)

Figure 7.7. Lower segment of Nahal Zipporim valley downstream of the gully head. Gravel covers the well-defined channel. *Pistacia atlantica* trees grow along the banks of the gravely channels at elevations above 700 m in Nahal Lotz, west of Mizpe Ramon. (Tree height as scale, approximately 10 m)

Figure 7.8. Nahal Sekher, northern headcut during a flood, December 23 1993. The headcut retreat during this event was more than 12 m

Figure 7.9. Nahal Sekher channel, downstream of the headcuts, showing a large collapse of the bank, caused by the undermining forces of the floodwater

The Negev Highlands climate

The Negev Highlands region, located within the hot Sahara-Arabian desert belt of the northern hemisphere, has a mean annual temperature of 17–19 °C, depending on topographic elevation. The average annual precipitation ranges from 120 mm in the northern part to <80 mm in the southern part (Figure 7.1); the annual evaporation range is 2000–2500 mm and the annual precipitation-to-potential-evapotranspiration ratio (P/EPT) is 0.07–0.04, which, according to the United Nations Educational, Scientific and Cultural Organization (UNESCO, 1979), classifies the Negev Highlands as an "arid zone." The rainy season lasts from October to April, but most of the rain falls during the winter (December to February) in low-intensity (<15 mm h^{-1}) rainstorms originating from the Mediterranean Sea (Sharon, 1972; Sharon and Kutiel, 1986). Occasionally, small-scale (10–50 km^2) short-lived (several minutes) but intense (30–120 mm h^{-1}) rainfalls occur (Sharon and Kutiel, 1986), originating mainly from rare tropical air masses passing across the Red Sea to the south. This synoptic system, known as the Red Sea Depression, yields occasional floods. Some of these floods are lost along the course of the drainage system (Yair, 1983, Yair and Kossovsky, 2002). Yair (1983) reported that the threshold amount of rainfall necessary to generate runoff on the rocky surfaces is on the order of 1–3 mm and for the soil-covered slopes it is higher: on the order of 3–5 mm. However, under normal conditions, only rain events that exceed 10 mm generate considerable floods in the drainage basins. This limit is lower in small rocky drainage basins

(Evenari et al., 1982; Yair, 1983, Yair and Kossovsky, 2002). A detailed description of the flood-generating mechanism in the Negev Highlands is given by Yair (1983) and Yair and Kossovsky (2002).

Figure 7.10 summarizes the annual precipitation recorded at the Sede Boqer station (for location, see Figure 7.1). This figure illustrates the multi-annual distribution of the precipitation in the region during the last 56 years, with an average of 92.5 mm. During the 17 years of monitoring (1990–2007) the average annual rainfall has dropped to 89 mm; 6 years were above the average and 11 were below it. The first 2 years (1990-1991 and 1991-1992) were significantly wet years (156–190% of the average rainfall), followed by two drought years.

This interval was repeated in the years 1996–2000, when two relatively wet years (1996-1997 and 1997-1998) were followed by two extremely dry years (1998-1999 and 1999-2000), in which precipitation was only 40–45% of the average rainfall. The following years (2001–2006) were characterized by six consecutive below-average rainy seasons, followed by the wettest rainy season ever reported (2006-2007), yielding 175.3 mm, nearly 200% of the average rainfall.

However, this unusually high amount of rainfall represents a relatively small area in the Negev Highlands in the vicinity of Sede Boqer. Other nearby regions received about average or even 75% of the average annual rainfall, demonstrating the highly spotty nature of rainfall in arid regions (Sharon, 1972; Sharon and Kutiel, 1986).

Figure 7.10. Plot of annual rainfall, Sede Boqer station, 1951–2007. The average rainfall during this period was 92.5 mm; during 1990–2007 it was 89.0 mm

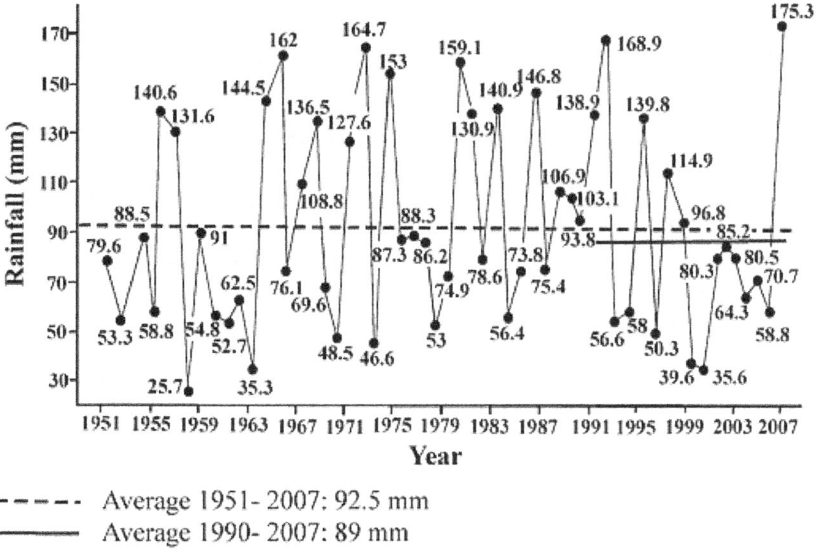

Present geomorphological features of the valleys in the Negev Highlands

In most of the valleys of the Negev Highlands, a major gully has developed at the bottom of the valley (Figures 7.5, 7.6 and 7.9). At each gully head, one or several headcuts 1.5–4.5 m high are incised into the valley floor, exposing the alluvium sequence down to a coarse conglomerate or to the bedrock. The gully heads divide the valleys in the region into two main segments:

The upper valley—upstream of the gully head up to the headwater of the drainage basin. In this segment the valley floors are 30–200 m wide and have a gradient of 1–3%. During floods, most of the valleys are irrigated by floodwaters, which penetrate 1–3 m into the soil section. Therefore, in this segment the natural flora is preserved and the ancient agricultural fields are still potentially valuable (e.g., Figures 7.3 and 7.4). According to a range survey carried out in the Negev Highlands by Seligman et al. (1962), the herbaceous pasture vegetation of this segment has the highest value in the entire region, in terms of food units and carrying capacity for grazing animals (sheep, goats and cattle).

The lower valley—downstream of the gully head as far as the outlet of the drainage basin. In this segment a relatively narrow gully (3-20 m wide) develops (Figures 7.7 and 7.9). The gully, 3–4 m deep with steep walls, channels the floodwater into itself, deep below the original valley floor. Within the gullies the water flows rapidly over the coarse gravel layer exposed at the bottom of the alluvial section or directly on top of the bedrock. As a result, floodwaters no longer irrigate most of the valley, and the area, which is outside the gullies, dries up rapidly, losing most of its vegetative cover within several years (Figure 7.7). The vegetation that does exist within the gullies is restricted to the narrow banks, 1–3 m wide, developed along the main channels. Occasionally, the increase in the velocity of a flood flowing within the channel damages the banks and the vegetation located along them (Figure 7.7). According to Seligman et al. (1962), the herbaceous pasture vegetation value of this segment for grazing animals is 83–99% less in comparison to that in the upper segment.

The dynamic backward migration of the gully heads

During flood events, the vertical headcuts gradually retreat up the valley (Figure 7. 8). As a result of the retreat, the lower segments of the valleys become longer, and the upper segments become shorter. This fundamental process dictates an ongoing dynamic geomorphic change, which advances up the valley, leading to a gradual ongoing loss in soil, biomass, and agricultural and grazing potential in the region. Rates of these parameters are strongly dependent on the velocity and long-term rates of the headcut retreat in each valley in the Negev Highlands.

Methods

A regional field survey was carried out in 1990 to locate representative drainage basins and headcut sites in the region between Beer-Sheva and Mizpe Ramon (Figure 7.1). The survey covered an area of approximately 1000 km^2 in the Negev Highlands. Four drainage basins were selected: the Zipporim (where preliminary observations were initiated in 1984), Revivim, Boqer and Sekher basins, each representing a unique geomorphic configuration. The mechanism and rate of gully retreat were studied in these basins over 17 years (1990–2007).

The following methods were used to monitor the headward erosion and to determine its rate:

1. In each site selected, several benchmark pins were installed around the headcut and the gully wall, to establish long-term measurement points. Similar techniques, summarized by Poesen et al. (2003), were implemented in several studies in a wide range of environments.
2. Regular measurements of the change in distance between the edge of the gully head or wall and the benchmark pins were conducted during the years 1990–2007. The measurements were taken several times a year and after major flood events. The measurements used metric strips; accuracy was in the range of ±10 cm.
3. The annual rate of retreat and the total loss of soil were calculated for each headcut site. The volume of the eroded soil was calculated by three-dimensional (3D) mapping of the headcut and given in cubic meters. Accuracy of 3D calculations was in the range of ±10%. Applying a value of 1500 ±250 kg m^{-3} for loess sediments (Ninari and Berliner, 2002) enabled soil erosion to be estimated in terms of weight. [It is important to note that this volume does not represent the total soil erosion in the entire basin under study. Therefore, these calculations are minimum erosion estimations for each drainage basin and do not include other soil erosion agents, such as rill and sheet erosion. According to the review presented by Poesen et al. (2003), the contribution of gully erosion to overall soil loss rates and sediment production rates on a global scale ranges widely between 10 and 94% (Poesen et al., 2003)].
4. An archeological survey was carried out in the Zipporim valley to assess the gully erosion rates in historical times. The survey mapped the agricultural installations and facilities (such as stone dams, agricultural terraces and irrigation channels) and their relations to the development of the Zipporim gully throughout the last few millennia. It is important to note that although the quantities estimated in this analysis are approximations, the trends through time are credible.

Results: headcut retreat rates

Nahal (wadi) Zipporim

The Nahal Zipporim basin is located in the central part of the Negev Highlands, 7 km southwest of Sede Boqer (Figure 7.1). It drains the northern edge of the Matred plateau to the Mediterranean Sea via Nahal Lavan. The present gully head is located in the upper Zipporim valley, and the basin above it occupies 9 km^2. The valley bottom near the gully head is 80–100 m wide, covered by an alluvial sequence composed of conglomerates and alluvial loess of late Pleistocene–Holocene age. The valley below the gully head is dissected by a prominent gully, 3–4 m deep with steep walls, which continues downstream for a distance of 9 km where it joins the major wadi of Nahal Lavan. No recovery effect was found anywhere along the lower Zipporim valley.

The gully head is composed of three distinct headcuts, which join together to form the Zipporim gully (Figure 7.11). The eastern and western headcuts are located on either side of the valley, while the main headcut is located in the center. Shallow channels, located along the valley margins above the gully head,

Figure 7.11. Map of Nahal Zipporim gully, showing the position of several headcuts each year between 1990 and 2007. The total volume of eroded soil from the gully head was 2000 m^3

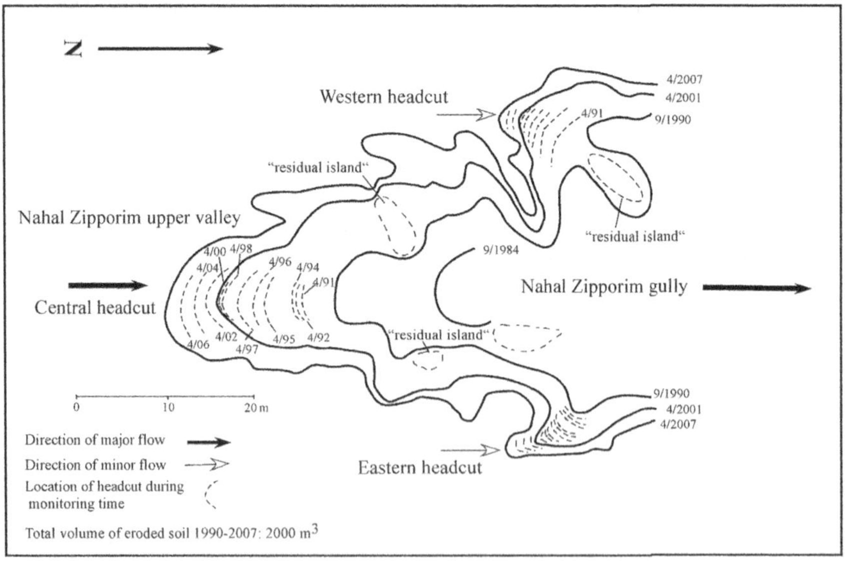

contribute floodwater to the two side headcuts, while the central headcut receives the flow running along the main valley. The exposed section in the central headcut, which is 3–4 m high (Figures 7.5 and 7.6), consists mainly of fine fluvial loess sediments. At the bottom of the eastern headcut, the bedrock (chalky limestone of the Maastrichtian–Ghareb Formation) is exposed, while the western headcut consists of 1.5–2 m of coarse conglomerate (limestone and chert pebbles), covered unconformably by 1–1.5 m of fluvial loess sediments. During floods, each headcut retreats separately according to its specific parameters; i.e., the erodibility of the alluvial section and its resistance to the flow stresses exerted by concentrated flow or the volume and velocity of flows running through the specific headcut. The different response of each headcut to the floods is evident in Table 7.1.

Table 7.1. Annual retreat rates in the Zipporim gully head (in cm)

Year	Total precipitation Sede Boqer station (in mm)	Eastern headcut (chalk covered by loess)	Central headcut (loess and fine conglomerate)	Western headcut (loess and coarse conglomerate)
1990-1991	138.9	30	360	160
1991-1992	168.9	20	55	140
1992-1993	56.6	0	0	0
1993-1994	58.0	30	30	60
1994-1995	139.8	105	290	50
1995-1996	50.3	5	110	50
1996-1997	114.9	40	180	30
1997-1998	96.8	60	170	10
1998-1999	39.6	20	0	0
1999-2000	35.6	50	25	0
2000-2001	80.3	50	15	0
2001-2002	85.2	40	150	10
2002-2003	80.5	180	60	0
2003-2004	64.3	70	70	30
2004-2005	70.7	10	10	10
2005-2006	58.8	40	70	30
2006-2007	175.3	10	170	170
Total retreat 1990–2007 (cm)		760	1780	770
Average retreat (cm yr^{-1})		44.7	104.7	45.3

Average rainfall: 89.0 mm yr^{-1}
Total length of gully retreat 1991–2007: 17.8 m
Total eroded soil 1991–2007: 2000 m^3
Average width of valley: 90 m
Total lost area of agricultural and range potential (1991–2007): 750 m^2

Table 7.2. Comparison of mean annual gully head retreats in the Negev Highlands

Site name	Catchment area upstream of headcut (km^2)	Headcut material	Total retreat 1990–2001 (m)	Average annual retreat ($m\ y^{-1}$)	Total eroded soil 1990–2001 (m^3) (t*)	Average annual eroded soil ($m^3\ y^{-1}$) ($t\ y^{-1}$)	Average annual eroded soil ($t\ km^{-2}$)	Total potential agricultural and range area lost 1991–2007 (m^2)	Average annual rate of lost area ($ha\ y^{-1}$)
Zipporim Eastern headcut	9	Chalk and loess	7.60	0.44					
Zipporim central headcut	9	Loess	17.80	1.04					
Zipporim western headcut	9	Loess and conglomerate	7.70	0.45					
Total for Zipporim gully head	9		17.80	1.04	2000 3000	117 176.4	19.6	750	0.004
Revivim southern headcut	15	Loess	27.1	1.6					
Revivim northern headcut	15	Loess and conglomerate	7.2	0.45					
Total for Revivim gully head	15		27.1	1.6	1000 1500	58.8 88.2	5.8	800	0.005
Boqer southern headcut	100	Loess	49.9	3.11					
Boqer northern headcut	100	Loess	45.2	2.82					
Total for Boqer gully head	100	Loess	49.9	3.1	2000 3000	117.6 176.4	1.7	1000	0.006
Sekher southern headcut	100	Sandy loess	196.2	17.8					
Sekher northern headcut	100	Sandy loess	57.1	11.42					
Total for Sekher gully head	100		250	16.6	9000 13500	600 900	9.6	8750	0.080

(*) Calculation based on 1500 kg per 1000 m^3 (after Ninari and Berliner, 2002).

In the wet years recorded during the 1990–2007 monitoring interval (1990-91, 1991-92, 1994-95, 1996-97, 2006-07), the three headcuts were active and their relative retreat rates reflected the dominance of the central headcut relative to the headcuts on both sides of the valley. However, in drought years (1992-93, 1993-94, 1998-99, 1999-2000, 2003-04, 2005-06), the central headcut was almost inactive and considerable retreat was recorded along the side headcuts. In addition, in the rainiest years (1991-92 and 2006-07) with 168.9 and 175.3 mm, respectively, the retreat of the Zipporim central headcut was minor or almost equal to the retreats recorded in less rainy years, such as 1996-97 (114.9 mm) and 1997-98 (96.8 mm). This data hints at indirect relations between annual precipitation and retreat rates of the gully system.

In general, the backward retreat measurements presented in Tables 7.1 and 7.2 indicate that between 1991 and 2007, the Zipporim gully migrated upstream 17.8 m, at an average annual rate of approximately 1.04 m y^{-1}. In the same period, the side headcuts migrated 7.6–7.7 m at an annual rate of 0.44–0.45 m y^{-1}, which is less than half of the migration rate of the central headcut. The 3D calculations of the volumes eroded (Table 7.2) shows that between 1991 and 2007, approximately 2000 m^3 (3000 tons) of soil and alluvial sediments were eroded from this gully head, at an annual average rate of 117 m^3 y^{-1} (176.4 t y^{-1}). During this period, approximately 750 m^2 of land lost their high agricultural and range potential due to gullying and soil loss, at an average annual rate of 0.004 ha y^{-1}.

A special archeological survey was conducted in Nahal Zipporim during the present study (see below) to reconstruct the location and the history of the agricultural installations relative to the present Zipporim gully (Figure 7.12). A significant topographic threshold was found within the Zipporim valley, approximately 650 m below the present headcut. The agricultural installations are located above this threshold in an open valley more than 100 m wide, spaced 25 to 40 m apart, and perpendicular to the valley axis. Below the threshold, the valley is dissected by the prominent Zipporim gully, which divides the valley into an upper level and the incised gully bottom. The agricultural installations are located mainly within the upper level, on both sides of the present gully. They are distributed sporadically in much smaller plots and are parallel to the gully banks.

In addition, several water conduits were constructed in the lower segment below the topographic threshold, which diverted the flow of water from the Zipporim channel to several small agriculture plots located on the western edge of the dissected valley.

This configuration indicates that the topographic knick point marks the approximate location of a paleo-headcut. The archeological evidence (such as water-harvesting installations, pottery remains and the ruins of a nearby farm) indicates that most of these agricultural installations were built during the Byzantine period (Avni, forthcoming). The location of a paleo-gully head 650 m downstream relative to the present one enables the historical backward migration rate of the Zipporim gully since the Early Islamic period (1100 y BP) to be calculated. This rate is approximately 0.60 m y^{-1}, almost half of the present monitored retreat rate of the Zipporim gully central headcut (see Table 7.2 and Discussion).

Figure 7.12. Map of Nahal Zipporim valley downstream of headcut, showing configuration of Byzantine terraces relative to Zipporim gully. The topographic threshold is located 650 m below the active gully head. For explanation, see text

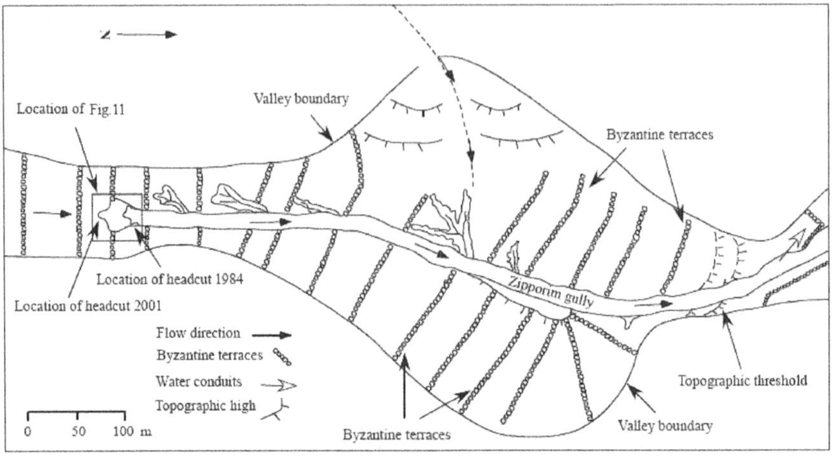

Nahal (wadi) Revivim

The Nahal Revivim drainage basin is located in the central part of the Negev Highlands region, 9 km northeast of Sede Boqer (Figure 7.1). The gully head is composed of two distinct headcuts, which join together to form the Revivim gully, which dissects the lower part of the valley. The area above the headcut drains 15 km^2. Most of the valleys above the headcuts are 100 to 120 m wide, covered with a fluvial section composed of redeposited loess sediments alternating with coarse conglomerate lenses of late Pleistocene–Holocene age. Below the headcut, a prominent gully developed, continuing downstream toward the Yeroham reservoir located 5 km to the north. The alluvial sequence exposed in the headcuts is composed of 1 m of basal conglomerate (20–30 cm limestone and chert pebbles), covered with a 2–3 m thickness of fluvial loess. The location of the site, close to the main road (Yeroham–Sede-Boqer), provided a good testing site for the relationship between the gully formation and migration, and the road system located along the valleys. Since the road was built in the early 1950s, it has been incised several times by gully migration in the vicinity of the Revivim gully head (Figure 7.13).

A detailed description of the headcut migration was presented by Avni (2005). In this chapter, the results are presented in Table 7.2.

Nahal (wadi) Boqer

The Boqer drainage basin is located 3.5 km northwest of Sede Boqer (Figure 7.1). It drains the central part of the Negev Highlands to the Mediterranean Sea via Nahal Besor, the major drainage basin in the region. The gully head is located in the lower part of the basin and the drainage basin above the gully occupies about 100 km^2. The headcut site is located in a flat valley 100–150 m wide, composed of chalk rocks of upper Cretaceous age (the Santonian Menuha Formation; Sneh et al., 1998). Two major headcuts, 3–4 m high, formed in this site on the central part of the valley. The results of the headcut migrations are presented in Table 7.2.

Nahal (wadi) Sekher

The Sekher drainage basin is located in the northern part of the Negev Highlands region, about 25 km north of Sede Boqer and 18 km south of Beer-Sheva (Figure 7.1). It drains the northern part of the Negev Highlands to the Mediterranean Sea via Nahal Besor, the major drainage basin in the region. The gully head is located in the central part of the basin and the drainage basin above the gully occupies about 100 km^2. The headcut site is located in a flat valley 30 to 40 m wide, composed of chalk rocks of Lower Tertiary age (the Eocene Adulam Formation; Sneh et al., 1998). This segment of the Sekher valley is a transition zone between the upper valley, covered by fluvial loess and conglomerates, and the lower valley, covered mainly by sandy loess. The sandy component within the alluvial section was contributed by the nearby dune-field that covers hundreds of square kilometers to the north and northwest of the site.

Two major headcuts, 2–3 m high, formed in this site on both sides of the valley. A detailed description of the headcut migration was presented by Avni (2005). In this chapter, the results are presented in Table 7.2.

Discussion

The mechanism of gully retreat

Observations at erosion sites during floods or a short time (a few hours to a few days) after flood events indicate that the mechanism of headcut migration is related mainly to direct erosion caused by flow shear stresses exerted by concentrated flow (Figure 7.8). The high erodibility of the alluvial section in the Negev Highlands is due to the high proportion of loessial soils and redeposited loess within the alluvial section that originated from the late Pleistocene loess deposits widespread in the region (Bruins, 1986; Bowman et al., 1986; Zilberman, 1992).

Figure 7.13. The Yeroham–Sede-Boqer road after the March 1991 floods near the Revivim headcut. The road was incised by a side bank headcut connected to the Revivim gully

The relationship between erodibility and gully retreat is well expressed in comparisons between the high retreat rates recorded for the Nahal Sekher headcut (17.8 m y^{-1}), composed of highly erodible sandy loess, and the moderate retreat rates recorded for the other headcuts (0.45–3.1 m y^{-1}), which are composed of less erodible redeposited loess and coarse alluvial material. The lowest retreat rate (0.44 m y^{-1}) was recorded from the eastern headcut of the Zipporim gully, which is composed mainly of chalky rocks. As the penetration of the floodwater into the loessial section in the valleys is limited to a depth of 1–3 m (Yair, 1983; 1987), the lower part of the alluvial section exposed in the headcut remains almost dry. The absence of less permeable soil horizons within the alluvial sections also lowers the possibility that the headcut retreat is related to seepage processes during or after floods.

During the retreat process, the gully head forms vertical headcuts. The height of the headcuts in most cases is equal to the thickness of the fine alluvial cover in the valley, exposing the bedrock or a highly resistant conglomerate unit at its base (Figure 7.6). Occasionally, some massive collapse of the gully banks occurs below the headcut and downstream along the gully as a result of the erosive undermining of the loess section by the force of the running water along the gully banks (Figure 7.9). The floodwater transports the eroded sediments downstream through the narrow channel for distances of up to several kilometers. Under natural conditions (e.g., without artificial digging), almost no local deposition of the loess sediments is found near the erosion sites, and the fine particles from the eroded alluvial section are transported downstream to the outlet of the drainage system, to the Mediterranean coast or to the Dead Sea basin. On the other hand, coarse alluvial material, (e.g., pebbles and chunks of rock), are deposited along the stream channels, forming shallow bar structures (Figure 7.7).

After heavy rainfall, the upper loessial section develops tension cracks, caused by the wetting and drying effect. The average thickness of the soil section separated by the cracks is 10–20 cm and they developed at the end of the wet season and during dry intervals within the wet season. This phenomenon is accompanied by soil topple, set in motion by gravity. In some cases the deep cracks led to a local collapse of the steep banks along the gullies. The contribution of this phenomenon to the total soil erosion in the gully head vicinity has not been investigated in detail. However, this process impacts on the increasing width of the gully banks below the gully head and on the total erosion rates along the gullies.

Initiation of the gullies in the Negev Highlands

Geological and archeological evidence clearly show that well-dissected gullies and alluvial channels have existed in the Negev Highlands region since the late Pleistocene OIS 2 and that gully activity intensified during the early Holocene and continued to do so throughout historical time (Evenari et al., 1982; Zilberman, 1992; Avni and Porat, 2002; Avni et al., 2006). In addition, the backward migration mechanism of headcuts implies that the location of the initial embryonic headcut must be downstream of the present headcut. As indicated above, the length of the gully channel below the headcut is several kilometers, hinting at the long time required for migration of a headcut from its initial location. Bull (1997) presented a concept connecting the initiation of gullies to topographical thresholds caused by the accumulation of alluvial deposits along the drainage network. Following this concept, the location of the embryonic gully head in the Negev Highlands should be located at the place where the alluvium sediments drifted from the high elevated terrain of the Negev Highlands and accumulated as alluvial fans along the Negev Highlands foothills (Zilberman, 1992). On the other hand, some local deposition may occur along the drainage network within the uplifted terrain of the Negev Highlands due to variations of valley floor gradients (Bull, 1997). Another simple explanation for headcut initiation links the younger headcut formation to a sudden collapse of one of the many stone dams, built in historical times by the ancient farmers. However, most of the well-developed gullies predate the agricultural activities in the region, as indicated above (Avni and Porat, 2002; Avni et al., 2006).

Another question is the importance of the vegetation cover in the valleys as an agent preventing erosion. Bull (1997), Fanning (1999) and Poesen et al. (2003) strongly stress the importance of the depletion of vegetation as the main cause for gully initiation. Observations made in the Negev Highlands region indicate that significant soil erosion also occurred in highly vegetated valleys. This implies that the scarce desert vegetation cannot totally prevent erosion during powerful floods resulting from high-intensity rainstorms, due to the high erodibility of the alluvial section. However, heavy grazing of desert vegetation can contribute to increasing retreat rates of the gullies due to the increase in soil erodibility. The exact location and timing of gully initiation in the Negev Highlands will require further research.

Rates of gully headcut retreat

The rates of headcut retreat presented in Tables 7.1 and 7.2 and in Figure 7.11 show great variability in annual values, both within the sites and between neighboring sites. These differences are summarized in Table 7.2.

Differences in annual rates at single sites and between sites depend on several factors, such as variations in flood discharge reaching each section along the cross-section of the valley due to micro-topographic differences, the specific location of the headcut within the valley floor (central versus side location), and the composition of the alluvial sequence and it erodibility. An example of these differences is reflected in the Zipporim headcut (Table 7.1), where the annual rates are normally higher in the central headcut (approximately 1 m y^{-1}), with several exceptions in which the western and eastern headcuts, located on both sides of the valley, retreat more because of local conditions. These differences become much stronger during drought years.

The importance of the section composition in the headcut retreat is reflected by the fact that the eastern headcut in the Zipporim gully, whose basal section is composed of chalky rocks, normally expresses the lower retreat rates (0.44 m $y\text{-}^{1}$) relative to the other headcuts composed of coarse conglomerate and redeposited loess, because of its relatively high resistance to erosion. This observation is not valid in several instances, especially in drought years, as a result of channeling of low floods into the shallow channel feeding the eastern gully of the Zipporim headcut. Similar differences are found between the two headcuts in Nahal Revivim in which the annual and the multi-annual rates of the southern headcut, composed mainly of erodible redeposited fluvial loess, are higher compared to those of the northern headcut composed of conglomerate and loess, which are less erodible (Avni, 2005). These differences in the local and annual scale become clearer in the multi-annual and regional scale presented in Table 7.2. The most important result reflected in Table 7.2 is that the mean retreat rates in Nahal Sekher are 5 to 40 times greater than those of the other sites, and the total volume of eroded soil from the Sekher site is 4.5 to 9 times greater than that from the others. This can be explained not only by the combined impact of the highly erodible alluvial component (sandy loess) but also by other factors, such as the relative northern location of the site (the mean annual precipitation in the upper basin is approximately 120 mm) and the size of the basin (100 km^2), which generates more powerful flood events under extreme and regional synoptic conditions (Sharon, 1972; Poesen et al., 2003).

In addition, the indirect relations and even inverse relations between the annual retreat rates and the annual rainfall, as presented in Table 7.1, hints that the actual annual and multi-annual retreat rates in each site are controlled by local conditions such as synoptic, geologic, geomorphologic, biotitic and anthropogenic. This phenomenon characterizes the arid and semi-arid environments worldwide (Bull, 1997; Fanning, 1999; Yair and Kossovsky, 2002; Poesen et al., 2003).

The general multi-annual rates reported from the Negev Highlands region are in accordance with the wide variety of retreat rates presented in other studies around the world (e.g., Bull, 1997; Fanning, 1999; Vandekerckhove et al., 2001, 2003; Poesen et al., 2003).

Hydrological significance of gully retreat

In Nahal Zipporim, the upstream valley above the headcut is 60–100 m wide, covered with a thick section of redeposited loess. Down the headcut, a narrow channel developed, 5–10 m wide, exposing the bedrock or a coarse conglomerate unit deposited on top of it. Downstream, gravel covers the channel bottom. In Nahal Revivim, the 100-m irrigated valley upstream was incised by the Revivim gully, 15–20 m wide. Coarse gravel covers the gully. A similar configuration appears in Nahal Sekher, where the upper valley, 30–40 m wide, is covered by redeposited sandy loess and dissected by a deep gully, 8–10 m wide, exposing the local chalk rocks on its bottom, and downstream is covered by coarse gravel. All these examples indicate that a general geomorphological change is taking place in the Negev Highlands region that is causing a narrowing of the valley floors into channels and gullies; these are changing the alluvial composition of the valley bottoms from thick sections of redeposited loess in the upper valley to coarse gravel within the gullies. This process is restricting the irrigation potential of the floods downstream of the headcuts to less than 5–25% of the original width of the valley and increasing the velocity and the flow power of the floods flowing within the narrow channels. These hydraulic parameters cause damage to the banks and the vegetation along them, and make attempts to control the water flowing within the gullies for irrigation operations more difficult.

On the wider scale, the headcut retreat process contributes to the reduction of the loessial segments within the valleys and to the parallel increase in the dimensions of segments whose bottoms are covered by rocks and gravel. In addition, the erosion of the highly erodible materials from the drainage basins, slopes and valleys (especially loess and redeposited loess sediments) causes the increase in the distribution and size of the bedrock outcrops in the entire region. This process generates a positive feedback effect that increases the runoff efficiency (due to ongoing change in the rock/soil ratio dictating the runoff coefficient; Yair, 1983, 1987), the power of the floods and their erosive impact. The gradual increase in floodwater volume in the arid and semi-arid zone can trigger better conditions for trapping more runoff water in reservoirs. In addition, the highly porous coarse alluvial sediments deposited along the gully floors (Figures 7.6 and 7.7) have the potential of contributing to the hydrologic recharge beneath the valley floor. This conclusion is hinted at by the occurrence of giant

Pistacia atlantica trees located along the gravelly channels in the Negev Highlands (Danin, 1983; Avni, Danin and Amit, 2003; Figures 7.7 and 7.14).

Effect of gully retreat on biomass, range and the agricultural potential

Change in the irrigation pattern in the valleys causes a sharp drop in the vegetative biomass. There is some 70–90% less floral biomass below the gully heads than above them (Avni, 2005; Figures 7.5 and 7.7). This phenomenon continues downstream to the lower segments of the valleys, almost without recovery. The drop in the biomass in the lower segments is caused by the combined effect of the limited vegetation located within the narrow gullies, and the gradual decrease in the biomass along the valley floor incised by the gullies, which increases parallel to the increasing distance downstream from the headcut. This process results from the ongoing increase in salinity in loessial soils left without floodwaters to leach-down the salts that accumulate in them following typically shallow penetration of rain into these soils is a region with such a high rate of evaporation (Yair, 1983; 1987).

The decrease in the biomass down the gullies is highly reflected in the range survey of the Negev Highlands (Seligman et al., 1962), which mapped the biomass in terms of grazing value and carrying capacity for grazing livestock. The upper segment of the valley above the headcut had the highest carrying capacity in the region, with rich herbaceous pasture vegetation. On the other hand, the ongoing development of the gullied lower segment caused an 83–99% reduction in the grazing value (Seligman et al., 1962). Simultaneously, the agricultural potential was severely reduced as a result of the stripping of the valley floor and the development of a dissected topography, thereby preventing cultivation (Figure 7.15). These observations are fundamental in the concept linking the contribution of the headcut migration to the natural desertification processes in the region (see below).

Figure 7.14. A *Pistacia atlantica* tree growing within the gravelly channel of Nahal Lotz in the Negev Highlands west of Mizpe Ramon, due to the lateral migration of the channel and the erosion of the banks

Figure 7.15. An agricultural terrace abandoned due to gully incision and soil erosion, preventing cultivation, in Nahal Zipporim

Changes in the ecological environment

The retreat of the gully heads along the valley axis causes a significant environmental change, as reflected in the composition of the vegetative assemblage. The upper valley vegetation, characterized by rich herbaceous vegetation accompanied by shrubs such as *Thymelaea hirsuta, Retama raetam, Moricandia nitens, Atriplex halimus* and others, is replaced within the gully by low density shrubs that can resist the high velocity flow running along the gully, such as *Achillea santolina*. Some of the shrubs, which are widespread in the upper valley, become rare within the narrow gully. The presence of herbaceous vegetation is almost negligible along the gully. On the other hand, isolated *Acacia raddiana* trees found along the gullies below 700 m and groups of *Pistacia atlantica* above 700 m seem to reflect the ecological change in which the annuals, subsided on the loessial soils, are replaced by low density trees subsided along the gravelly gullies (Danin, 1983). A recent study (Avni et al., 2003) indicated that there is a link between the distribution of young *Pistacia atlantica* trees in the Negev Highlands and the increased dimensions and distribution of the gravelly covered gullies in the region (Figures 7.7 and 7.14).

Some observations made in the region indicate that the species richness has increased within the gullies (Ward et al., 2001) as a result of the creation of new ecological niches and relative increase in the water availability along the gullies. A similar phenomenon is known from the rocky slopes in the desert, in which the increase in available water, resulting from the concentration of runoff on the exposed rock outcrops, enables the penetration of several Mediterranean floras to the Negev Highlands (Yair and Danin, 1980; Yair and Berkowicz, 1989).

Influence of gully migration on modern infrastructures

The gully head migration up-valley and instability in gully width over time are damaging modern infrastructures established in the region during the last 60 years, such as roads, pipelines and power lines (Figure 7.13). Most of this damage occurs because these constructions are predominantly located along the valleys, which serve as the most convenient transportation corridors in the rocky-hilly terrain of the Negev Highlands. In some cases, modern construction increases erosion and gully formation as a result of land disturbance caused by heavy machinery used in these activities. The parallel influences of gully development on infrastructures and vice versa are well known around the world (Bull, 1997; Fanning, 1999; Poesen et al., 2003).

Historical implications

Archeological surveys carried out in the Negev Highlands during the past two decades (Lender, 1990; Haiman, 1992; Avni, 1996) indicate that the first agricultural activity in the region was initiated during the Early and Middle

Bronze Age (5200–4000 BP; Lender, 1990; Avni, 1996). During these periods, cultivation activities were concentrated in the wide loessial valleys with no special installations (Lender, 1990). The first documented evidence of runoff harvesting installations (such as agricultural terraces) is from the Iron Age (3000–2700 BP; Bruins, 1986), indicating that during this time, soil erosion and downslope transportation of loessial sediments were well-known features. The high number of cisterns and large water channels found in the region indicate that long rocky slopes were already exposed in the region during this time. The agricultural activities were intensified during the Byzantine and Early Islamic period, (1700–1100 BP), in which an intensive effort toward increasing the area of arable land was made in the Negev (Rubin, 1990; Rosen and Avni, 1993; Avni, 1996). The Byzantine population terraced the wadis and leveled the valley floors. These activities greatly reduced the erosion of the wadi beds and enabled accumulation of the sediments transported from the slopes behind the terraces.

The archeological survey done in the Zipporim valley (Figure 7.12) indicated that human settlement in the vicinity of the Zipporim gully began during the Iron Age. The main effort to analyze the relations between development of the Zipporim gully and the agricultural installations built along the valley focused on the reconstruction of the Zipporim valley during the Byzantine period. The main conclusions from this survey are:

1. A large number of agricultural terraces were incised by the retreat of the Zipporim gully (Figure 7.15). Most of the agricultural terraces were attributed to the Byzantine period.
2. A significant topographic threshold, which marks the approximate location of a paleo-headcut during the early Byzantine period, has been reconstructed within the Zipporim valley, approximately 650 m below the present headcut (Figure 7.12).
3. The configuration of the agricultural installations described above and below the topographic threshold indicates that the Zipporim gully was already formed below it during the Byzantine period.

These observations and conclusions indicate that gully initiation in the Zipporim valley pre-dates the Byzantine period. The location of a paleo-gully head 650 m downstream relative to the present one enables the historical backward migration rate of the Zipporim gully to be calculated since the agricultural activity was terminated some 1100 years ago during the Late Byzantine-Early Islamic period. This rate is approximately 0.6 m y^{-1}, which is almost half of the present monitored retreat rate of the Zipporim gully central headcut (Table 7.2). Since its abandonment, the Zipporim gully has became entrenched along approximately 10% of the total length of the upper Zipporim valley. Similar rates and geomorphic significance were presented by Avni et al. (2006) in the Nahal Zena basin, 9 km south of Nahal Zipporim.

Examination of the positions of archaeological sites and agricultural installations relative to the location of gullies in the region indicates that gully formation and headcut retreat was a well-known feature during the Byzantine period. Besides the stone dams built by the ancient farmers to distribute floodwater in the agricultural fields, they built stone walls and dams parallel to the major gullies to protect the fields from damage caused by gully incision. In addition, water cisterns were constructed during the Byzantine period within dissected gullies utilizing the concentrated flow (e.g., Nahal Sekher, approximately 1.4 km below the present headcut). The nature and design of these constructions indicate that erosive processes and gully migration were active before and during the Byzantine period, following a pattern that seems to be similar to the present one.

The calculated rate of headcut retreat since historical times is 0.6 m y^{-1}, while present rates recorded from most of the monitored headcuts in the region are much higher (Table 7.2). These rates hint at accelerated erosion at present, generated by a positive feedback effect that is inherent in the erosion process (see below).

The concept of natural desertification

The parameters discussed above, i.e., the long process of soil erosion and gullying, the ongoing loss of biomass and richness of species, and the decreases in range and agriculture potential, indicate that the Negev Highlands region is undergoing a complex environmental change. Most of these parameters coincide with the definition of desertification processes, as established by the United Nations Conference on Environment and Development (UNCED, 1992). Therefore, these environmental indicators can be described in terms of ongoing increasing natural desertification of the region, which is advancing proportionately to the rate of the headcut retreat in each basin. During the monitoring period (1990–2007), the linear gully retreat in the study area ranged from 17.8 to 250 m, at an average annual rate of 1.04–17.8 m y^{-1} for each gully head. The total soil loss in these sites ranged between 1000 and 9000 m^3, at an average rate of 58.8–600 m^3 y^{-1} for each gully head. During the monitoring period, approximately 0.07–0.87 ha lost their agricultural and range value in each basin under study (Table 7.2). Since the Early Islamic period (1100 BP), approximately 6.5 ha, representing 10% of the high-value agricultural and rangeland in the Zipporim valley, were lost due to gully erosion (Figure 7.15). Similar rates were reported from other drainage basins in the region (Avni et al., 2006). It is important to remember that these calculations are based only on soil erosion monitored in the headcuts and do not represent soil erosion rates in the entire drainage basin caused by other erosion processes, e.g., sheet and rill erosion; nor do they take into account the possible existence of other headcuts operating simultaneously within the upper segments of the drainage basins in the region. Therefore, traditional calculations of soil loss rates, in terms of t km^{-2} y^{-1} (ranging in the study basins from 1.7 to 19.6, see Table 7.2) serve only as minimum estimations. Parallel to the soil loss, the biomass was depleted by 70–90% and the range value by 83–99% (Seligman et al., 1962).

However, several parameters seem to contradict the overall trend of natural desertification and can be described in terms of de-desertification processes. These processes include the increase in the species richness within some of the gullied segments (Ward et al., 2001), the increase in the number of young *Pistacia* trees germinated along the gullies (Avni et al., 2003) and the increases in floodwater volume and their potential contribution to shallow aquifers within the arid and semi-arid regions. Similar processes resulting from the erosion of loessial sediments along the hill slopes generate the appearance of several Mediterranean species in rocky niches within the desert environment (Yair and Danin, 1980; Yair and Berkowicz, 1989). Most of these parameters are a function of the gradual increase in rock outcrops in the region, as described in detail by Yair (1983; 1987), Yair and Berkowicz (1989) and Yair and Kossovsky (2002). These processes are active simultaneously with the above-mentioned natural desertification processes and have to be evaluated as part of the total environmental change.

Figure 7.16. A rocky basin in the Negev Highlands west of Mizpe Ramon with very limited vegetation cover, range or agricultural potential—an example of the end product of soil erosion processes

Combining the above-mentioned parameters facilitates total formulation of the non-linear and complex environmental changes acting in the region. Most of the parameters indicate that the environmental changes are leading toward increasing natural desertification of the region. The expected continuation of these processes in the future leads to the expectation of rocky desert regions with very limited vegetation cover, very limited biomass and very limited range and agricultural potential (Figure 7.16). Because of the slow and ongoing nature of these environmental processes, each drainage basin and region will achieve critical natural desertification status within its own individual time-scale, as a function of its own unique environmental parameters (Bull, 1997; Fanning, 1999; Poesen et al., 2003).

Model of Soil degradation and desertification for the Negev Highlands

The above-mentioned observations indicate that ongoing environmental change has been active in the Negev Highlands for the past several millennia. The reconstruction of this long-term environmental change leading toward natural desertification was formulated in terms of a simple conceptual model (Figure 7.17), with the following stages:

1. Geological and climatological manifestations (Issar and Bruins, 1983; Bowman et al., 1986; Avni, 1991; Zilberman, 1992; Offer et al., 1992; Avni et al., 2006) indicate that during the last glacial phase in the late Pleistocene (OIS 4 and 3), the Negev Highlands were covered by a thick section of loessial sediments, causing a reduction in the flow gradients. During this stage, almost no runoff could be generated in the region due to its thick soil cover (Yair, 1983; 1987).
2. Following OIS 2, and more intensively subsequent to the post-glacial late Pleistocene major climatic shift, these sediments were eroded from the slopes and accumulated as redeposited loess in the valleys (Bruins, 1986; Zilberman, 1992; Avni et al, 2006). The gradual exposure of the rocky slopes triggered an ongoing increase in the runoff coefficient (Yair, 1983, 1987; Yair and Kossovsky, 2002), and the flow from the slopes reached the valleys.
3. As a result of the floods running down the valleys, embryonic gullies formed in topographic threshold locations (Bull, 1997). From their initial locations the headcuts gradually retreated upstream.
4. Agricultural activity, initiated in the region during the Bronze Age (Lender, 1990; Avni, 1996), and reaching its zenith during the Byzantine and Early Islamic period (Rosen and Avni, 1993; Patrich, 1995; Avni, forthcoming), was concentrated mainly in the non-eroded upper segments of the valleys and entailed fields irrigated from floods and runoff generated by the nearby rocky slopes (Yair, 1983, 1987). During the time that these installations were utilized, the suspended sediments in the floodwater were trapped behind the stone dams and accumulated to depths of 1.5–2 m (Avni et al., 2006). As part of the construction of the agricultural installations, much effort was invested in protection strategies against soil erosion and penetration of headcuts into the fields (Evenari et al., 1982). Simultaneously, water cisterns were dug along the dissected gullies. At this stage, irrigation water and land suited for cultivation were widespread in the region and the ancient farmers developed a dynamic equilibrium for some time between natural soil erosion and land conservation activities.

5. As the process advanced, agricultural soil was eroded mainly by gullying. As a result of the ongoing increase in rock exposure, the runoff increased, generating powerful floods in a positive feedback effect, with no relation to climatic change. At this stage, agricultural activities were transferred upstream to less-eroded segments, and the region suffered escalating natural desertification.
6. As these processes continued to their final stage, the region was transformed into a totally rocky desert with a very limited biomass, range and agricultural potential (Figure 7.16).

Figure 7.17. A conceptual model presenting the relations between soil degradation, runoff and agricultural potential in the Negev Highlands. For explanations see text

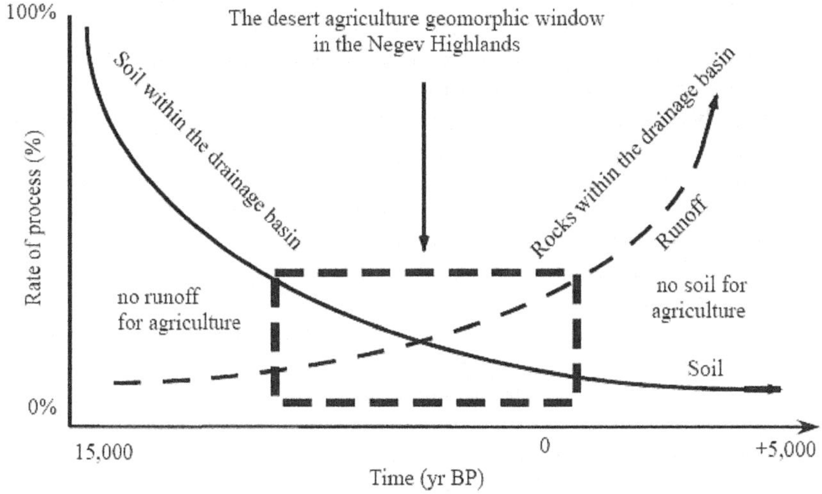

As reflected in the model, the environmental change was generated mainly by the natural dynamic change in the soil/rock ratio evolving within the drainage basins since the post-glacial late-Pleistocene major climatic shift. This phenomenon can be related to a long-term readjustment of the geomorphologic system to the Holocene climate. As a result, the late Pleistocene sediments, originally deposited within the drainage basins, are slowly being removed, causing loss of agricultural potential and a rapid decrease in the vegetation biomass (Figures 7.5 and 7.7).

As mentioned above, the present retreat rates in most of the gully heads are greater than those calculated for historical times (0.6 m y^{-1}), hinting at accelerated erosion. This phenomenon is a result of the positive feedback effect between the increasing portion of rocks in the drainage basins and the runoff generating soil erosion (Figure 7.17). A similar feedback effect was presented by Bull (1997) from the Arroyo phenomenon in the American Southwest and by

Fanning (1999) from gully retreat in southeast Australia. It can be concluded that if the erosion process continues to follow the pattern described herein, most of the valleys in the Negev Highlands will be transformed into rocky valleys, losing their agricultural potential, within a few thousand years (Figure 7.15, Figure 7.16). As desert agriculture is based on runoff harvesting techniques, the delicate relations between runoff efficiency and agricultural soils within the drainage basins are critical for the success of agriculture. These delicate relations are pictured in Figure 7.17 as "the desert agriculture geomorphic window." Therefore, the dynamic change evolving within the drainage basins endangers the agricultural activities and reduces it potential over time.

The long-term erosion process presented in the model that seems to be almost linear is triggered by the fact that the fine loessial sediments in the Negev Highlands, most of them brought to the region during the last glacial phase (Issar and Bruins, 1983; Bowman et al., 1986; Zilberman, 1992), are limited in quantity and of non-renewable origin. The fundamental geomorphic processes causing the land degradation and natural desertification, such as the ongoing change in the soil/rock ratio in the basins, will seemingly continue in the same trend in the future. After reaching a certain state of soil erosion, the incision process in the region becomes irreversible, as do the desertification processes relating to it. The irreversible nature of the erosion process occurring in the Negev Highlands is evident in the geological record, which shows very limited remains of early and middle Pleistocene sediments in the drainage basins (Avni, 1991; Zilberman, 1992), indicating that the erosive processes active in the region during the interglacial phases were aggressive and most likely washed the fine sediments from the drainage basins.

The Negev Highlands model for soil erosion presented here differs from the concept of the Arroyos cyclic development presented by Bull (1997), claiming that about 100 years are needed to create a prominent gully and 500 to 2000 years are required for complete aggradation within it up to a complete recovery. It seems that the main difference between the two models is related to the balance of the fine sediments in the basins. The large resources of the fine sediments in America's southwest (Bull, 1997) or in Australia (Fanning, 1999) make some of the land degradation processes in these regions reversible, especially after the recovery of the vegetation (Bull, 1997; Fanning, 1999). Another difference between the two models is the importance attributed to human intervention in environmental processes, relative to the natural processes. Human intervention seems to be very important especially in southeast Australia (Fanning, 1999), where no significant gully erosion is reported before the European settlement, some 200 years ago. On the other hand, in the Negev Highlands desert, the anthropological effect on the environment plays a much more complex role, which is superimposed on the long term natural trend (Avni et al., 2006).

Combating gully erosion

Measures were taken for combating soil erosion in the Negev Highlands region almost parallel to the initiation of the agricultural activities in the region. Land conservation techniques (such as terrace construction) were implemented in the region during historical times. The Byzantine-early Islamic cultural intervention was sustained for at least 500 years, indicating that ancient farmers practiced land management, which established a dynamic equilibrium between ongoing natural erosion and human activities. Once the land was abandoned, natural processes were reestablished, following the natural Holocene trend of erosion.

As a result, some of the Byzantine terraces collapsed and the fields were stripped of their soil. At other locations, the stone works still exist, representing an excellent example of long-impact land conservation (Figure 7.18). Some of them managed to stop soil erosion for more than 3000 years and they serve as evidence that human intervention in the environment can decrease natural soil erosion and counteract natural desertification processes. Today the region under study is a marginal area with very limited agricultural importance. Most of the modern agricultural activities are limited to plots irrigated by pipes. The local Bedouin living in the vicinity of the gullies use the preserved upper valleys as preferred rangeland, with no attempt to prevent or to combat the escalating erosion (Portnov and Safriel, 2004). However, a few headcuts, located near major roads, were restored by construction of cemented or stone-built spillways. Other examples of combating soil erosion and gully retreat—by the planting of trees—are widespread in the northern Negev, to the north of the Negev Highlands region (Portnov and Safriel, 2004).

Conclusions

The geomorphological, hydrological, biological and environmental parameters described above indicate that a natural desertification process is ongoing in the desert ecosystem of the Negev Highlands. This process is proceeding at a rate that is proportional to the rate of headcut retreat of the gullies in the region. The gully-retreat process is accompanied by erosion of soil that has high agricultural and range value. Since the Early Islamic period (1100 BP), 10% of the high-value agricultural and range land in the Zipporim valley has been lost due to gully erosion, accompanied by a sharp drop in biomass, and by a reduction in range value in the eroded segments of the valleys. Similar processes were observed in most of the valleys in this region. If the erosion processes described herein continue, most of the drainage basins in the Negev Highlands are likely to be transformed into rocky basins, losing their agricultural potential within a few millennia.

This major environmental change is the result of a long-term natural ongoing process, which seems to follow the long-term readjustment of the geomorphologic system to the Holocene climate, causing the ongoing degradation of the soil and biomass, and leading to the increasing natural desertification of the region. This process is not related to any modern anthropogenic activity or short-term global climatic change.

However, historical and archeological data indicate that the ancient inhabitants of the region had already implemented successful long-term land conservation techniques 3000 BP, thus implying that a sustainable land management policy can be adapted to the Negev Highlands, and to other semi-arid regions in the Middle East. Similar natural desertification processes are operating in most of the arid and semi-arid regions around the world, but their specific parameters have to be characterized and evaluated regarding the local conditions in each region.

Figure 7.18. An agricultural terrace in the Negev Highlands south of Sede-Boqer, serving as an example of long-term land conservation—constructed during the Byzantine Period (1400–1600 BP), it is currently in use by the local Bedouin population

Acknowledgements

Financial support for this research was provided by the Geological Survey of Israel, by the Ramon Science Center, The Desert Research Institute, Mizpe Ramon, and by the Ministry of Science and Technology of Israel. The archeological survey of Nahal Zipporim region was carried out by G. Avni from the Israel Antiquities Authority. N. Porat from the Geological Survey of Israel provided the OSL age determinations of loessic sediments and soils widely used in this study. J. Plakht from the Ramon Science Center is most appreciated for the fruitful discussions and for the fieldwork.

References

Arkin, Y., & Braun, M. (1965) Type sections of Upper Cretaceous formations in the Northern Negev (southern Israel). *Israel Geological Survey*. Stratigraphic Section no. 2a., Jerusalem.

Avni, G. (1996) *Nomads, farmers and town-dwellers: pastoralist-sedentist interaction in the Negev Highlands, sixth–eighth centuries CE*. Israel Antiquities Authority, Jerusalem.

Avni, G. forthcoming. The Byzantine-Islamic transition in the Negev—an archeological perspective. *Jerusalem studies in Arabic and Islam*.

Avni, Y. (1991) The geology, paleogeography and the landscape evolution of the central Negev Highlands and the western Ramon structure. *Israel Geological Survey*, Report GSI/6/91, Jerusalem (in Hebrew, with English abstract).

Avni, Y. (2002) Gully incision as the main cause for ongoing aridization of the Negev Highlands. *Israel Geological Survey Current Research*, 13:19–27.

Avni, Y. (2005) Gully incision as a key in desertification in an arid environment, the Negev Highlands, Israel. *Catena*, 63:185–220.

Avni, Y., Danin, A., & Amit, R. (2003) The *Pistacia atlantica* in the Negev Highlands. Unpublished report, submitted to the Ministry of Science in Israel.

Avni, Y., & Porat, N. (2002) Environmental effects related to the degradation and erosion of Pleistocene sediments in the Negev Highlands. *Israel Geological Survey Report*, GSI/1/2002. Jerusalem (in Hebrew).

Avni, Y., Porat, N., Plakht, J., & Avni, G. (2006) Geomorphic changes leading to natural desertification versus anthropogenic land conservation in an arid environment, the Negev Highlands, Israel. *Geomorphology*, 82:177–200.

Billi, P., & Dramis, F. (2003) Geomorphological investigation on gully erosion in the Rift Valley and the Northern Highlands of Ethiopia. *Catena*, 50:353–368.

Botha, G. A. (1996) Cyclical colluvial accretion on bedrock pediments during the late quaternary in northern KwaZulu-Natal, South Africa. *Zeitschrift fur Geomorphologie N.F.*, 103:85–102.

Botha, G. A., Wintle, A. G., & Vogel, J. C. (1994) Episodic late quaternary paleogully erosion in northern KwaZulu-Natal, South Africa. *Catena*, 23:337–340.

Bowman, D., Karnieli, A., Issar, A., & Bruins, H. J. (1986) Residual colluvio-aeolian aprons in the Negev Highlands (Israel) as a paleo-climatic indicator. *Paleogeography, Paleoclimatology, Paleoecology*, 56:89–101.

Braun, M. (1967) Type sections of Avedat group, Eocene formations in the Negev (southern Israel). *Israel Geological Survey*, Stratigraphic Section no. 4, Jerusalem.

Bruins, H. J. (1986) Desert environment and agriculture in the central Negev and Kadesh Barnea during historical times. Published Ph.D. dissertation, The Agricultural University of Wageningen, the Netherlands. The Netherlands: Midbar Foundation, Nijkerk.

Bruins, H. J., & Berliner, P. R. (1998) Aridity, climatic variability and desertification. In H. J. Bruins, & H. Lithwick (Eds.), *The arid frontier: Interactive management of environment and development* (pp. 97–116). Dordrecht: Kluwer Academic Publishers.

Bruins, H. J., & Lithwick, H. (Eds.) (1998) *The arid frontier: Interactive management of environment and development*. Dordrecht: Kluwer Academic Publishers.

Bruins, H. J., & Yaalon, D. H. (1992) Parallel advance of slopes in aeolian loess deposits of the northern Negev, Israel. *Israel Journal of Earth Sciences*, 41:189–199.

Bull, W. B. (1997) Discontinuous ephemeral streams. *Geomorphology*, 19:227–276.

Clarke, M. L., Vogel, J. C., Botha, G. A., & Wintle, A. G. (2003) Late quaternary hillslope evolution recorded in eastern South African colluvial badlands. *Palaeogeography, Palaeoclimatology, Palaeoecology*, 197:199–312.

Danin, A. (1983) *Desert vegetation of Israel and Sinai.* Jerusalem, Israel: Cana Press.

Eriksson, M. G. (1998) Landscape and Soil History in Central Tanzania. Ph.D. Dissertation, Stockholm University, Stockholm, Sweden.

Evenari, M., Shanan, L., & Tadmor, N. (1982) *The Negev–the challenge of the desert.* Cambridge, MA: Harvard University Press.

Fanning, P. C. (1999) Recent landscape history in arid western New South Wales, Australia: a model of regional change. *Geomorphology*, 29:191–209.

Haiman, M. (1992) Early settlements patterns of the Negev Highlands. Ph.D. Thesis, Hebrew University of Jerusalem, Israel (in Hebrew, with English abstract).

Issar, A., & Bruins, H. J. (1983) Special climatological conditions in the deserts of Sinai and the Negev during the latest Pleistocene. *Palaeogeography, Paleoclimatology, Palaeoecology*, 43:63–72.

Lender, Y. (1990) Map of Har Nafha (196) 12-01. Jerusalem: Israel Antiquities Authority.

Mainguet, M. (1991) Desertification: natural background and human mismanagement. *Springer Series in Physical Environments*, vol. 9. Heidelberg: Springer Verlag.

Mainguet, M., & Le'tolle, R. (1998) Human-made desertification in the Aral Sea Basin: planning and management failures. In H. J. Bruins & H. Lithwick (Eds.), *The arid frontier: Interactive management of environment and development* (pp. 129–142). Dordrecht: Kluwer Academic Publishers.

Ninari, N., & Berliner, P. R. (2002) The role of dew in the water and heat balance of bare loess soil in the Negev desert: quantifying the actual dew deposition on the soil surface. *Atmospheric Research*, 64:323–334.

Offer, Z. Y., Zangvil, A., & Azmon, E. (1992) Characterization of airborne dust in the Sede Boker area. *Israel Journal of Earth Sciences*, 41:239–245.

Oostwoud Wijdenes, D., Poesen, J., Vandekerckhove, L., Nachtergaele, J., & De Baerdemaeker, J. (1999) Gully-head morphology and implications for gully development on abandoned fields in a semi-arid environment, Sierra De Gata, Southeast Spain. *Earth Surface Processes and Landforms*, 24:585–603.

Patrich, J. (1995) Church, state and the transformation of Palestine: the Byzantine period. In T. E. Levi (Ed.), *The archaeology of society in the Holy Land.* London: Leicester University Press.

Poesen, J., Nachtergaele, J., Verstraeten, G., & Valentin, C. (2003) Gully erosion and environmental change: importance and research needs. *Catena*, 50:91–133.

Poesen, J., Vandekerckhove, L., Nachtergaele, J., Oostwoud Wijdenes, D., Verstraeten, G., & Van Wesemael, B. (2002) Gully erosion in dryland environments. In L. J. Bull, & M. J. Kirkby (Eds.), *Dryland rivers: Hydrology and geomorphology of semi-arid channels* (pp. 229–262). John Wiley and Sons.

Portnov, B. A., & Safriel, U. N. (2004) Combating desertification in the Negev: dryland agriculture vs. dryland urbanization. *Journal of Arid Environments*, 56:659–680.

Rosen, S. A., & Avni, G. (1993) The edge of the empire: the archaeology of pastoral nomads in the southern Negev Highlands in late antiquity. *Biblical Archaeology*, 56:189–199.

Rubin, R. (1990) *The Negev as settled land: Urbanization and settlement in the desert in the Byzantine period.* Jerusalem: Yad Ben Zvi Press (in Hebrew).

Seligman, N., Tadmor, N., & Raz, Z. (1962) Range survey of the central Negev. The National and University Institute of Agriculture. Bulletin 67 (in Hebrew with English abstract).

Sharon, D. (1972) The spottiness of rainfall in desert areas. *Journal of Hydrology*, 17:161–175.

Sharon, D., & Kutiel, H. (1986) The distribution of rainfall intensity in Israel, its regional and seasonal variations and its climatological evaluation. *Journal of Climatology*, 6:277–291.

Shi, C., Dian, Z., & You, L. (2002) Changes in sediment yield of the Yellow River basin of China during the Holocene. *Geomorphology*, 46:267–283.

Sneh, A., Bartov, Y., Weissbrod, T., & Rosensaft, M. (1998) Geological map of Israel, 1:200,000, sheets 1–4. Jerusalem: Israel Geological Survey.

UNCED (1992) United Nations Conference on Environment and Development. Rio de Janeiro. June 1–12.

UNESCO (1979) Map of the world distribution of arid regions: explanatory note. Man and Biosphere (MAB) Technical Notes no. 7. Paris: UNESCO.

Vandekerckhove, L., Nachtergaele, J., Oostwoud Wijdenes, D., Verstraeten, G., & Van Wesemael, B. (2002) Gully erosion in dryland. In L. J. Bull & M. J. Kirkby (Eds.) *Dryland rivers: Hydrology and geomorphology of semi-arid channels* (pp. 229–262). John Wiley and Sons.

Vandekerckhove, L., Poesen, J., & Govers, G. (2003) Medium-term gully headcut retreat rates in southeast Spain determined from aerial photographs and ground measurements. *Catena*, 50:329–352.

Vandekerckhove, L., Poesen, J., Oostwoud Wijdenes, D., & Gyssels, G. (2001) Short-term bank gully retreat rates in Mediterranean environments. *Catena*, 44:133–161.

Vandekerckhove, L., Poesen, J., Oostwoud Wijdenes, D., Nachtergaele, J., Kosmas, C., Roxo, M. J., & De Figueiredo, T. (2000) Thresholds for gully initiation and sedimentation in Mediterranean Europe. *Earth Surface Processes and Landforms*, 25:1201–1220.

Ward, D., Feldman, K., & Avni, Y. (2001) The effect of loess erosion on soil nutrients, plant diversity and plant quality in the Negev desert wadis. *Journal of Arid Environments*, 48:461–473.

Yair, A. (1983) Hillslope hydrology, water harvesting and areal distribution of some ancient agriculture systems in the northern Negev desert. *Journal of Arid Environments*, 6:283–301.

Yair, A. (1987) Environmental effects of loess penetration into the northern Negev desert. *Journal of Arid Environments*, 13:9–14.

Yair, A., & Berkowicz, S. M. (1989) Climatic and non-climatic controls of aridity: the case of the Northern Negev of Israel. *Catena Supplement*, 14:145–158.

Yair, A., & Danin, A. (1980) Spatial variations in vegetation as related to the soil moisture regime over an arid limestone hillside, northern Negev, Israel. *Oecologia* (Berlin) 47:83–88.

Yair, A., & Kossovsky, A. (2002) Climate and surface properties: hydrological response of small arid and semi-arid watersheds. *Geomorphology*, 42:43–57.

Zilberman, E. (1981) The geology of the central Sinai-Negev shear zone, central Negev: Part A. The Saad-Nafha lineament. *Israel Geological Survey Report* Hydro/1/81, Jerusalem.

Zilberman, E. (1991) Landscape evolution in the central, northern and northwestern Negev during the Neogene and the Quaternary. Ph.D. Thesis, Hebrew University of Jerusalem. *Israel Geological Survey Report* GSI 45/90. Jerusalem (in Hebrew, with English abstract).

Zilberman, E., (1992) The late Pleistocene sequence of the northwestern Negev flood plains: a key to reconstructing the paleoclimate of southern Israel in the last glacial. *Israel Journal of Earth Sciences*, 41:155–167.

Chapter 8

Livestock Production and Fodder Improvement

A. Allan Degen

My research is focused on livestock production and improvement of fodder availability. In this chapter I describe four projects:

- Examination of livestock production by Bedouin of the Negev Desert in Israel (Degen, Benjamin and Hoorweg, 2000);
- Dietary selection by grazing Karakul sheep in southern Kazakhstan (Degen, Benjamin, Abdraimov and Sarbasov, 2002);
- Selection of fodder trees and intercropping for forest rejuvenation and goat production in the mid-hills of Nepal; and
- Livestock production in drought prone Afghanistan (Degen and Weisbrod, 2004).

Livestock production by Bedouin of the Negev Desert in Israel*

Introduction

There are approximately 170,000 Bedouin in the Negev Desert of Israel—about 50% living in urban communities and 50% in spontaneous hamlets (Statistical Yearbook of the Negev, 2004). About 10,000 Bedouin or about 1000 families depend on raising small ruminants (mainly sheep) as their main source of livelihood. Grazing lands allotted to them are insufficient to meet year-round flock maintenance, yet flock movement to available grazing areas is often restricted by government rules and regulations. Today, flock owners are perceived by government authorities and, indeed, by the general public, as living on both the geographic and economic margins of the Negev. They practice pastoralism on uncultivable lands or on fallow and aftermath fields, and only about 10% of Bedouin depend on sheep for their livelihood. It is surprising then, that in the late 1980s—when many *kibbutzim* and *moshavim* abandoned sheep raising because of rising input costs and falling sheep prices—Bedouin continued to keep their flocks. There are indications that they may even have increased their holdings. How were the Bedouin able to support their households by raising sheep in a changing physical, economic and socio-political environment not conducive to this enterprise and how are they able to do so today?

Findings

The changes that have taken place since 1948 have turned the Bedouin pastoralists of pre-Israel times into marginal pastoralists today. Two important characteristics of current Bedouin pastoral activity in the Negev are evident: 1) pastoralism can be practiced mainly in the margins of other agricultural activities, on fallow and aftermath fields and in uncultivable areas; and 2) pastoralism has become a marginal occupation for the Bedouin population, as only about 10% (one-thousand families) derive their livelihood from raising sheep. The grazing sources allotted to them are insufficient to meet year round flock maintenance, yet flock movement to available grazing areas is strictly limited by rules and regulations pertaining to their herding practices.

---

* Reprinted with permission from Nomadic Peoples, with deletion of two tables, from "Bedouin households and sheep production in the Negev desert, Israel," A. Allan Degen et al., Nomadic Peoples NS (2000) (4)1: 125–147.

In spite of these difficulties, there appears to have been a considerable increase over the years in the number of sheep that are raised. In 1961, there were approximately 70,000 sheep in the Negev, today there are up to 300,000 sheep. Officially, 150,000 are registered with the Ministry of Agriculture's Veterinary Services, but the Ministry has estimated that this is approximately 50% of the actual number of sheep owned by Bedouin.

Flocks are officially recognized as being owned by household heads. However, it may not mean their outright possession, but rather their responsibility for the decision making in the management of the flock. Each household is usually related to others in a complex of alliances, obligations, reciprocal relations and family status. Within the flock, some members of the household (wives) and extended family (sons) may own some animals and these are considered their property (Abu-Rabia, 1994). Then again, household heads often register some of the animals in the names of other family members and/or deliberately understate the number of animals in their flocks to the government veterinary authority (compulsory for vaccination against contagious diseases). It also appears that total holdings are always understated and not registered with the authorities. Each household is considered a patriarchal unit where the household head or elder male makes all the decisions regarding flock management for all members of the family who may have ownership of some of the animals in the flock. As a family-operated enterprise, wages are not paid (except where shepherds are hired) nor are expenses and income shared amongst family members.

Marketing

To maintain their flocks and cultivate crops, Bedouin must have access to grazing areas, supplementary feeds (straw, bran, hay, grain), seeds and agricultural contractors for cultivation, harvesting and transport. They must also have outlets for their produce, and access to retail markets for essential human foodstuffs. There are no official marketing channels, such as those available for other agricultural enterprises in Israel (e.g., milk, poultry, fruit, citrus marketing boards), for Bedouin livestock raisers. In addition, because they cultivate land below the 220 mm isohyte, they are not eligible for drought compensation. The absence of any organized marketing venue and lack of any drought compensation effectively prevent the Bedouin from financial assistance such as credits and guaranteed prices. They are rarely able to negotiate bank loans or overdraft accounts. In addition, Bedouin buy their agricultural inputs individually. Large processing mills and factories, as well as Israeli agricultural cooperatives (*kibbutzim* and *moshavim*), however, prefer to sell their products either to large traders or through marketing boards.

Bedouin pastoralists, from roughly the mid-1950s, changed relatively quickly from an essentially subsistent and self-contained economy into the money-dependent market economy that exists today. Nearly all transactions take place on a purely cash and carry basis, that is, the ability to buy and sell depends on the ready availability of cash money. This situation appears to have arisen out of the desire of the Bedouin to hide their income sources from tax and other government authorities. Hence money received from all sources of income is kept in cash by household heads and by those women eligible for maternity and child benefits.

Personal connections are important in acquiring grazing sites. Some Bedouin have managed to establish long-standing, friendly relations with *kibbutzim* and *moshavim,* the main growers of small-grain winter crops, which enable them to lease aftermath grazing areas long before the summer grazing season begins and often at favorable rentals. These relationships, with time, have become solidly based on mutual trust and benefit. Others have managed to establish similar favorable relations with official authorities responsible for issuing permits for livestock grazing movement, such as the Ministry of Agriculture's Bedouin section, the army and the Jewish National Fund's forestry service. The ability to establish these relationships provides needed security for a grazing system marked by natural drought and changing governmental policies that constitute important factors in livestock management. Apart from the above input sources and their own cultivated areas, inputs such as straw, hay and grains are purchased from large-scale traders who can also deliver purchases to the homestead.

Livestock is often sold from the households, some to other Bedouin, including members of the extended family, for traditional purposes such as religious holidays, weddings, births and circumcisions, and some to traders, often Palestinians. These sales eliminate transport costs and keep the transaction secretive as veterinary supervision (apart from a transport permit) is eliminated. Prices of sheep increase considerably at holidays. Sheep are also sold at weekly markets (Kressel and Ben-David, 1995). A market is held in Beer Sheva on Thursday morning and smaller markets are held in Tel Sheva on Friday morning and in Rahat on Saturday morning. Most of the buyers are traders from the Palestinian administered towns of Dahariyya, Hebron and Gaza.

Flock size and management

After the establishment of Israel, Bedouin pastoral practices can be regarded as a response adaptation to government restraints. Noy-Meir (1975) identified major Bedouin pastoral systems within the Negev as determined by ecological conditions, flock movements and types of pasture, and the degree of feed supplementation. They practice the seasonal type of pastoral nomadism in the semi-arid to arid Negev, in that flocks are kept near the homestead in winter and are moved from the homestead for grazing in spring and summer.

Bedouin usually raise Awassi sheep, a fat-tailed breed raised in the Middle East that is well adapted to the desert (Degen, 1977). Bedouin also own black goats of mixed local origin for household purposes, and some donkeys for local transportation and herding. In addition, camels can be kept for milk, some chickens for eggs and meat, a cow or two (mainly by Bedouin of peasant origin) for milk and calves and an occasional horse for riding (prestige).

Bedouin living in municipalities and non-pastoralist Bedouin living in spontaneous hamlets also raise sheep. We estimated that 35% of the municipal families own about twelve sheep each for a total of 27,000 head, and that half of the non-pastoralist families in the spontaneous hamlets have about twenty-five sheep each for a total of 38,000 head. Dinero (1998) reported that 35% of municipal families in one town (Segev Shalom) owned livestock in 1993, which declined to 22% in 1996. The number of sheep held by pastoralists is about 85,000 to 235,000 head or 85 to 235 sheep per family.

In families living in spontaneous hamlets, unmarried daughters shepherd animals; wives and occasionally husbands and sons help. As more girls are now attending school since education is compulsory by Israeli law (Abu-Saad, 1996), the family labor force is being reduced. When needed, shepherds from the administered territories are hired. None of the sons plan to raise livestock as a livelihood, although many maintain small numbers of animals within the parental household flock or near their homes. Wives own animals and either keep them with the main flock or corralled at home. Husbands decide when and where to graze, when to corral the main flock at home and when to sell or buy sheep. Husbands are responsible for flock movement, either by foot or by truck, to distant pasture sources. They are also responsible for the everyday management of the flock, such as providing drinking water (by either piped water or tractor-drawn water tanks) and other essential supplies.

Each year, household heads apply to the Government Land Authority for grazing areas. Flock movement to these areas is limited by permits issued by the Bedouin Affairs Department of the Ministry of Agriculture to about nine months during spring and summer. Flocks are kept at the homestead for about three months during the late summer and winter. Black goats are restricted to the confines of the household because of the "black goat" law passed in 1977 (as grazing black goats were considered destructive). Permits are issued only after flocks have received mandatory veterinary vaccinations. Consequently, veterinary care involves a compulsory and important expense in the flock management budget.

Grazing permits allow flock movement from approximately mid-February to October. These permits are issued by the Ministry of Agriculture in conjunction with the Government Land Authority in Beer Sheva, and they stipulate the designated areas and exact dates for grazing. The grazing period allowed is dependent on the condition and availability of pasture. In spring, these are usually lands in the control of the Jewish National Fund (mainly forests), the army and the Land Authority. Payment for these sites is minimal and is based on flock numbers. Arrangements are also made to graze private land, such as those of *kibbutzim* or *moshavim.*

Some sheep and goats are milked by all families for household use. However, few families milk all the lactating ewes; labor shortage is given as the main reason for not milking. Milking is done by wives and/or daughters. In an average flock, milk yield is approximately 20 kg per day over a two-month period between February and March, for a total of 1200 kg. The milk is processed into yoghurt *(laban)*, butter *(samne)* and a hard, dry cheese *(afig),* which is stored for future consumption (Degen, 2003). Most of the *samne* and *afig* is made for home consumption, but occasionally some is given as presents; very little is sold by the women.

All sheep are hand-sheared prior to summer. Wool yield averages about 1.5 kg per sheep. Husbands, wives, children and shepherds do the shearing. There is little demand for the wool and it is used mainly for blanket and pillow filling. Almost no tents are made of wool and no weaving uses wool. Tents are made from burlap bags and carpets are hand-woven from colorful, synthetic material.

Beside occasional droughts, there are other problems in sheep management and marketing of animals by Bedouin. First, the Palestinian West Bank territories are often closed for security reasons, thus preventing buyers from the territories from entering Israel. This is a serious setback as these buyers purchase most of the Bedouin sheep and sell them in the territories. In addition, shepherds from the territories cannot continue to be employed, as, by law, they cannot remain overnight in Israel. Second, outbreaks of foot and mouth disease occur often, which prevents animal movement and limits sales. These factors, together with the droughts, result in lower sheep prices and higher grain and fodder prices.

Production results

There are approximately 130 breeding ewes per household and two to four rams. Lambing occurs mainly between November and March, but continues throughout the year as rams graze with the ewes. Of the lambing ewes, 2.5% lamb twins and produce a total of 120 lambs, of which 105 survive. Thus, the lambing rate is 0.93, with lamb mortality at 12.5%. The sheep mortality is about 3.0%. Of the surviving lambs, 68 (55–60%) are sold at three to five months of age, which mainly covers expenses. Of the rest, 15 ewe lambs are kept as replacements for ewes that either die, are sold, or are slaughtered, and 20 lambs are used as presents and for traditional slaughter. Sick animals are not slaughtered for home consumption, but sold whenever possible. The income from sheep averages US$ 9,394 per year, but the variation among individual households is considerable.

The expenses are fairly constant for individual cases, except the expenses for land rental, which varies considerably from year to year. Most of the expenditure is on animal feed (60%), followed by land rental (20%), wages (8%) and veterinary costs (7%). On average, expenses are US$ 7,470 and constitute about 80% of the incomes, although this varies from household to household.

Income from sheep raising averages US$ 1,924. This does not take into account lambs used for social purposes and for home consumption. Large annual fluctuations in income are common, as was also found by Ginguld (1994).

Discussion and conclusions

The average income is only US$ 1,924 per annum, and some Bedouin lose money. Grazing restrictions are imposed upon the Bedouin and land conflicts are common. Why then do Bedouin persist in raising sheep for their livelihood today?

The maintenance of their traditional lifestyle remains important. One wife expressed this quite convincingly when she stated that without the presence and smell of sheep, her life has no meaning. Economic difficulties and high unemployment in the wage labor market may also provide some of the answer. Retention of a few flocks may be a rational choice as a supplement for those who are financially stressed, providing families with milk and other dairy products for personal consumption. Thus, maintaining some sheep acts as a hedge against the risk of unemployment and, if sheep raising does become more profitable, it would be easy to start this enterprise (Dinero, 1996). Moreover, as pointed out by Ginguld et al. (1997), Bedouin have a decided advantage over other sectors in raising sheep, namely, cheap labor available in most households, cheap inputs in that marginal land is used and little investments in equipment.

A reason given by the married sons of the household heads was that their fathers had no other expertise and, because of their ages, no other occupational options. For their part, the married sons unanimously stated that they would not continue to raise sheep after their father's demise. This would indicate that the number of sheep should be declining. However, demographic logic and the official and unofficial statistics of long-term Bedouin sheep holdings do not appear to support this. Demographically, elders are dying and being replaced by some of their sons, but the number of sheep, according to official records, is increasing by 2–4% per annum. This would indicate that either sheep holdings are increasing per household or more households are raising fewer sheep. The Bedouin, however, insist that it is now more difficult to hide the actual number of sheep they possess from the authorities and that is the main reason for the increase.

Another reason for raising sheep given frequently by Bedouin elders, researchers, and officials is that continued agro-pastoralist activities enhance Bedouin claims to land ownership by their continued land occupancy. This inherent attitude of the Bedouin was reinforced by the Land Procurement Act of 1980, which resulted from a dispute over land expropriated from the Bedouin within the *siag* by the State for the building of a military airbase. To do so, the State forcibly evacuated 500 families from an area of 30,000 ha and settled them in Ksifa, a newly established town. This resulted, indirectly, in the earlier-than-scheduled establishment of Aro'er, due to Bedouin dissatisfaction with Ksifa (Fenster, 1995). Under the terms of the Land Procurement Act, the government acknowledged, for the first time, the need to compensate Bedouin for expropriated land. Furthermore, due primarily to Bedouin protest, the initial area in dispute contracted from 30,000 ha to only 6,500 ha (Meir, 1997), that is, Bedouin were given compensation for 23,500 ha. Meir (1997) states that:

> "The fact that the state began to recognize Bedouin as land claimers was significant in itself.... This did not imply that all Bedouin land disputes (about 6,000 claims for a total of 70,000 km^2) would be solved or were approaching solution. But the significance of the above evacuation case was that it became a land-mark in the gradual process of settlement of Bedouin land ownership issues either by allowing them formal ownership and residence on the land they claimed, or by evacuating them in return for compensation."

About 40% of the total Bedouin population lives in spontaneous hamlets. Although sons are reluctant to continue the pastoralism of their fathers, and 75% of the inhabitants of the hamlets obtain their livelihood from sources other than agro-pastoralism, these non-pastoralists are not relocating to existing municipal localities and, in fact, may be resisting urbanization. Dissatisfaction with the urbanization alternative may have arisen from the following factors existing in the municipalities: high unemployment rate and drug abuse among Bedouin youth, high crime rate, liberalization of women, low standard of public services (e.g., schools, medical clinics, sewerage that are marginally better than those available for hamlets), tribal mixing and low alternative employment opportunities (Abu-Saad, 1996).

Meir (1997) recognized that this might lead to permanent settlement in some hamlets:

> "Recently, demands for the legalization of many of these unrecognized settlements (i.e., spontaneous hamlets) have grown. In an attempt to contain the conflict with the Bedouin, the state has gradually begun to yield to these demands, a policy which has resulted in more selective court orders, recent tacit recognition of one of these settlements (Abu-Rabia, personal communication), and state readiness to consider recognizing more. Plans are underway for establishing an additional four towns and seven agricultural villages."

The integration of many spontaneous hamlets into the national water supply network further illustrates the recent changing state policy towards the spontaneous hamlets.

Furthermore, there has been a change in the state's previously negative attitude of reducing the Bedouin's agro-pastoral economy to one of slowly recognizing it as a positive national economic resource. This positive attitude may be a result of the Bedouin's persistent maintenance of livestock holdings—not only by the inhabitants of hamlets (agro-pastoralists and smallholders), but also by urban dwellers (Ben-David, 1988). It has been realized that these holdings are not a drain on the national economy in that they are not subsidized, but are in fact being maintained by market processes that today include those of the areas under the control of the Palestine Authority in the former occupied territories. As such, they are an important economic factor in the peace process presently under way. It is well known that the sharing of the region's water resources will entail radical changes in Israel's agricultural policy in the near future. Of paramount importance to this issue is Israel's blueprint to use recycled sewerage water for irrigating the Negev's industrial crops, and to restrict sweet water irrigation to edible crops grown under "protected agriculture" conditions (glass and polyethylene structures). This will result in an increase in the area used for dry-land (rain-fed) winter cereal crops and, with it, a corresponding huge increase in straw and aftermath availability that can be utilized by Bedouin livestock.

It would appear that household heads persist in raising sheep mainly for the following reasons: 1) maintenance of Bedouin traditional lifestyle; 2) a claim to the land; 3) some income, including household dairy products and meat; and 4) a means to hold their money. The future of most Bedouin would appear to be integrating into the Israel urban economy, while attempting to maintain cultural traditions. A relatively small but stable number of households will continue to practice agro-pastoralism as a means of livelihood.

Dietary intake of grazing Karakul sheep in deserts of southern Kazakhstan†

Introduction

Large tracts of land in the Central Asian Republics of Kazakhstan, Uzbekistan, Turkmenistan and Kyrgyzstan are desert and semi-desert. Approximately 60% of this land is used for grazing, making range-based livestock production important to the national economy of these countries as well as to the livelihood of a large portion of the rural population (Kerven, 1995; Kaliev, 1999; Brent, 2001). From a total area of 274 million ha in Kazakhstan, 166 ha are classified as desert and semi-desert rangeland, of which 112 million ha are particularly favorable for raising Karakul sheep. The low annual rainfall and low winter and high summer ambient temperatures support sparse stands of grasses, ephemerals, ephemeroids, shrubs, dwarf shrubs, semi-shrubs and some trees. Low winter temperatures suppress the growth of the herbaceous species until the spring months when they achieve their maximum growth. The shrub-like species are mainly photoperiodic; some shed their leaves in the summer, but all rejuvenate in late summer and autumn with the onset of rainfall (Abdraimov, 1988).

Extremes of rainfall and ambient temperatures, characteristic of the deserts and semi-deserts of Kazakhstan, cause frequent shortages in forage availability (Kaliev, 1999). In the absence of supplementary feeding and adequate animal shelters, severe sheep losses can occur, particularly during the winter but also in other seasons (Esenova and Dobson, 2000). Improvements in fodder availability in the rangelands of Kazakhstan have been confined mainly to the planting out of extensive stands of shrubs and trees (Abdraimov, 1987).

The phenolic content of tree and shrub foliage can vary with age of the plant, type of foliage (twigs or leaves), age of foliage, position in the canopy, reaction to browsing, ambient temperature, sunlight and season (Furstenburg and van Hoven, 1994). In general, phenolic content is higher in leaves than twigs (Hanley et al.,

---

† Reprinted with permission from the copyrighted article in the Journal of Agricultural Science, with deletion of one table, from "Browse selection by Karakul sheep in relation to plant composition and estimated metabolizable energy content," by A. Allan Degen et al.," Journal of Agricultural Science (2002) 139:353–358. Cambridge University Press.

1992), in young plants than old plants (Furstenburg and van Hoven, 1994) and in spring than the other seasons (Palo, Sunnerheim and Theander, 1985). Even when present in relatively low concentrations, phenolics can cause foliage to be unpalatable and severely restrict intake by most domesticated ruminants (Benjamin, 1995).

Consumption of foliage of trees and shrubs by sheep could provide much needed fodder in Kazakhstan. It was predicted that sheep would select plants with low phenolic contents, high metabolizable energy yield, high protein content and low fiber content. To test this prediction, the selective foraging of Karakul sheep, introduced into an undisturbed area where 15 woody species, natural to the region, were growing and being studied for their growth characteristics, were observed. Shrub foliage from these species was analyzed for chemical composition and in-vitro metabolizable energy yield and the results were related to the selective foraging of the plant species by the sheep. The study was carried out at the Kazakh Institute of Karakul Sheep Breeding, Buchtulan Dryland Experimental Station, situated 170 km northwest of Chimkent (the south Kazakhstan regional center) and 3 km from the Syr Darya River.

Karakul sheep and browsing procedure

Ewes have a mature body mass of 40–50 kg and generally produce a single lamb per lambing, but, as not all ewes lamb, produce less than one lamb per ewe in the flock. The natural breeding season is in October, with lambing in March. The breeding season appears to coincide with the late-summer browsing of shrubs and the onset of rainfall. While summer breeding is supported by the browsing of shrub-like species, the survival of the ewe and the ability to meet energy and nutrient demands for late pregnancy is dependent on green herbage availability in February–March. The most important product of Karakul sheep production is the fur-like wool pelts from lambs slaughtered shortly after birth. Lambs raised for meat are usually slaughtered as yearlings, mainly for local consumption. Karakul sheep are well adapted to the vegetation of the region. However, the carrying capacity of the range is low (10–20 ha per sheep) and without feed supplements, particularly in the severe winter, sheep mortality can be over 35% of the flock (Olcott, 1981). Attempts to reduce mortality by raising Karakuls in higher rainfall regions with higher forage availability and quality have resulted in poorer quality pelts and were thus discontinued (Abdraimov, 1991).

During May 1995, a flock of 20 mature Karakul ewes was introduced into a previously undisturbed 10 ha field that contained 15 shrub species that were present in relatively large numbers (at least 10 of each species in the field). The ewes were 4 years of age, had lambed in March and had previous access to the plant species. Lambs were slaughtered shortly after birth for pelts. The ewes were kept in the field from dawn to sunset (moved at mid-day for drinking) to graze out the more palatable herbaceous species in order to induce them to browse from the less palatable species. The sheep were to be kept in the field until net foraging time was less than 2 h per day, at which time the study was terminated so that the ewes would not lose significant body mass. During the day, two observers noted the plant species that were browsed without regard to degree of edibility.

Findings

The herbaceous species in the field were grazed out within 2 days and only from the third day did the sheep spend most of their foraging time browsing from the woody species. After 5 more days most of the edible shrubs, with foliage accessible to the sheep, had been defoliated. During the sixth day, the sheep spent about 2 h nibbling soft twigs from defoliated plants but were inactive most of the time and so the study was terminated. Of the 15 woody species, six had been defoliated and nine were not eaten by the sheep.

Crude protein of the plants selected by Karakul sheep ranged between 116 and 227 g kg^{-1} DM, metabolizable energy yield ranged between 4.26 and 9.99 MJ kg^{-1} DM and digestible organic matter ranged between 351 and 724 g kg^{-1} organic matter. There was no statistical difference between eaten and non-eaten plants in metabolizable energy yield, digestible organic matter, crude protein content, neutral detergent fiber, acid detergent fiber, acid detergent lignin and ash content.

One of the uneaten species, *Calligonum gracile*, had an exceptionally high total phenolic content of 352 g kg^{-1} DM; in other species, total phenolics ranged between 10.6 and 55.0 g kg^{-1} DM. Plants eaten by sheep had a significantly lower (Mann-Whitney U [9, 6] = 45; $P<0.05$) total phenolic content than non-eaten species: 17.7±6.3 versus 66.4±129 g kg^{-1} DM (30.7±15.6 g kg^{-1} DM excluding *Calligonum gracile*). Total tannins ranged between 0.30 and 29.2 g kg^{-1} DM (except for *Calligonum gracile*, with a content of 304 g kg^{-1} DM) and condensed tannins ranged between 0.0 and 45.3 g kg^{-1} DM (except for *Calligonum gracile*, which had a content of 54.8 g kg^{-1} DM). Plants eaten had a total tannins content of 4.5±3.5 g kg^{-1} DM and condensed tannins of 0.40±0.4 g kg^{-1} DM; for non-eaten plants, these values were 40.8±120.1 g kg^{-1} DM (7.9±11.7 g kg^{-1} DM without *Calligonum gracile*) and 11.2±26.0 g kg^{-1} DM (5.7±20.2 g kg^{-1} DM without *Calligonum gracile*), respectively.

Discussion and conclusions

Ruminants can be divided into three major classes: concentrate selectors, intermediate feeders and bulk/roughage feeders. Concentrate selectors (e.g. many ungulates) limit their diets to highly digestible parts and low fiber portions of plants in their diets. Intermediate feeders, typified by goats, are adapted to a variety of forages and select grasses, forbs and browse in that order. Grass and roughage eaters, such as cattle and sheep, are well adapted to fibrous forage rich in cellulose (Van Soest, 1982; Hofmann, 1989).

In general, herbivores do not graze at random, but have marked food preferences. The dietary choices of herbivores can be attributed to two main factors, namely, the nutrition and level of toxic constituents of the plants. As an example of this, grazing kudus selected a diet of higher digestible dry matter and crude protein but lower phenolics than the mean levels occurring in available plants (Owen-Smith and Cooper, 1989). Furthermore, when offered a choice, mule deer and black-tailed deer selected plant forages containing the highest digestible dry matter and lowest non-tannin phenolics. However, these deer preferred low energy plants when the forage contained high levels of non-tannin phenolics. Tannins had little effect on dietary choice in these deer (McArthur et al., 1993).

The Karakul sheep browsed mainly from shrubs whose foliage had a total phenolic content below 20 g kg^{-1} DM. This was the only chemical component significantly different ($P<0.05$) between the eaten and non-eaten plant species. Both total and condensed tannins were relatively low in all plant species (except for *Calligonum gracile*), indicating that most of the phenolics were non-tannins. Plant phenolics can be divided into tannins and non-tannins, two functional groups that have different effects on the grazing animal. Tannins reduce consumption and digestibility of forages and can be toxic to herbivores (Murdiati, McSweeney and Lowry, 1992). They form precipitates with proteins, resulting in the formation of indigestible tannin-protein complexes (Makkar, 1993). The non-tannin phenolic component of plants consists of low molecular weight compounds. The small phenolics and/or their products are metabolized by gut microflora and absorbed from the gut. They can function as metabolic toxins and can be fatal to the animal (McArthur et al., 1993).

Explaining the reasons why animals may reject forage can be problematic. A reason may be palatability; however, it is often difficult to determine whether palatability or a physiological reason has caused feed rejection (Van Soest, 1982). Palatability should be distinguished from selection, which is a source of variation in determining intake under grazing and/or browsing conditions. For example, when animals are browsing from a particular plant species, more palatable parts are eaten first and the less palatable later. Thus with time, selection and intake will vary as the forage supply decreases. A similar selection and intake process occurs when animals are browsing from a variety of woody species. The more palatable species will be browsed first and only when they are completely defoliated will less palatable species be browsed. However, palatability both within and among species is a comparative characteristic and can vary with seasonal growth, plant development and differentiation, and age and rejuvenation after defoliation (Van Soest, 1982). It is also recognized that some plant species are never eaten because they contain high levels of toxic material to animals, especially to domesticated ruminants (Coley, Bryant and Chapin, 1985).

The current study was carried out at one time of the year. On-site investigators, experienced in raising sheep, reported that a number of plants species were not eaten at this time and it was important to know why this was the case. Indeed, some woody species were avoided by the sheep despite their availability, whereas other plant species were defoliated, particularly in the absence of herbaceous species. It is clear that sheep avoidance from browsing the inedible plant species was not arbitrary, but caused by the anti-nutritional qualities of the phenolics in the foliage. Furthermore, the composition of the plant species could change with the phenological changes that occur during the year and, consequently, change the edibility of the plants.

Rejuvenation of forests with fodder trees for goat production in the hills of Nepal‡

The overall goal of this ongoing project is to rejuvenate communal forest and grassland areas in the mid-hill regions of Nepal that have either been denuded of lower branches or degraded to dwarf-shrub and/or barren land by over-utilization and poor management by small holders. It endeavors to establish a sound sustainable economic base for subsistence farmers who rely on goats for their potential income by using integrated agro-silvi-pasture techniques. This goal is being achieved by focusing on the following:

- Determining the ease of establishment, development, growth rate and re-growth after harvesting of promising indigenous fodder trees and shrubs when cultivated between rows with annual legume crops and grasses to maintain soil fertility.
- Evaluating nutritional value of selected trees, shrubs, legumes and grasses to sustain improved goat production.
- Developing a management strategy that will prevent further degradation of existing forests, new plantations and rejuvenated grasslands based on animal choice.

Based on the results of the above, the following objectives are being achieved:

- Planting out species that meet the criteria in both degraded communal areas and in and around individual farms.
- Using existing farms as pilot farms, within a hill country village complex that has access to both partially degraded communal forest and barren dwarf-shrub areas.
- Using computer models to test whether the generated results meet the socio-economic criteria set by farmers and national authorities.

The studies benefit Nepal because they integrate research carried out at three levels: research institutes, on-site farms, and communal areas, while taking into account socio-economic criteria and priorities set by farmers and national authorities.

---

‡ Reprinted, with some deletions, from "Semi-Annual Progress Report" for January, 2007 on Nepal by A. A. Degen.

Accomplishments

Feeding trials to determine food preference of male and female goats and to determine protein and energy requirements were conducted at the Bandipur research station, within the mid-hill ecological zone of Nepal. Following these trials, the participating farmers were offered field workshops during which time agricultural methods for animal husbandry and nutritional aspects of livestock production, including veterinary treatment, were studied.

Many studies that primarily intend to help local farmers often fail to achieve practical means because their academic findings remain theoretical. As our study was addressing basic questions at the household level, we found it vital at this stage to receive direct feedback from the participating farmers. The Nepalese and Israeli investigators invited all the farmers participating in the study to a general meeting at Katteldada village. Responses were excellent; the farmers were most enthusiastic and proud of their achievements thus far, and urged us to continue with the program. They emphasized that the number of goats and income from them had increased substantially as a result of the project.

Chemical composition and energy yield of the foliage of fodder trees in the communal forest was monitored seasonally. In addition, at the "Buck Service Center" (created during the 2003–2004 breeding season), the breeding service register was continuously updated and farmers using the service recorded doe kidding performances and kid growth parameters. Twelve collaborating farmers also kept quantitative and qualitative records of the various feeds offered to their goats. These farmers, active members of the Katteldada goat raiser's cooperative group, are also involved in a goat feeding trial that began in December 2004.

Project Impact

The Nepalese investigators, together with regional livestock extension officers, are effectively coordinating and implementing the program with the Katteldada farmers. The farmers are the main benefactors of the program, as they carry out all the project's field activities. In addition, the regional extension officers benefit from their active collaboration in the project, and they, in turn, disseminate the project results to other villages. In recognition of their achievements, due to the project, the village goat-raising committee has been upgraded to the district level, to become the "District Goat-Raising Committee of Gorkha." The participating farmers are most enthusiastic about the project and have reported tangible achievements, including a substantial increase in goat numbers. To demonstrate the Katteldada project's training activities and agricultural achievements, TV and radio broadcasts have been promoted and a two-hour video film was made. Currently, two copies of the film (in the Nepali language) are available at NARC, Nepal and at Ben-Gurion University of the Negev, Israel. English subtitles are planned, to enable this film to be used by the international community. In addition, a booklet on goat production and fodder management, based on this project, was published for the farmers in the Nepali language [(Vol. 3, Issue 1, November 2006) Fodder Production, Management and Utilization for Forest Rejuvenation. Animal Feed Technology, Nepal. 24 pages].

Nepalese laboratories and investigators have been helped by the project as they implement up-to-date recording methods. Most importantly, the project has encouraged them to conduct applied agricultural research and development studies at farm sites with collaborating farmers rather than at research stations with technicians. This way, there is an immediate and direct transfer of knowledge, training and material benefits from the research institution to the target farmers, resulting in cooperative solving of problems as they occur. The methodologies that are being introduced will provide a model for future activities by the Nepal Agricultural Research Council at the village level.

Livestock production by pastoral nomads in the drought-prone Registan Desert of southern Afghanistan

Goal of study

In Afghanistan, there are several million nomadic people known as *Kuchis*, a generic term used there mainly by non-nomads to describe pastoral nomads. A severe drought in 1998 caused the displacement of approximately 100,000 *Kuchis* from the Registan desert in southern Afghanistan. Most of these people are now living in temporary settlements between the Arghandab and Helmand Rivers and Registan. Approximately 57,000 are being supported by the UN in camps for internally displaced persons (IDPs) in Maywand and Panjwayi districts of Kandahar Province; the rest are living in unsupported settlements.

The UN is currently seeking strategies to return the *Kuchis* to their traditional livelihood of raising livestock in Registan and end UN support of the IDP camps. This also appears to be the desire of most *Kuchis*, at least the elders whom we interviewed, although some *Kuchis* believe that the nomadic lifestyle is coming to an end (Amani, 2003). How can this be achieved and sustained? We start with the premise that the return of *Kuchis* to their traditional nomadic lifestyle is possible if carefully planned, seeing that it was done successfully by *Kuchis* themselves for many centuries.

Our goal was to examine possibilities for the return of *Kuchis* to Registan. We reviewed reports published to date, visited several well clusters in Registan, interviewed *Kuchis* from the IDP camps and non-IDP settlements and had discussions with personnel from a number of organizations working with *Kuchis*, including the United Nations High Commission on Refugees (UNHCR), United Nations Assistance Mission in Afghanistan (UNAMA), Voluntary Association for Rehabilitation of Afghanistan (VARA) and Central Asian Development

Group (CADG). Reports included recent independent studies that described the past and present situation of *Kuchis*, the state of vegetation, soils and well clusters and water catchments (*nawars*) in Registan. Registan is situated in southern Afghanistan, bordered by the sand dunes south of the Arghandab River in the north and by the Pakistan border in the south. It is part of the African-Indian Desert region with a total area of about 24,000 km^2 that includes part of the Kandahar, Helmand and Nimroz districts. It has a Mediterranean arid climate characterized by low winter rainfall from November to May, high summer air temperatures (above 40°C in August) and a high rate of potential evaporation (about 3000 mm annually). This region is used by nomads and semi-nomads who, until recently, depended solely on the raising of livestock for their livelihood.

Background

Approximately 10% (2–2.5 million people) of the population of Afghanistan consists of *Kuchis*. They comprise a number of ethnic communities, the Pashtuns being the majority group (UNCHA, 2004). In Registan there are only two main ethnic groups: Baluch, comprising about 80%, and Pashtuns about 20% of the nomadic population. They raise mainly Baluchi sheep for meat, milk and wool. In addition, camels are raised for transport, milk and drawing water, goats for milk, hair and meat and donkeys for transport.

Wells and *nawars* are the sources of water for *Kuchis* and their animals in Registan. *Nawars* are dug-out depressions that act as collecting reservoirs for runoff water. Only Baluch *Kuchis* have wells; Pashtun *Kuchis* are dependent on *nawars*. Some Baluch *Kuchis* having no access to wells are dependent on *nawars* as well. Livestock in Registan grazes only natural vegetation; apparently, no supplementary feed is offered. According to the *Kuchis* we interviewed, two methods of livestock raising are practiced:

- Grazing animals in Registan during winter/spring months of the year and then migrating to near the Arghandab River or Helmand River areas for the rest of the year. This was practiced by all Pashtun *Kuchis* as the *nawars* dried out by the summer, and by some Baluch *Kuchis*.
- Grazing in Registan all year was practiced by some Baluch *Kuchis* as long as drinking water and fodder were available. In a report of the Food and Agricultural Organization (FAO 2003) it was stated that some *Kuchis* from Registan migrated across the Pakistan border for the summer months. None of the *Kuchis* whom we interviewed mentioned this migration, however.

Drought of 1998–2002

In 1998, an extreme drought brought disaster to Afghanistan. Combined with many years of war, landmines and devastated livestock numbers it caused the death of many Afghans. The average number of sheep per family decreased from 21.9 to 2.9 (FAO, 2003). Registan *Kuchis*, in particular, were affected greatly.

The Taliban, who were in power at the time, evacuated them with their remaining livestock to refugee camps at Maywand and Panjwaji near the Arghandab River, where they have remained until today (Frontier Post, 2000). Other *Kuchis* crossed into Pakistan where refugee camps were established. Three years of drought followed and over the entire period of drought over 90% of their sheep and about 40% of their camels perished. About 60% of *Kuchi* households lost all livestock. The Taliban Government was unable to support these people and the UN has been providing food and clothing for the IDPs in the camps.

Four IDP camps were established near the Arghandab River in Kandahar province: one in the Maywand district (Qala-e-Shamir) and three in the Panjwaji district (Marghar, Talukan and Moshan). *Kuchis* from these four camps were evacuees from Registan in the Kandahar province and the Garmser well cluster in Helmand province. The study was directed towards the IDPs in these four camps. The number of animals owned per household in the IDP camps is generally very low. Three methods of raising sheep are presently being practiced:

- Zero grazing: seasonally available fodder, including straw, grasses and grape leaves, is collected from irrigated land near the river and transported back to the camp, usually by camel.
- Aftermath grazing: animals graze on rented aftermaths of mainly wheat, but also other crops such as maize, for all or part of the year; zero grazing is practiced when aftermaths are not available.
- Migration to Registan: sheep from a number of families are combined into a flock of 100–300 head and shepherded by one or two families to Registan. Water is provided from wells or from tanks that are brought out by tractor. Sheep are maintained in Registan in winter/spring for as long as possible and then returned to IDP camps where each family provides for its sheep by one of the above two methods.

Payment for fodder collected or for land grazed can be in cash but usually is in exchange for work done by the sheep owner for the landowner.

Livestock

Kuchis raise mainly sheep but also have goats, camels and donkeys. As the restoration of sheep numbers is the main consideration, only sheep and their management will be described in detail in this report.

The only sheep breed raised by *Kuchis* in Registan is the fat-tailed Baluchi. Like all fat-tailed sheep, they are well adapted to desert conditions, are good walkers and grazers, require relatively little water and are not prolific breeders. They are mated in summer during August, so that lambing coincides with the emergence of ephemerals after the first rains during January. Usually one ram per ten to twenty females is used for mating. The rate of twinning is low, averaging about 5% of the ewes lambing. In the total flock, ninety to ninety-five lambs are obtained from one hundred ewes under a regime of good nutrition, but this can be reduced to only fifty lambs per hundred ewes under poor nutritional conditions. Until they are weaned, lamb mortality can range from 5% with good nutrition to

25% with poor nutrition. Lambs are weaned at five to six months of age and weigh 20–25 kg. The ram-lambs are sold in September/October for about 3,500 Afghanis (AFA 50 = US$ 1), mainly to feedlot owners for fattening. Some ram-lambs are kept for a year or more until they weigh about 40–50 kg and can be sold for AFA 8,000–10,000. The number and timing of the sale of ram-lambs depend on cash requirements and in practice, *Kuchis* prefer to hold on to their sheep for as long as possible.

Ewes that are milked produce about 0.5 liters of milk per day and 100 liters of milk per lactation period. Milk is mainly for family consumption and some is processed into butter, ghee and dried yoghurt. Any surplus is sold. Sheep produce between 2.0 and 2.5 kg of carpet wool per adult animal per year. One kilogram of white wool is sold for AFA 50 and one kilogram of black wool at AFA 15–20. Baluch *Kuchis* eat old, dying sheep and sick lambs, slaughtering them just before death, but Pashtun *Kuchis* do not.

Veterinary services are not used by the *Kuchis*. Livestock are neither vaccinated nor dewormed. However, according to Dr Ali Khan, a VARA veterinarian, there is a veterinary clinic at Qala-e-Shamir, which services the four IDP camps. In addition, a mobile clinic visits the camps once a week. The most prevalent diseases are anthrax and enterotoxemia; brucellosis and foot and mouth are not problems. No vaccines are offered, but deworming is available for free once a year. The *Kuchis* we interviewed were unaware of the above veterinary services and claimed that their livestock was never treated for parasites. They informed us that they purchase medicines for livestock, often in Pakistan, when the animals are ill.

Wells and **nawars**

The CADG, lead by Abdul Ali (Jan Agha), surveyed wells in the twelve clusters in Registan. Of seventy-six wells examined, eight were functioning, fifty-nine had collapsed and nine were dry. The last well dug by *Kuchis* was thirteen years ago at Jahanan, while the UN dug one well at Jahanan and two wells at Taband five years ago. Wells collapsed or dried up following the three to four years of drought. In general, wells are about 1–1.5 m square and 90–140 m deep, but can be as shallow as 21 m. They are hand dug by lowering a well digger with ropes secured under the thighs, using camels to lower and raise the diggers. Typically, the upper part of the well (5–15 m) is of sandy soil that is naturally unstable and, therefore, reinforced with wood that requires maintenance every year or two, but can be reinforced with bricks and cement that require very little, if any, maintenance.

A large number of *nawars* exist in Registan, their volume ranging from several cubic meters to thousands of cubic meters. Many of them are partially or completely clogged with sediments that migrated with floods or with eolic sand from nearby sand dunes. Since the *nawars* have not been maintained since 1998, their capacity to hold water has been reduced over the years. Although *nawars* appear to be a major source of water for *Kuchis*, we saw only one large *nawar* (perhaps the largest in Registan), which was in disrepair, during our trip. This *nawar* was approximately 100 m long by 30 m wide. We were told that in good rainfall years the water was up to 3 m deep which is equivalent to about 9,000 m^3 of water. Although the *nawar* was dry, the soil underneath was moist, indicating that there had been run-off water this year. It should be remembered that *nawars* are unreliable sources of water as they are totally dependent on seasonal rain. In drought years, *nawars* remain empty while wells still produce water. However, in rainy years, *nawars* could provide large volumes of high quality water.

Discussion

Population

In all, there are approximately 9,600 families in the four IDP camps or about 57,600 people (assuming six persons per family, according to UNHCR), which is about 60% of the families that inhabited the above area. The other 40% are living in unsupported, spontaneous settlements mainly situated between the Arghandab and Helmand Rivers and Registan, but not in IDP camps. Thus, in all, there are about 16,000 displaced families or 96,000 people. Many of the men have become farm laborers in areas such as Helmand and Kandahar where crop irrigation is practiced (UNCHA, 2004). The great majority of these *Kuchis* are illiterate, so that apart from the census carried out in Afghanistan on a district basis in 1995, 1998 and 2003 (FAO, 2003), there are no verifiable vital statistics available. Even the 2003 census is equivocal, because there is no information available to distinguish the *Kuchi* inhabitants of Kandahar Registan from the general population of the Kandahar and Helmand provinces.

Table 8.1 shows the results of a family count carried out during 2003 by VARA. The total number of families is 26% greater than the 9,600 families quoted by UNHCR. Heads and/or chiefs of the various sections in the camps would tend to exaggerate rather than lower the number of families in their section in order to receive more humanitarian aid.

Table 8.1. Numbers of families in IDP camps, 2003 (van Engelen, 2003a)

Camp	Sections	Families
Marghar	9	1,957
Qala-e-Shamir	10	1,440
Talukan	27	5,225
Moshan	18	3,470
Total	64	12,092

Area and climate of Registan

The estimated area of Registan that was utilized by the 16,000 families was approximately 10,000 km^2. The area is considered to be a desert, having an arid climate characterized by low Mediterranean-type rainfall, high summer ambient temperatures and relatively cold continental winters. Average annual rainfall is up to 90 mm. The ambient temperatures recorded at Kandahar are summer highs of well over 40 °C and winter lows of slightly above zero. Although *Kuchis* cannot quantify rainfall, they do claim that they had never experienced a drought as severe as that of 1998, nor experienced continuous drought years like those between 1998 and 2002. Although this was never stated by *Kuchis* we interviewed, such an arid climate indicates an ecosystem subject to fluctuations within and between years in terms of soil water and fodder availability that would necessitate nomadic practices for survival.

Soils and vegetation

According to recent soil surveys carried out independently by Toderich (2004) and Ayubi (2004) for CADG, "vast areas are covered by immature soils derived from unconsolidated materials, such as gravel, rocks and sand" (Toderich, 2004). They found almost all the varieties of geomorphic and desert type soils: hammadas, regs, ergs, nebkhas, takyrs and playas. In layman terms, the most important characteristics are the presence of mobile sand dunes, separated by hammadas (soils covered with a top layer of small stones, with varying slopes) that protect relatively stable sandy and loess soils. Hammadas are important as being relatively free of vegetation and are conducive to water run-off for potential water collection when there are rainfall events of sufficient intensity. Sandy and loess type soils are those that support the more important (for livestock) desert-type vegetation. For this discussion, the most important vegetation types are perennial and annual psammophytic species, and annual ephemeral vegetation.

Perennial and annual psammophytic species predominate on sand dunes and sandy fixed areas. Tree and shrub perennials are relatively unpalatable to sheep but more palatable to camels and donkeys. Perennials are important for protecting sandy fixed areas and well clusters from sand encroachment, for stabilizing sand dunes, for shade, for fuel and for *Kuchi* dwellings and well construction. During the drought, many shrubs were collected by *Kuchis*, who travel into Registan by tractor, and the shrubs were sold for firewood in Lashkar Gah and other centers.

Annual ephemeral vegetation predominates on sandy-loams and loess soils. It is highly palatable and constitutes the main fodder intake of small ruminants grazing in Registan. This vegetation usually and universally rejuvenates following a significant rainfall event, even after a drought year and particularly after consecutive drought years. The reason is because the soil seed-bank is replenished in the absence of grazers of the few plants that do germinate and seed. The sustainable annual carrying capacity of Registan is that of the total production system; i.e., the number of animals maintained by *Kuchis* that not only graze in Registan but also migrate to other areas. In these terms, it is best to refer to the length of time spent in Registan by *Kuchis* as a point of reference. Time spent in Registan during any particular year is determined by soil water availability for plant growth (edible biomass) and drinking water availability from wells and *nawars*. Interdependent relationships between vegetation- and drinking-water availabilities for livestock determine the transhumance of *Kuchis* from Registan to the semi-arid and irrigated regions adjacent to the Helmand and Arghandab rivers.

Livestock

A reliable estimate of the number of animals held by the 16,000 *Kuchi* families before 1998 is not available from the 1995 census conducted by the FAO for the same reasons as discussed above for determining population numbers. The statistics most often quoted by the various investigating agencies were that 90% and 60% of the sheep and camel populations, respectively, perished following the disastrous droughts of 1998. Any estimate is thus dependent on the estimate of present sheep and camel populations. Table 8.2 shows the results of an animal count carried out in the four IDP camps during 2003 by van Engelen (2003b).

Table 8.2. Estimated sheep and camel populations in IDP camps, 2003 (van Engelen, 2003b)

Camp	Sections	Families	Camels	Sheep/Goats
Marghar	9	1,957	NA	NA
Qala-e-Shamir	10	1,440	996	4,693
Talukan	27	5,225	NA	NA
Moshan	18	3,470	981	10,289
Totals counted	64	12,092	1,977	14,982
Animals calculated per family			0.40	3.05
Estimated total animals in camps			4,837	36,881

However, if we assume that the animal head count is correct, but that there are 26% fewer families in the IDP camps (as discussed above), there could be 26% more camels and sheep per family. On this basis there are perhaps 0.50 camels and 3.84 sheep/goats per family. Thus the estimated total animal numbers could have been approximately five camels and thirty-five sheep/goats per family before 1998.

In a survey of ten representative *Kuchi* families out of 6,400 that once lived at different well clusters in Registan, but are now living in unsupported spontaneous settlements, Ayubi (2004) reported that, on average, they owned thirty-one sheep and one camel per family. According to these families, before the drought they owned approximately 263 sheep and 21 camels per family. The existing numbers of sheep and camels per family are 11.8% and 4.8%, respectively, of the sheep and camels they previously possessed. While the percentage of remaining sheep per family is relatively consistent with that generally accepted, the remaining camels are very much less than the 40% that is commonly quoted. The reason for this discrepancy is unclear. Possibly more camels survived the drought but were subsequently sold for any number of reasons. What is of importance from Ayubi's report is that families possessing, on average, thirty-one sheep and one camel were subsisting unsupported.

Water resources

In southeastern Registan, there are twelve well-clusters spread unevenly over the area. According to a recent survey undertaken by VARA to map these clusters, 380 wells have been identified and mapped (van Engelen, 2003a). Of these identified wells, few were functioning and water was drawn from them from depths as great as 140 m (a consequence of low water tables). In general, the functioning wells were those that were brick-lined in their upper reaches, with surfaces protected by concrete. According to van Engelen (2003b), there are 30–40 traditional well diggers/maintenance specialists in the IDP camps who are currently unemployed.

No research has been done to explore the aquifer unit underlying the different well clusters in Registan. Therefore, the potential of each well to provide more water, the type of aquifer, its hydraulic properties and its response to droughts are unknown. Local hand-dug wells penetrate only a few meters below the water table (due to technical limitations) and the volume of water pumped from each well is very small (limited by the time that the bucket takes to be lowered into the well). In addition, there are *nawars* associated with these well clusters. These are depressions where run-off water is collected from sloping hammada areas when rainfall is of sufficient intensity to cause run-off. After 4–5 years of neglect, most of these *nawars*, like most of the wells, have been filled and covered with sand. It is probable that the *nawars* were conveniently utilized by grazing animals until depleted and only then was water labor-intensively drawn for the animals from wells.

There are no formal owners of wells, although some *Kuchi* chiefs interviewed claimed they had deeds of ownership. However, it appears that a well cluster was controlled by a group of related families by virtue of what had become their traditional occupancy of the pasture areas surrounding the well cluster. Probably the extent of control may have been a radius of up to 10–15 km from a well cluster, a shepherding distance that could allow easy watering of livestock in the presence of water in *nawars* but would be more difficult in their absence or after their depletion. Traditionally, it would appear that it was customary to allow families from other groups to draw and transport water to their animals but they could not bring their animals to within walking distance of the wells. In this way pasture and browse resources were reserved for livestock of resident families.

Water and pasture inter-relationships

In any particular year, the length of time spent in the Registan region before eventual migration was restricted either by pasture availability or the drinking water available and possibly, but rarely, by both (when dependent on *nawars*). However, when wells were functioning, it is probable that pasture availability was the main constraint limiting the time spent in Registan. The beginning of the time spent in Registan would have had to be when vegetative growth, following a significant rainfall event(s), had progressed sufficiently to allow animals to forage enough fodder to approach their maintenance requirements. Because the number of animals foraging was a function of the stocking rate of the year-round nomadic system and not the actual carrying capacity of Registan fodder resources, overgrazing would have been the rule rather than the exception. That is, grazing/browsing continued until, inevitably, all edible biomass had been consumed after a few months. If foraging began too early, animal intake rates would have been greater than plant growth rates, which would have led to an early migration. The initiation of foraging would have had to take into consideration the availability (purchase and/or renting) of fodder resources on the migratory route to the semiarid and irrigated regions adjacent to the rivers. This reasoning was reinforced by interviewed *Kuchis* who, each having had access to a functioning well, had returned during the winter/spring seasons of 2002/2003 and 2003/2004 to Registan. They reported that even though their flocks were only a tenth of their former size, they had returned to the IDP camps after only two to three months. Possibly, however, they may have also returned because of work prospects in harvesting wheat and aftermath availability at the irrigated areas.

Socio-economic factors

It would appear that *Kuchis* practiced a well-defined transhumance form of nomadism in which they always returned during the winter/spring seasons to grazing/browsing areas that were centered around maintained wells and *nawars* to which they had traditional rights of use. The dramatic change caused by the disastrous years of drought and the evacuation of *Kuchis* with few animals to IDP camps is that the IDP camps have replaced their traditional areas in Registan as their home territory to which they return if they migrate. This is because few of the wells and few, if any, of the *nawars* are functioning in Registan, whereas at the IDP camps, water and humanitarian aid is readily available at no charge. In addition, for those *Kuchis* with animals, fodder is available albeit at a price. Self-interest guides them to exploit the resources currently available to them.

The 1998 evacuation to IDP camps of Registan *Kuchis* by the then-ruling Taliban Government was for economic reasons. *Kuchis* had lost so many of their animals that their means of livelihood no longer existed and their main means of livelihood became the humanitarian aid provided by outside agencies, mainly the UN. Unlike in many other war-affected countries, where refugee camps have become and are becoming permanent fixtures maintained by aid agencies, *Kuchis* are not prevented from returning to their former unoccupied territories by threats of violence from opposing forces, but remain only because of the lack of economic incentives. Because they do not have the earning capacity to purchase sufficient numbers of animals needed for their traditional form of livelihood and in the absence of any other means, they are fated, if only by self-interest, to remain in IDP camps as long as humanitarian aid is forthcoming. Unless the present conditions of self-interest for *Kuchis* are changed, the IDP camps are liable to become a self-perpetuating and permanent financial burden on a reluctant UN. For the UN, the solution in the short term is to continue to fund the camps for humanitarian reasons and to fund consulting NGOs in an endeavor to find long-term solutions to the problem of how to provide the economic incentives for *Kuchis* to return to their traditional way of livelihood.

General comments

Two of the projects are still in progress: livestock production by the Negev Bedouin and fodder trees and goat production in Nepal; whereas, two of the projects have been completed: dietary intake of grazing sheep in Kazakhstan and livestock production in Afghanistan. On the whole, cooperation from the participating groups has been excellent and results very promising.

The livestock study of Negev Bedouin has been well accepted by the Bedouin population. The Bedouin often visit the research station for advice, mainly on veterinary treatments, including vaccinations and disease problems. They also enquire about alternative feeds for their livestock and about estrous synchronization, which is being used more commonly with their sheep and goat flocks. A Bedouin was accepted as a Ph.D. student in the project and he completed his degree in 2007 on dietary selection of free-grazing sheep and goats in the Negev. He is the first Ph.D. recipient residing in Tel Sheva and was recently appointed as director of Research and Development for Negev Bedouin.

The study in Nepal is being met with great enthusiasm among the villagers where the project is being carried out. The villagers claim that they have greatly increased goat production as a result of the improved fodder. They are encouraging us to continue and to carry out a similar project with water buffalo, an animal extremely important for their milk supply. Most of the fodder and statistical analyses have been done in Israel; however, the principal investigators from Nepal have visited Israel on two occasions thus far and there is an M.Sc. student enrolled at the University who is part of the project. This student received his degree in 2008. The principal investigators and this student will establish research methodologies in Nepal.

The study in Kazakhstan was a four-year study that was completed in 1997. During this project, prolificacy of Karakul sheep was increased with the use of hormone treatment and fodder production was improved by using run-off water and planting out preferred and nutritious plants. Most of the analyses of plant composition and energy values were done in Israel; however, the principal investigators visited the labs in Israel on three occasions. In addition a post-doctorate candidate from Kazakhstan spent two years in Israel and two trainees from Kazakhstan spent three months each in Israel. The Kazakh students were able to learn methodologies in Israel and apply them at the research station in Kazakhstan.

The project in Afghanistan endeavored to suggest methods for increasing livestock production and improving fodder availability for nomads (*Kutchi*) raising livestock in the desert. Four years of drought in the Registan Desert had decimated livestock numbers and, as a result, the *Kutchi* were relegated to IDP camps, where they depend on UN assistance for survival. The goal was to return the *Kutchi* to their nomadic ways in the desert and to be independent of outside help. A hydrologist and I traveled to the Registan Desert to understand the system being used for livestock production and to offer recommendations on what could be done to improve it. Our main aims were to establish fodder reserves that could buffer droughts and to provide water more efficiently to the livestock. The *Kutchi* were very helpful and cooperative once they learned the reason for our visit and welcomed our initiative. We wrote up an initial report, which was to be followed by further visits. However, the security situation in Afghanistan did not allow further visits.

Acknowledgements

"Livestock production by Bedouin of the Negev Desert in Israel" was funded by the Netherlands-Israel Development Research Programme (NIRP), under contract NIRP-97-145-7, in collaboration with Nuffic, The Hague and HAIGUD, Jerusalem.

"Dietary intake of grazing Karakul sheep in deserts of southern Kazakhstan" was funded by the US-Israel Cooperative Development Research Program, Office of the Science Advisor, United States Agency for International Development (USAID), under Grant Numbers TA-MOU-CA-13-051 and TA-MOU-CA-13-061.

"Rejuvenation of forests with fodder trees for goat production in the hills of Nepal" was funded by the US-Israel Cooperative Development Research Program, Office of the Science Advisor, United States Agency for International Development (USAID), under Grant Number TA-MOU-00-C17-014.

References

Abdraimov, S. A. (1987) *Arid fodder production in Kazakhstan*. Alma-Aty, Kazakhstan: Academy of Agricultural Sciences (in Kazakh).

Abdraimov, S. A. (1988) *The Kazakhstan arid pastures*. Alma-Aty, Kazakhstan: Academy of Agricultural Sciences (in Kazakh).

Abdraimov, S. A. (1991) *Recommendations on radical improvement and efficient utilization of desert pastures*. Alma-Aty, Kazakhstan: Academy of Agricultural Sciences (in Kazakh).

Abu-Rabia, A. (1994) *The Negev Bedouin and livestock rearing: Social, economic and political aspects*. Oxford: Berg Publishers Limited.

Abu-Saad, I. (1996) Provision of educational services and access to higher education among the Negev Bedouin Arabs in Israel. *Journal of Education Policy*, 2:527–41.

Amani, W. (2003) *Kuchis*' way of life ending. *The statesman*, http://www.statesman.com.pk/Feature/Feature12.htm.

Ayubi, A. G. (2004) Work plan for Registan water/pasture assessment. Report submitted to the Central Asian Development Group (CADG). Singapore.

Ben-David, Y. (1988) *Bedouin agriculture in the Negev - Policy proposals*. Jerusalem: The Jerusalem Institute for Israel Research.

Benjamin, R. W. (1995) Potential and actual contribution of fodder trees and shrubs to ruminant nutrition in drought prone regions. In K. Becker, P. Lawrence & B. Ørskov (Eds.), *Sustainable small-scale ruminant production in semi-arid and sub-humid tropical areas* (pp. 55–81). Stuttgart: Institute for Animal Production in the Tropics and Sub-tropics, University of Hohenheim.

Brent, L. (2001) Food production, consumption, and marketing in Kazakhstan. Central Asian Livestock Sector in Transition Series, Report No. 9, International Agricultural Browse selection by Karakul sheep 357. Madison, Wisconsin: Program at the University of Wisconsin-Madison.

Coley, P. D., Bryant, J. P., & Chapin, F. S. (1985) Resource availability and plant antiherbivore defense. *Science*, 230:895–899.

Degen, A. A. (1977) Responses to dehydration in native fat-tailed awassi and imported German mutton merino sheep. *Physiological Zoology*, 50:284–293.

Degen, A. A. (2003) Roles of urbanized Negev Bedouin women within their households. *Nomadic Peoples*, 7:108–116.

Degen, A. A., Benjamin, R. W., Abdraimov, S. A., & Sarbasov, T. I. (2002) Browse selection by Karakul sheep in relation to plant composition and estimated metabolizable energy content. *Journal of Agricultural Science, Cambridge*, 22:170–174.

Degen, A. A., Benjamin, R. W., & Hoorweg, J. C. (2000) Bedouin households and sheep production in the Negev Desert, Israel. *Nomadic Peoples*, 4(1):125–147.

Degen, A. A., & Weisbrod, N. (2004) Can pastoral nomads return to their traditional livelihood of raising livestock in the Registan Desert of southern Afghanistan? *Nomadic Peoples*, *8*(2):214–229.

Dinero, S. C. (1996) Resettlement and modernization in post-nomadic Bedouin society: The case of Segev Shalom, Israel. *The Journal of Planning Education and Research*, 15:105–116.

Dinero, S. C. (1998) Planning in a post-nomadic Bedouin town: An update and follow-up assessment of development and change in Segev Shalom, Israel. *Scottish Geographical Magazine*, 114:66–75.

Esenova, S., & Dobson, W. D. (2000) Changing patterns of livestock, meat, and dairy marketing in post-communist Kazakhstan. *Central Asian Livestock Sector in Transition Series, Report No. 3.* Madison, Wisconsin: International Agricultural Program at the University of Wisconsin - Madison.

Fenster, T. (1995) Participation as a political process in enforced resettlement projects: The Bedouin in Israel. *Geographic Research Forum*, 15:33–48.

Food and Agriculture Organization (FAO) (2003) First-ever livestock census in Afghanistan. http://www.fao.org/english/newsroom/news/2003/25511–en.html.

Frontier Post (2000) Drought reduces life to a torture in Afghanistan. http://rawa.fancymaking.net/drought3.htm.

Furstenburg, D., & Van Hoven, W. (1994) Condensed tannin as anti-defoliate agent against browsing by giraffe (*Giraffa camelopardalis*) in the Kruger National Park. *Comparative Biochemistry and Physiology*, 107A:425–431.

Ginguld, M. (1994) Managing herds and households: Management practices and livelihood strategies of sheep-owning Bedouin households in the Negev region of Israel. M.A. thesis, The Institute of Social Studies, The Hague, The Netherlands.

Ginguld, M., Perevolotsky, A., & Unger, E. D. (1997) Living on the margins: Livelihood strategies of Bedouin herd-owners in the northern Negev, Israel. *Human Ecology*, 25:567–589.

Hanley, T. A., Robbins, C. T., Hagerman, A. E., & McArthur, C. (1992) Predicting digestible protein and digestible dry matter in tannin-containing forages consumed by ruminants. *Ecology*, 73:537–541.

Hofmann, R. R. (1989) Evolutionary steps of ecophysiological adaptation and diversification of ruminants: a comparative view of their digestive system. *Oecologia*, 78:443–457.

Kaliev, G. A. (1999) The economic aspects of the development of sheep breeding in Kazakhstan. In A. Khazanov, V. Naumkin, K. Shapiro & D. Thomas (Eds.),*The present state of cattle breeding and animal husbandry in Kazakhstan and the prospects of their development* (pp. 88–92). Moscow: Russian Center for Strategic Research and International Studies (in Russian with English summary).

Kerven, C. (1995) Review of planning and policies on extensive livestock development in central Asia. Set 38. London: Pastoral Development Network, Overseas Development Institute.

Kressel, G. M., & Ben-David, J. (1995) The Bedouin market - Corner stone for the Founding of Be'er Sheva: Bedouin traditions about the development of the Negev capital in the Ottoman period. *Nomadic Peoples*, 36/37:119–144.

Makkar, H. P. S. (1993) Antinutritional factors in foods for livestock. In M. Gill, E. Owen, G. E. Pollott, & T. L. J. Lawrence (Eds.), *Animal production in developing countries* (pp. 69–85). Occasional publication, British Society of Animal Production, no. 16.

McArthur, C., Robbins, C. T., Hagerman, A. E., & Hanley, T. A. (1993) Diet selection by a ruminant generalist browser in relation to plant chemistry. *Canadian Journal of Zoology*, 71:2236–2243.

Meir, A. (1997) *As nomadism ends: The Israeli Bedouin of the Negev*. Boulder: Westview.

Murdiati, T. B., McSweeney, C. S., & Lowry, J. B. (1992) Metabolism in sheep of gallic acid, tannic acid and hydrolysable tannin from *Terminalia oblongata. Australian Journal of Agricultural Research*, 43:1307–1319.

Noy-Meir, I. (1975) Primary and secondary production in sedentary and nomadic grazing systems in the semi-arid region: Analysis and modeling. Research Report submitted to Ford Foundation. Department of Botany, Hebrew University, Jerusalem, Israel.

Olcott, M. B. (1981) The settlement of the Kazakh nomads. *Nomadic Peoples*, 8:12–23.

Owen-Smith, N., & Cooper, S. M. (1989) Nutritional ecology of a browsing ruminant, the kudu, through the seasonal cycle. *Journal of Zoology, London*, 219:29–43.

Palo, R. T., Sunnerheim, K. & Theander, O. (1985) Seasonal variation of phenols, crude protein and cell wall content of birch (*Betula pendula* Roth.) in relation to ruminant *in vitro* digestibility. *Oecologia*, 65:314–318.

Statistical yearbook of the Negev (2004) Negev Development Authority, Ben-Gurion University of the Negev, Beer Sheva, Israel.

Toderich, K. (2004) Water/pasture assessment of Registan desert. Report submitted to the Central Asia Development Group (CADG), Singapore.

UN Office for the Coordination of Humanitarian Affairs (UNCHA) (2004) Afghanistan: Focus on Nomads and the drought. irrinnews.org, http://irinnews.org/print.asp?Report ID = 26549.

van Engelen, A. (2003a) An assessment of economic opportunities for Kuchi internally displaced persons in Maywand and Panjwaji districts. Report submitted to Voluntary Association for Rehabilitation of Afghanistan (VARA), Zeist, The Netherlands.

van Engelen, A. (2003b) Return of Kuchi to Registan: A phased approach. Report submitted to RDP International B.V. The Netherlands.

Van Soest, P. J. (1982) *Nutritional ecology of the ruminant.* Corvallis, Oregon: O & B Books

Chapter 9

Fish in the Desert

Samuel Appelbaum

Fish farming in the desert may at first sound like an anomaly, but in Israel over the last decade a scientific hunch has turned into a bustling business.

Israeli scientists realized they had made an important discovery when they found that brackish water drilled from underground desert aquifers hundreds of feet deep could be used to rear warm-water fish. The geothermal water, ten-fold less saline than seawater, free of pollutants and a toasty 37° C (98° F) on average, proved an ideal match.

"It wasn't easy convincing people that growing fish in the desert makes economic sense," said Samuel Appelbaum, a professor and fish biologist at the Jacob Blaustein Institutes for Desert Research at the Sede Boqer campus of Ben-Gurion University of the Negev.

"It is important to dispel the fallacy that arid land is non-fertile, useless land." said Professor Appelbaum, who pioneered the concept of desert aquaculture in Israel in the late 1980s. "We should consider arid land where subsurface water exists as land that offers great opportunities, especially in food production, because of the low level of competition for this land and because it provides opportunities to its relatively sparse population."

The next step in this country, where water is scarce and expensive, was to show farmers that water utilized for cultivating fish could later be used to irrigate their crops in a double-usage system. The organic waste produced by the cultured fish makes the water especially useful, because it can serve as fertilizer for their crops.

Fields watered by brackish water dot Israel's Negev and Arava Deserts in the south of the country, where they spread out like green blankets against a landscape of sand dunes and rocky outcrops. At Kibbutz Mashabbe Sade in the Negev, the recycled water from the fishponds is used to irrigate acres of olive and jojoba groves. Elsewhere, it is also used for irrigating date palms and alfalfa.

The chain of multiple users for the water is potentially a model that can be applied elsewhere, especially in arid third-world countries where farmers struggle to produce crops. Israeli scientists have recently sought to generate interest abroad in their ideas.

People living in the dry lands that cover about 40 percent of the planet are often among the poorest in the world. Israeli scientists are working to share desert aquaculture technology that they have developed with Tanzania, India, Australia and China, among others. Similar methods of fish farming are also being used in the Sonoran Desert of Arizona.

The history of Israeli fish farming in non-desert areas, mostly in the Galilee region, dates back to the late 1920s, before Israel was established as a state. At the time, the country was extremely poor and meat was considered a luxury. But fish was a cheap food source, so fish farms were set up on several kibbutzim in the Galilee. Eventually, these farms expanded to rearing many varieties of fish, including tilapia, carp, striped bass and mullet, as well as ornamental fish.

The past decade has seen the establishment of about 20 fish farms producing both edible and ornamental fish in the Negev and Arava Deserts.

Fish farming, meanwhile, has become more lucrative worldwide as people include more fish in their diet for better health, and ocean fisheries are increasingly becoming depleted.

The practice is not without critics, who say it can harm the environment and the fish. In Israel, a decision was made by the government to stop fish farming in the Red Sea near the southern city of Eilat by 2008 because it was deemed damaging to nearby coral reefs.

Some also argue that the industry is not sustainable in the long term because most of the fish that are farmed are carnivorous and must be fed a protein-rich diet of other fish, usually caught in the wild. Another criticism is that large numbers of fish are kept in relatively small areas, leading to a higher risk of disease.

Advocators of desert aquaculture contend that the controversy surrounding fish farming in ocean areas does not apply to desert aquaculture, which is in an isolated, controlled area, with much less competition for resources.

In Kibbutz Mashabbe Sade, in the middle of the Negev desert, about 15,000 fish are reared at a time. Up to 500,000 m^3 of water from the fishponds are used for field irrigation every year, creating a chain of users for the same water.

Figure 9.1. Open fishpond in the desert

Figure 9.2. Feeding experimental fish

Figure 9.3. Mother barramundi grown in the desert

Figure 9.4. Indoor fish farm in the desert

Figure 9.5. Aeration of desert fishpond

Figure 9.6. Farming fish fingerlings in the desert

Figure 9.7. Harvesting fish in the desert

Figure 9.8. Experimental systems

Figure 9.9. Intensive fish farming in the desert

Figure 9.10. Brood stock for the production of juvenile fish

Figure 9.11. System for growing ornamental fish

Figure 9.12. Wellhead for the subsurface brackish geothermal water

Figure 9.13. Covered fishpond in desert fish farm

Figure 9.14. Tapping a desert aquifer

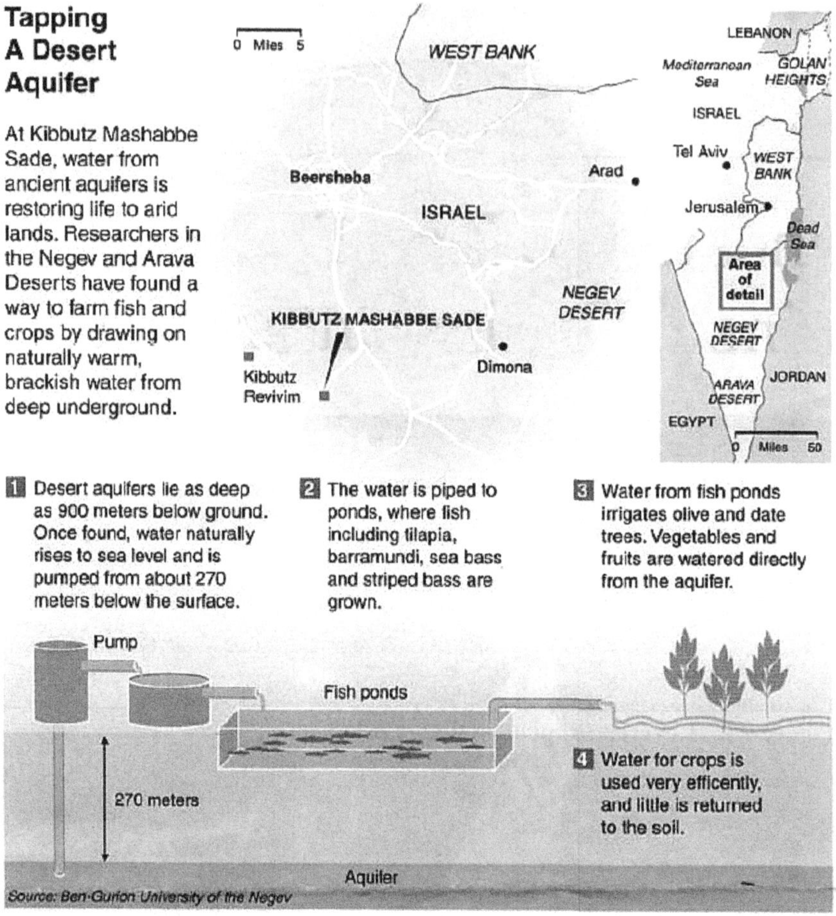

Chapter 10

Dust on Solar Collectors

Sergey Biryukov

Solar energy research—the field in which I work—has grown in importance because solar energy is an unlimited natural resource that is harmless to the local and global environment. A very important part of my research relates to the development of new solar concentrators—particularly to precise measurement of their optical properties. Optical quality is a very important factor in solar power production, because reflectance of solar-collector and concentrator mirrors and coating-transparency of glass photovoltaic panels directly influence the conversion efficiency of solar photons into useful thermal and electrical power.

In this chapter I would like to concentrate on my research in the optical quality of solar collectors, which is very important to future large-scale solar power production. Its significance becomes even more important as development of highly concentrated solar energy progresses. The problem I will discuss pertains to the negative impact of the environment on the efficiency of solar energy conversion.

The principal aspect of any solar device is that it must perform its job outdoors while being irradiated by photon flux from the sun. Correspondingly, any nontransparent physical body positioned between the sun and the solar collector (receiver) blocks part of this flux, thus reducing the amount of useful power produced. Similarly, airborne dust contamination on solar collector surfaces leads to reduced conversion efficiency.

Let us consider this problem. Arid and semi-arid regions of the planet are attractive zones for solar-electricity generation, due to their high average yearly solar-radiation levels and to the availability of large tracts of inexpensive land associated with these regions. A major obstacle, however, is the correspondingly high concentration of atmospheric dust prevailing in these same regions. These two circumstances are clearly illustrated by the global distribution map of bright sunshine (Figure 10.1) and the corresponding global distribution map of atmospheric dust production and frequency of dust storms (Figure 10.2).

Figure 10.1. Average yearly total number of hours of bright sunshine

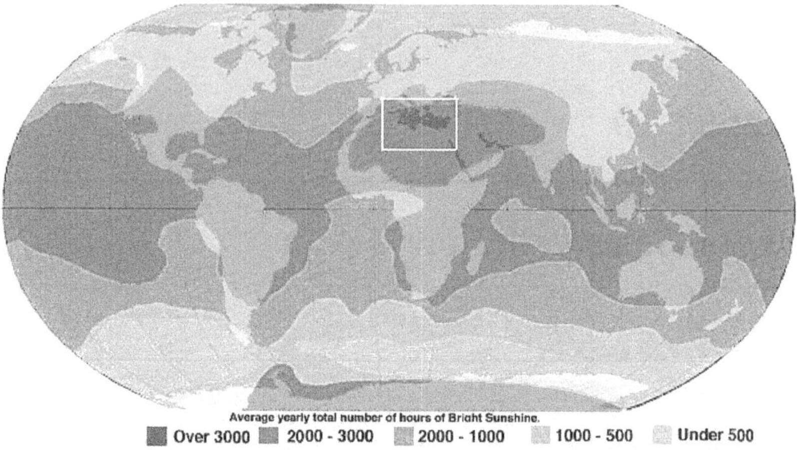

Figure 10.2. Distribution of dust producing regions on earth

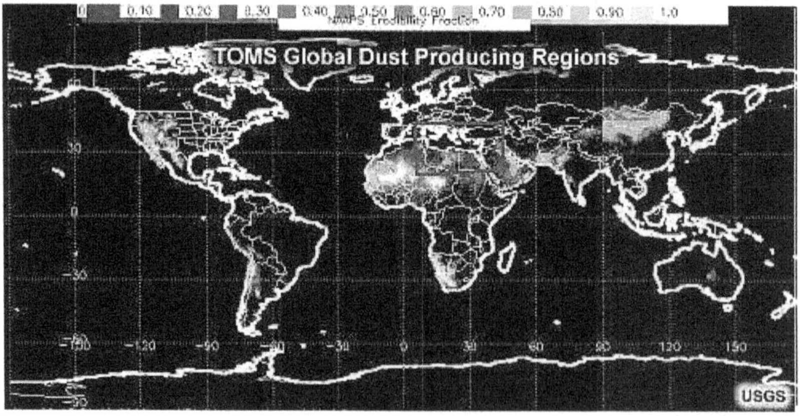

Deposition of airborne dust on solar collector surfaces leads to degradation of their optical properties, with corresponding loss of efficiency, resulting in negative economic consequences. In desert climates, rain is scarce and thus not capable of maintaining acceptable levels of collector cleanliness. The usual method for cleaning solar collectors is to wash them with deionized water under pressure (Figure 10.3).

Figure 10.3. Simultaneous cleaning of two parallel strings of parabolic trough collectors

However, the general scarcity of water in such regions—of which Israel is a typical example—makes this commodity very problematic for cleaning very large-scale fields of solar concentrator mirrors and/or photovoltaic modules. This is especially true when these cover an area as large as the solar power plant in California seen in Figure 10.4, which is capable of generating about 150 MW of electrical power from its >1 million m^2 of 94%-reflectivity, high-precision parabolic mirrors.

Fig 10.4. Aerial view of the five 30-MW parabolic trough plants at Kramer Junction in the Mojave Desert of California

Taking into account that even 150 MW is not a large quantity of power when compared to competing "conventional" power plants (a single block of a nuclear power plant typically produces 600–800 MW), cleaning costs to maintain the high optical quality of square kilometers of expensive perfect mirrors must be kept to a minimum. Clearly, an alternative method for keeping solar collector surfaces clean is very desirable.

In fact, surface cleanliness is not specifically a solar energy problem—many other fields require a solution to this same problem. Efficient solar energy collection imposes stringent limitations on the physical methods that may be considered: spending more than 1% of the energy produced by the plant or blocking an equivalent area (in terms of monetary loss) of the aperture to solve this problem would obviously not be acceptable. So, if this problem can be solved for the solar energy industry, the same approach could solve similar problems in several other fields.

Faced with this problem, my first inspiration came to me as I wrote my name with my finger on a dusty solar concentrator mirror—perhaps application of an *electric field* might prevent long-term accumulation of dust particles on a surface. A device known as a "coherer" (Figure 10.5)—which utilized the action of an electromagnetic field on small particles to make the invention of radio possible—has been around for more than a century: Thomas Edison, Nichola Tesla and Jagadish Chandra Bose are just a few of the great inventors, engineers and physicists who have worked with this device. The physical principles behind some of the modifications of this device are still not perfectly understood. However, the "beauty of physics" did not let me down, the device actually worked, and its successful application for the purpose I had in mind seemed promising.

Application of an electric field also offered the advantage of keeping surfaces clean regardless of their location—solar collectors are frequently positioned rather high above the ground or are of geometric shapes that make water-cleaning applications difficult.

Figure 10.5. Marconi "coherer-receiver" (three examples)

From basic physics we know that electric interactions between two bodies can be attractive or repulsive when they are charged (but not only under such circumstances). This particle-attraction effect is widely used in a broad variety of filters. It is well known that particle attraction to and accumulation on cleaning elements necessitates periodic cleaning of these elements (e.g., electrostatic precipitators) or periodic removal and replacement of these filters (e.g., car engines). For this reason I decided from the very beginning to concentrate on repulsive interactions to develop a system capable of keeping dust from accumulating on surfaces or to periodically removing dust from such surfaces; in other words—to use "house-keeping" terminology—to create an "electrical broom" rather than the analogy of a vacuum-cleaner.

Physics history aside, my initial confidence that the chosen solution—application of an electric field—could solve the problem was based on the high sensitivity of my fingers. I can sometimes tell if an electrical cable is plugged into a socket just by touching its insulation. Through my skin I can sense the mechanical vibrations of the ponderomotive forces caused by polarization of the dielectric. If this effect is strong enough to produce sensible deformation of skin, why couldn't it be strong enough to remove (i.e., shake off) a dust particle? I put a small amount of dust on a thin wire with thin insulation, and plugged the wire into an electric socket. A significant amount of dust was shaken off the wire!

I systematically searched the scientific literature for sources of information about repulsive effects between an electric field and small dielectric particles. Searching for this information in sources related to air filtration, where repulsive interactions leading to parasitic effects should be found, shortened my search considerably.

Following this direction, I encountered the phenomenon of dust reentrainment from the collecting electrodes of electrostatic precipitators. This is caused by a growing electric charge in the dust layer, followed by what I would call a "coulombic explosion." Figure 10.6, taken from Rose and Wood (1966), demonstrates this phenomenon.

My initial inspiration to apply this repulsive interaction for removal of dust particles, nevertheless, was not related to Coulomb's Law (the most obvious approach). I knew that the natural charge on dust particles was small and widely distributed, and that it even included both negative and positive charges, so Coulomb forces on such particles would be small and even of different signs.

Universality of force direction for all particles was definitely a necessary condition, so I began to consider polarization of particles in an electric field gradient. A polarized particle in such a gradient experiences a force that is directed toward the higher strength of the field, independently of the field direction. Thus, the direction remains the same, not only for static electric fields, but also for alternating electric fields.

In a preliminary experiment, a thin copper wire stretched taut in the vicinity of a dusty surface and parallel to it, served as the source of such a gradient. The wire successfully attracted dust to it, as can be seen in Figure 10.7. This phenomenon, known as dielectrophoresis (Pohl, 1978), is used in biology and nanotechnology, usually in liquids but not in air.

Figure 10.6. Re-entrainment of dust in electrostatic precipitators, having coulombic origin

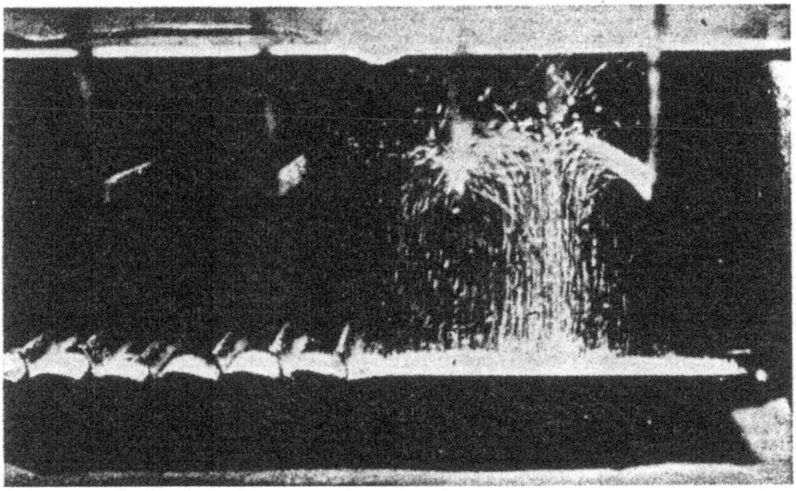

Figure 10.7. Attraction of dust to a wire electrode, demonstrating the phenomenon of dielectrophoresis, due to the polarization effect

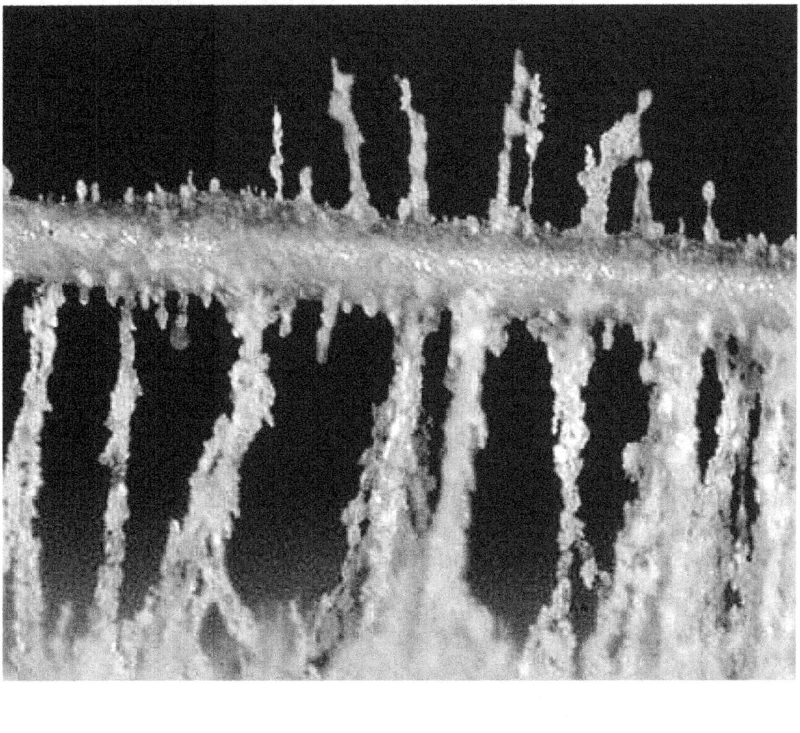

Such an electrode could attract dust to itself, like a vacuum cleaner or the elements of a filter, and would be capable of keeping nearby surfaces clean. But my goal was repulsion. My original idea was to use a high-frequency external electric field to induce an internal-field phase lag, or negative phase shift, in the dust particles, relative to the external field, which would convert the attraction effect into a repulsion effect, thus achieving negative dielectrophoresis.

Through further experimentation I also demonstrated repulsion of dust from the same electrode, creating conditions that were rather similar to those responsible for the dust–re-entrainment effect demonstrated in Figure 10.6. This effect was repulsive and worked at greater distances from the electrode. After further optimization of conditions through experimentation, I can say that my problem-solving approach is based on applying an electric field in the vicinity of the surface of interest: to create a repulsive interaction within the pre-surface interface between the field and the dust particle to be removed—or to prevent the particle from even settling. The class of phenomena I studied and used may be defined by a single term: "electrohydrodynamic interaction," which includes dielectrophoresis, electrophoresis, and the hydrodynamic phenomenon known as "ionic-"- or "electric-wind."

From the very beginning I realized that particle size d_p was of crucial importance to the problem I was attempting to solve. According to Stokes' Law, particle deposition velocity v (terminal velocity under gravity in air) is proportional to the square of a particle's size ($v \sim d_p^2$) but only linearly proportional to its density; thus particle density was of lesser importance.

The ability of any physical interaction to remove deposited particles from a surface also depends on the size of the particles to be removed, because two important size-dependent forces must be overcome, i.e., force of adhesion (particle adhesion to the surface) and force of gravity (particle weight).

The last statement may be illustrated by the following diagram (Figure 10.8), where:

f_{adh}	=	force of adhesion between the particle and the surface,
$f_{gravity}$	=	force of gravity (weight of the particle)
f_{int}	=	force of interaction between the particle and an external field, (e.g., when $f_{int} \sim d_p^2$);
$n(d_p)$	=	numerical concentration of particles with size d_p.

The particles of interest are those that influence the optical properties of surfaces. By order of magnitude, such particles are >0.1 μm in size; i.e., not much smaller than the wavelengths of light that comprise a major part of the solar spectrum. However, from aerosol physics it is known that in this size range, only particles ≥1 μm in size have effective natural deposition mechanisms (i.e., gravity and inertia). Because they are larger than the wavelengths of visible light, these particles can be measured with ordinary optical microscopy.

Figure 10.8. Diagram of forces acting on particles deposited on a surface, superimposed upon a hypothetical particle size distribution.

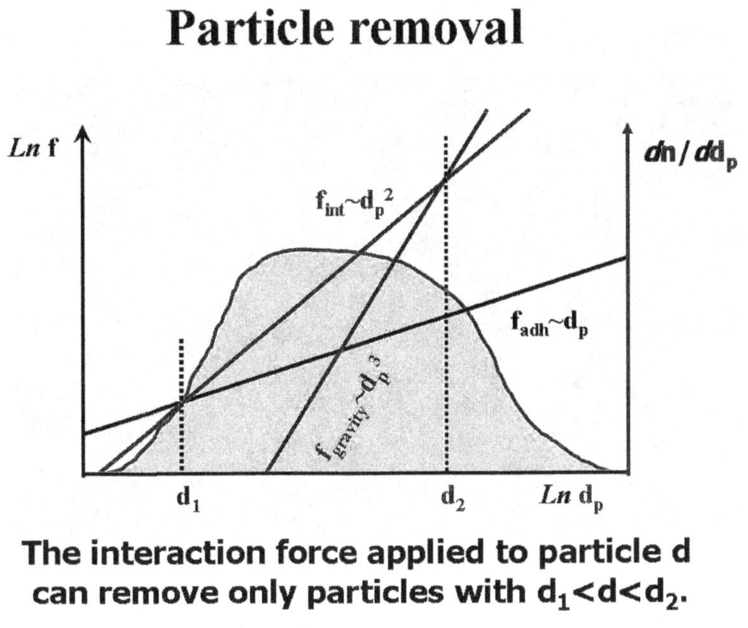

Taking this into consideration and supported by my previous experience in nuclear physics, I was inspired to develop corresponding software tools, which my colleague, Dr. A. Goldfeld, and I turned into a program package. We then designed and built a computerized optical microscope (Biryukov, Faiman and Goldfeld, 1999) that was capable of collecting images of dusty surfaces (Figure 10.9), recognize the particles, separate them from the background, and produce a particle size distribution (PSD) for individual frames or for sets of multiple frames in order to improve statistical accuracy (Figure 10.10).

Very importantly, the distance between our microscope's objective and its sample stage is large, to facilitate real-time "live" viewing on a monitor screen of particle interactions with electric fields (Figure 10.11a, b, c).

Figure 10.9. Image frame of a dusty mirror, taken by our computerized optical microscope

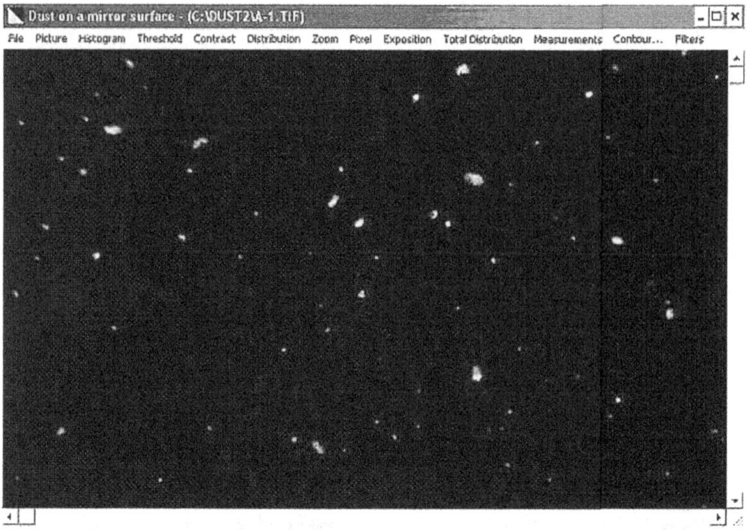

Figure 10.10. Histogram of particle size distribution, produced by our software

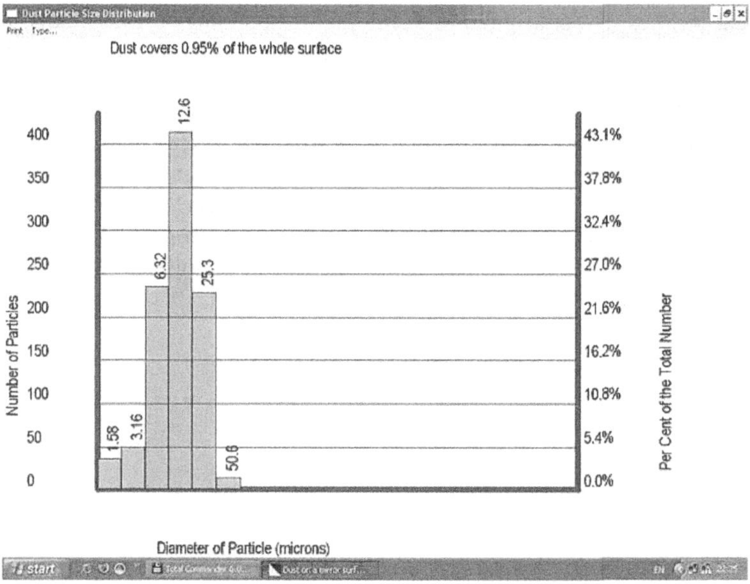

Figure 10.11a. Aggregated dust particle under microscope

Figure 10.11b. Beginning of particle splitting, caused by field charging

Figure 10.11c. Result of coulombic explosion of initial aggregate (Biryukov, 1997)

The computerized-microscopy technique we developed enabled me to investigate dust-deposition mechanisms (Biryukov, 1998), real particle-size distributions and dust-deposition rates (Figure 10.12). It was successfully applied to measuring photovoltaic-panel performance degradation (Biryukov, Faiman and Karki, 2004) and dust-removal efficiency from solar collector surfaces by means of an electric field.

Figure 10.12. Differentiated and cumulative size distribution and deposition rates of dust particles on a solar collector surface

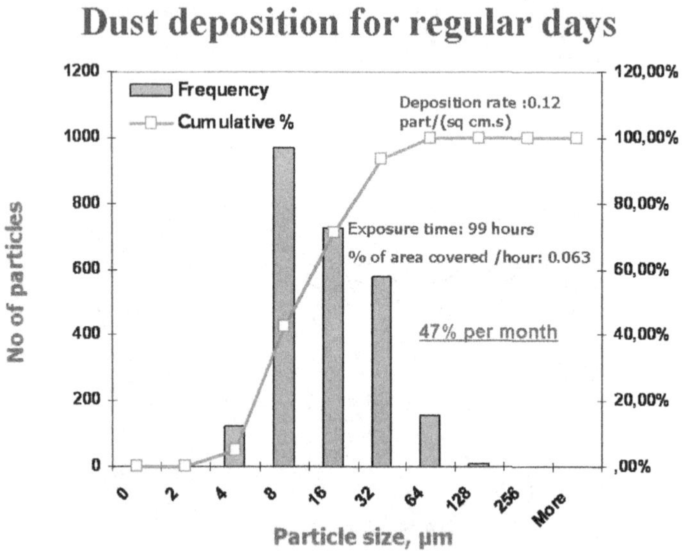

As a result of this research, several devices have been designed and built. The technique developed fulfills all the specific needs of solar energy applications; i.e., it is inexpensive, transparent to light, and has negligible power consumption. It is capable of keeping surfaces of solar collectors, including mirrors, photovoltaic panels and other important parts of solar converting devices free of dust.

The computerized microscopy method mentioned above was employed to quantitatively characterize the performance efficiency of these devices.

Tests of the device I call the "Kinnor" (lyre in Hebrew), because of its similarity to the musical instrument (Figures 10.13 and 10.14), carried out during a recent period of dust storms (April-May 2007) demonstrated its high efficiency (>>90%).

Figure 10.13. "Kinnor-2" prototype device for dry electrohydrodynamic cleaning of a 400-mm wide photovoltaic panel

Figure 10.14. "Kinnor-2" – dust removal in progress

Acknowledgement

This work was supported by grants from the Israel Ministry of Infrastructure. Development of photovoltaic panels with dust-protecting systems based on these results will be continued in collaboration with the Fraunhofer Institute for Solar Energy Systems in the framework of a Bi-National Germany-Israel research grant for Cooperation in Energy Research.

References

Biryukov, S. (1997) Coulombic splitting of loess aggregates on surface. *Journal of Aerosol Science*, 28 (Suppl. 1): 515–516. "Abstracts of the 1997 European Aerosol Conference," held in Hamburg, Germany, 15–19 September 1997.

Biryukov, S. (1998) An experimental study of the dry deposition mechanism for airborne dust. *Journal of Aerosol Science*, 29:129–139.

Biryukov, S., Faiman, D. & Goldfeld, A. (1999) An optical system for the quantitative study of particulate contamination on solar collector surfaces. *Solar Energy*, 66:371–378.

Biryukov, S., Faiman, D. & Karki, I. (2004) The cumulative effect of desert dust on photovoltaic modules. *Proceedings of the 19th European Photovoltaic Solar Energy Conference*, held in Paris, France, 7–11 June, 2004, II: 2117–2120.

Pohl, H. A. (1978) *Dielectrophoresis: The behavior of neutral matter in nonuniform electric fields*. Cambridge: Cambridge University Press.

Rose, H. E., & Wood, A. J. (1966) *An introduction to electrostatic precipitation in theory and practice* (2nd Ed.). London: Constable and Company.

Chapter 11

Ecological Sanitation: An Alternative for Sustainable Dryland Development

Amit Gross

In the modern world, the use of natural resources and production of wastes and contaminated effluents have significantly increased and now pose severe health and environmental risks in many regions—especially in arid regions. There is an urgent need to remedy already contaminated sites and find means of minimizing these trends. A new field of research, currently called Ecological Sanitation (ECOSAN), is a modern alternative to conventional sanitation techniques. The objective is to protect human health and the environment. Unlike traditional sanitation methods, ecological sanitation processes human waste to recover nutrients that would otherwise be discarded.

Issues, such as the following, are the main goals of the "EcoSan" field:

- Prevention of disease
- Protection of the environment: by means of waste product treatment, clean water conservation and pollution prevention of natural bodies of water.
- Recovery of nutrients: nitrogen and phosphorus remaining in feces and urine should be recycled to reduce dependence on artificially created fertilizers for agriculture

The means to achieve the EcoSan goals rely on simple designs that are easily constructed and maintained. The systems should meet the needs of all household members regardless of gender, age, or social status. They should be affordable, low-cost systems that enable people everywhere, even in developing countries, to implement ecological sanitation practices. Last, but not least, cultural acceptability must be considered. Planners should be sensitive to regional cultural and social values regarding the handling of human wastes. Examples of such technologies may be found in the two articles that follow.

Advanced technologies for greywater treatment and reuse*

The quantity of freshwater available worldwide is declining, bringing with it a pressing need for more efficient use. One method of conserving water is by recycling greywater for irrigation, a practice that can reduce potable water use in arid areas by up to 50% (Friedler, 2004). Greywater is all domestic wastewater, excluding toilet effluents. Moreover, use of greywater reduces the need to add fertilizers to the soil due to their presence in such effluents, and ensures a supply of water for irrigation even in times of drought. The need to recycle such water is greatest in small isolated settlements that are situated far from central wastewater treatment systems. Several studies carried out at the Zuckerberg Institute for Water Research of the Jacob Blaustein Institutes for Desert Research, Ben-Gurion University of the Negev, have shown that raw greywater is often contaminated and its use without pre-treatment can cause environmental harm and pose health risks (Casanova, Gerba and Karpiscak, 2001; Al-Jayyousi, 2002; Gross et al., 2005; Wiel-Shafran et al., 2005; Wiel-Shafran et al., 2006; Gross, Kaplan and Baker, 2007a). A number of commercial systems were tested and failed to meet acceptable standards for treating effluents, mainly because they were not adapted for the specific characteristics of greywater in arid areas (Gross et al., 2006).

As a result of these findings, our laboratory has developed a Recycled Vertical Flow Constructed Wetland (RVFCW) to provide an economically sound, low-tech and easily maintainable treatment system that will facilitate safe and sustainable use of greywater for landscape irrigation in small communities and households (Gross et al., 2007b).

The RVFCW system (Figure 11.1) has two components: 1) a three-layer bed consisting of planted organic soil over an upper layer of filtering media (i.e. tuff or beads) and a lower layer of limestone pebbles, and 2) a reservoir located beneath the bed. Wastewater flows directly into the plant root zone and trickles down through the filter into the reservoir, allowing passive aeration. From the reservoir, the water is recycled back to the filter for several hours, until the desired purification is achieved. The recirculation pump keeps the wetland constantly wet and functioning, even when no water is added to the system.

The RVFCW combines the principles of a trickling filter with that of traditional constructed wetlands. In this way performance is improved and the treatment system can be adapted to deal with the various problems that may be associated with a water treatment plant in a private garden, i.e., the need for a small

---

* From Gross, A. (2006) *A Voice from the desert*. Bulletin of Jacob Blaustein Institutes for Desert Research, Ben-Gurion University of the Negev. November 2006.

plant, capable of treating water of highly variable quantity and quality—possibly containing high concentrations of cleaning detergents—and supplying this treated water to a regime with varying demands. Moreover, such a system must be affordable to the average family, it must be reliable and it must be suitable for maintenance by unskilled operators. Environmental nuisances such as bad odors and mosquitoes were also taken into consideration when designing the system.

Figure 11.1. Diagram of a RVFCW system. Wastewater flows directly into the upper layer, trickles down into the lower reservoir, and is recycled back to the filter. After several hours of recycling treatment the effluent flows out to the irrigation system

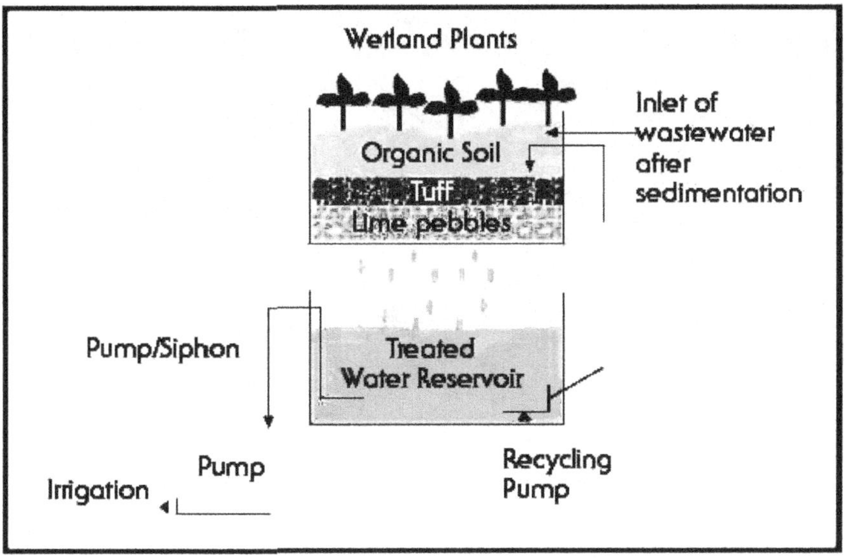

The recirculating pump between the lower reservoir and the upper filter allows the wastewater to undergo treatment several times (according to the owner's specifications) before being released to the garden, so that the quality of the water can be closely controlled and the size of the unit reduced. The recirculation mixes the untreated water with the treated water and reduces the load on the treatment system as a whole. The wastewater flows directly into the root zone and this prevents any contact between the raw wastewater and the environment.

The reservoir and the recirculating pump allow the system to keep going and keep the upper filter active without any new input of water for several days (for example when on vacation). The trickling of water from the upper to the lower containers via the root zone oxidizes the water and prevents unpleasant odors. The system itself can be partially buried beneath ground level, and the planted upper part provides an aesthetic bush of greenery that blends in with any garden view (see Figure 11.2).

Figure 11.2. Example of a RVFCW system installed in a private household four years ago and successfully being used to irrigate a garden in the Negev

Beyond the environmental and educational aspects of water reuse, a feasibility analysis of the small unit indicated a return over investment after less than three years for an average 5-person family (Based on Israeli water prices). Pilot systems have already been installed in a number of homes and are functioning successfully. The researchers running the project believe that the invention of technical solutions is not enough—support from institutions and the media is also vital to disseminating information and increasing public awareness regarding water saving and reuse, setting of acceptable standards for greywater reuse, and professional monitoring. The synergy between technology and such raised awareness will lead to more efficient and safer use of this vital resource. If these aspects are ignored, the unsafe practice of reusing greywater for irrigation without any treatment will continue, and even increase as the availability of fresh water decreases, leading to an unnecessary rise in environmental damage and health risks.

This system is expected to be implemented in Australia in the future through B. G. Negev Technologies.

The article above demonstrated how EcoSan technology can be applied to treatment of liquid waste i.e., greywater. Below is an example of an EcoSan technology that is applicable to treatment of solid organic waste.

Nutrient recycling from organic waste for the production of organic fertilizers and protection of the environment

By Amit Gross, Ofer Guy, Pinchas Fine and Ali Nejidat [1]

Background, technology and rationale

Organic wastes, specifically manures, pose an environmental concern as they are produced in large quantities and their disposal is problematic and expensive. The main environmental issues associated with manure wastes are the spread of diseases, release of substances to the atmosphere (such as greenhouse gases), formation of bad odors, and leaching of materials into the groundwater—mainly nitrogen, but also other less-monitored components (such as hormones and antibiotics). Recycling of manure for compost, biogas, and solid fertilizers has been suggested as a green solution to mitigate these problems and has been practiced for many years (Hansen, Alroe and Kristensen, 2001; Hue and Ikawa, 2001), but with limited results. Most of these practices have more than one of the following disadvantages: environmental impacts, insufficient quality, nutrient losses, high costs, low efficiency in recovering nutrients, and only limited success with pathogen removal (Hadas and Rosenberg, 1992; Hadas, 1997; Gross et al., 2008).

We have recently developed a novel concept for the production of nitrate-rich liquid organic fertilizer (Figure 11.3). This involves extraction of nitrogen from the manure, followed by separation (via ammonia volatilization) of the nitrogen from the extract and its subsequent conversion to nitrate (Arusi, et al., 2004; Guy, 2007). This concept mitigates most of the environmental concerns mentioned above and has several advantages over the traditional production of liquid fertilizers for organic agriculture. These include higher recovery of nitrogen, reduction of the extraction period, ability to recycle bio-solids that are not suitable for use in current practices (i.e., various sludges and marine algae) and lower costs. In addition, the method is more environmentally friendly as it recycles the bio-waste, recovers most of its nitrogen, supplies a higher value product, and ensures pathogen removal (see details in expected outcome below). The novel concept was successfully tested in pilot-scale experiments, producing several hundred liters of nitrate-rich (~ 1% N for guano) organic liquid fertilizer from guano and broiler manures. Doubling the concentration is also achievable in these systems. Moreover, analysis of plants grown with fertilizers produced by this model demonstrated better quality and shelf life compared with the regular organic fertilizers in use in Israel.

Figure 11.3. Lab-scale reactor for the production of nitrate-rich liquid fertilizer for fertigation in organic agriculture

Commercial potential

The commercial potential of such a fertilizer is tremendous and includes:

- Industries that produce large quantities of sludge, animal manure or other N-rich bio-solids that need to be disposed of properly;
- Organic farmers.

Development and research

The aim of this research is to develop farm- and regional-scale devices for the extraction of nitrogen from organic wastes in general and manures in particular. The systems should produce, with good efficiency, high-quality liquid fertilizer suitable for organic agriculture, and reduce environmental pollution related to leaching of chemicals, release of gases to the atmosphere, spread of odors, and spread of diseases from manure.

Specifically, the goals of the research are as follows:

1. To determine suitable extraction protocols for different bio-waste materials (i.e. different types of manure).
2. To develop a mathematical model that will describe the chemical and biological processes in the system.
3. To establish reliable design parameters for the device, while operating under different environmental conditions, with different bio-resources and with different system volumes.
4. To scale up our pilot systems on a farm, utilizing several tons of manure for the production of fertilizers; based on results from the farm experiment and the model we intend to scale up the system to a regional scale.
5. To perform a cost analysis for the device.
6. To establish efficient protocols for recycling the remaining bio-waste.
7. To test the system's ability to reduce emission of greenhouse gases and leaching of materials into the groundwater, as compared to prevailing practices.
8. To establish deliverable bio-starters for the extraction and nitrification processes occurring during fertilizer production.

Expected outcome

We expect to significantly reduce environmental pollution associated with bio-waste in general, and manures in particular, by recycling and stabilizing manures. Specifically we expect the following outcomes:

1. Efficient production of nitrate-rich organic liquid fertilizer for nitrogen top-dressing via the drip irrigation system. Nitrogen fertigation during the growing season is essential in intensive organic practice, but to date there are no sufficient means for the production of high quality nitrogen-rich (1.5–2% N) liquid fertilizers. All current practices are beset by one or more of the following problems: a) low recovery of nitrogen from the manures (i.e. high losses); b) ammonia is the main inorganic N form (it can cause plant deformations and other non-microbial diseases; c) liquid fertilizer clogs the irrigation systems (due to high levels of suspended solids and enhancement of bio-film growth). The proposed fertilizer will eliminate these problems.
2. The remaining digested bio-solids are "pathogen-free" and we expect that in most cases they can be used directly to improve soil nutrition (except for N) and soil properties in terms of texture and structure, and possibly to suppress diseases (Welztein, 1986; Rangarajan, 2001; Welke, 2001). The bio-waste can also be introduced into a compost pile. In any case, the remaining bio-waste is fairly stable and almost nitrogen free and therefore could be disposed of regularly without any foreseen environmental harm.
3. Reduced emission of nitrogenous greenhouse gases—and possibly others —by entrapment of the released gases in a bio-reactor (nitrification unit), where they are absorbed and, in the case of nitrogen, mostly converted to nitrate.

4. Reduction of bad odors as the digestion takes place in a closed system, and the last stage is aerobic and does not produce bad odors.

The device and its concept were submitted for a patent in the USA in November 2005.

Note

1. Amit Gross, Ofer Guy, and Ali Nejidat are affiliated with the Zuckerberg Institute for Water Research, Jacob Blaustein Institutes for Desert Research, Ben-Gurion University of the Negev, Midreshet Ben-Gurion, Israel. Pinchas Fine is affiliated with the Institute of Soils, Water and the Environment, ARO, Volcani Center, Bet Dagan, Israel.

References

Al-Jayyousi, O. R. (2002) Focused environmental analysis for GW reuse in Jordan. *Environmental Engineering and Policy*, 3(1-2): 67–73.

Arusi, R., Gross, A., Fine, P. Ronen, Z. & Nejidat, A. 2004 Nitrogen extraction from fowl manures for fertigation in organic agriculture. *Proceedings of the 9th European Biosolids and Biowaste Conference*, Wakefield, UK. 13(40): 1-15.

Casanova, L. M., Gerba, C. P., & Karpiscak, M. (2001) Chemical and microbial characterization of household graywater. *Journal of Environmental Science and Health, Part A Toxic/Hazardous Substances and Environmental Engineering*, 36(4): 395–401.

Friedler, E. (2004) Quality of individual domestic greywater streams and its implication on on-site treatment and reuse possibilities. *Environmental Technology*, 25(9): 997–1008.

Gross, A. (2006) *A voice from the desert*. Bulletin of Jacob Blaustein Institutes for Desert Research, Ben-Gurion University of the Negev. November 2006: 6–8.

Gross, A., Arusi, R., Fine, P., & Nejidat, A. (2008) Assessment of extraction methods with fowl manure for the production of liquid organic fertilizers. *Bioresource Technology*, 99:327–334

Gross, A., Azulai, N., Oron, G., Ronen, Z., Arnold, M., & Nejidat, A. (2005) Environmental impact and health risks associated with greywater irrigation—a case study. *Water Science and Technology*, 52(8): 161–169.

Gross, A., Kaplan, D., & Baker, K. H. (2007a) Removal of chemical and microbiological contaminants from domestic greywater using a Recirculating Vertical Flow Bioreactor. *Ecological Engineering*, 31:107–114.

Gross, A., Shmueli, O., Ronen, Z., & Raveh, E. (2007b) Recycled Vertical Flow Constructed Wetland (RVFCW)—a novel method of recycling greywater for landscape irrigation in small communities and households. *Chemosphere*, 66:916–923.

Gross, A., Wiel-Shafran, A., Bondarenko, N., & Ronen, Z. (2006) Reliability of small commercial greywater treatment systems: A case study. *International Conference on Protection and Restoration of the Environment VIII,* Chania, Greece. 8 pp. Not paginated.

Guy, O. (2007) Entrapment of Nitrogen from Fowl Manure Extracts for Fertigation in Organic Agriculture. M. Sc. thesis, Ben-Gurion University of the Negev, Israel.

Hadas, A. (1997) Guano and organic fertilizer alternatives. *The Field*, 77:18–20.

Hadas, A., & Rosenberg, R. (1992) Guano as a nitrogen source for fertigation in organic farming. *Fertilizer Research*, 31:209–214.

Hansen, B., Alroe, H. F., & Kristensen, E. S. (2001) Approaches to assess the environmental impact of organic farming with particular regard to Denmark. *Agriculture, Ecosystems and Environment*, 83:11–26.

Hue, V. N., & Ikawa, H. (2001) Organic soil amendments for sustainable agriculture. *Rx for soils and crops*. College of Tropical Agriculture and Human Resources, University of Hawaii, Manoa, p. 14.

Rangarajan, A. (2001) Impact of disease suppressive composts on organic vegetable quality, composition and yields. *Organic Farming Research Foundation - Information Bulletin*, 9:21–25.

Welke, S. (2001) Effectiveness of compost tea extracts as disease suppressants in fresh market crops. *Organic Farming Research Foundation - Information Bulletin*, 9:16–20.

Welztein, H. C. (1986) The effects of compost extracts on plant health. *Global Perspective on Soil Agricultural Systems*, 2:551–553.

Wiel-Shafran, A., Gross, A., Ronen, Z., Weisbrod. N., & Adar, E. (2005) Effects of surfactants originating from reuse of greywater on capillary rise in the soil. *Water Science and Technology*, 52(10-11): 157–166.

Wiel-Shafran, A., Ronen, Z., Weisbrod, N., Adar, E., & Gross. A. (2006) Changes in soil properties following irrigation with surfactant-rich greywater. *Ecological Engineering*, 26:348–354.

Chapter 12

Groundwater Pollution: The Deep Vadose Zone Monitoring System[1]

Ofer Dahan, Yaara Rimon,
Boaz Tatarsky and Ruth Talby[2]

Introduction

The mechanisms controlling water flow and contaminant transport from land surfaces through the vadose zone have been extensively investigated for several decades. It has been recognized that the vadose zone plays a major role in controlling groundwater quality and recharge quantity. The scientific efforts to study flow and transport processes through the vadose zone are well documented in a large array of published studies, from small-scale laboratory experiments (Van Geel and Sykes, 1994; Glass, Nicholl and Tidwell, 1995; Walser, Illangasekare and Corey, 1999; Heilig et al., 2003) to superfunded sites with large experimental field operations (Dahan et al., 1999; Wolfsberg et al., 1999; Doughty, Salve and Wang, 2000; Tseng et al., 2003). Although the basic information that allows adequate characterization of the flow process in a well-defined medium is relatively well known, the processes occurring in a natural heterogeneous environment are still far from being clearly described or properly predicted. While theoretical models and computation tools have significantly improved in recent years (Philip, 1957; Kool, Parker and van Genuchten, 1987; Verburg, Ross and Bristow, 1996; Šimůnek , Šejna and van Genuchten, 1998; Flint et al., 2002), the ability to provide reliable field observations to support the simulated process is still being developed.

The scientific efforts to provide better monitoring tools for the study of flow and transport through the vadose zone have led to remarkable developments, from the construction of huge experimental monitoring test sites, e.g. in Nevada, Ohio, Denmark and Israel (Sidle et al., 1998; Nativ, Adar and Becker, 1999;

Wolfsberg et al., 1999; Liu et al., 2004), to the implementation of new technologies, such as airborne penetrating radar (Kowalsky, Finsterle and Rubin, 2004). However, the most important monitoring tools traditionally used for the study of water flow and solute transport in unsaturated conditions are water-content sensors, such as time-domain reflectometry (TDR) probes, tensiometers for water-tension measurements, and suction cups for pore-water sampling (Looney and Falta, 2000). These tools were originally designed to be implemented in shallow soils for agricultural purposes, and their implementation in deep sections of the vadose zone is usually not straightforward. Traditional TDR probes are made of two or three parallel metal rods (2–6 mm in diameter and 10–30 cm long). The maximum depth for implementation of the traditional probes is usually limited by the ability to "push" the probe into the sediments.

The need for accurate, real-time measurements of water content in the deep vadose zone has spawned several technical developments. Yokuda and Smith (1993) developed TDR sensors embedded in a rigid rod that can be pushed into the ground using a vibratory drill to depths of up to 4.5 m. Although this increased the depth of installation, its use in consolidated sediment/rock or gravelly formations is still very limited. Measurements of water-content profiles in deep sections of the vadose zone may also be obtained through a permanently installed access borehole using neutron thermalization (Hillel, 1980, Flint and Flint, 2000), or the TRIME® TDR system (www.mesasystemsco.com/trime_index.asp) [http://www.mesasystemsco.com/pdf/IMKO_catalog_bw.pdf].

Application of tensiometers and suction cups in deep horizons of the vadose zone faces similar technical problems. The porous tip, which is usually made of brittle ceramic material, needs to be pushed into the ground or installed in a small-diameter hole specially drilled for that purpose. To overcome the limited installation depth of both TDR probes and tensiometers, Zilberbrand and Gvirtzman (1996) installed soil sensors through the walls of a wide-diameter vertical shaft. This concept was further developed by Murdoch and others (2000), who developed an instrumentation technique that allows installation of soil sensors as TDR probes through borehole side walls. Haimerl (2004) used TDR probes and tensiometers in deep vertical boreholes (up to 20 m) drilled in a coarse alluvial formation. In his setup, the sensors were placed in the deepest part of a borehole that was filled with fine sand and then sealed to the surface. The measurements were conducted in a relatively large volume of sand embedded to the desired depth, assuming that the hydraulic conditions in sand reflect those in the surrounding formation.

The limited application depth for standard soil sensors and the need for a reliable monitoring system that can provide continuous real-time measurements under relatively undisturbed sediment conditions necessitated the development of an alternative monitoring setup. The alternative setup allows installation of multiple TDR probes, tensiometers and pore-water sampling ports at any depth from land surface to the groundwater. Although our current experience in a full-scale field experiment is to a depth of 21 m in the vadose zone, there is no technical limitation, and the same system could be applied to even greater depths. This chapter presents the monitoring setup and preliminary results from several experiments that have been conducted with the vadose-zone monitoring system. The results are presented here in order to demonstrate the system's performance, rather than to provide a thorough discussion of the infiltration process that was observed in each experiment. Accordingly, only a "sample" of the data from each experiment is presented.

Methods

Methodological concept

The vadose-zone monitoring system allows attachment of customized probes to the upper side of an uncased small-diameter slanted borehole (Figure 12.1). A flexible sleeve made of thin PVC or rubber liner, hosting several probes along its length in the desired distribution, is lowered into an uncased borehole (Figure 12.2). The probes are attached to the sleeve's outer side (Figure 12.3). In the borehole, the probes are aligned along its upper side wall facing the undisturbed sediment column extending from the probe to the surface (Figures 12.1 and 12.2). Immediately after insertion of the sleeve into the borehole, the sleeve is filled with liquid filling material that solidifies in the mold (such as two-component urethane or concrete). The hydrostatic pressure generated by the filling material within the sleeve causes its expansion. As a result, the probes are pushed against the borehole's upper side wall with sufficient force to achieve good contact between the probe and the sediments on the upper side wall of the borehole. After curing of the filling material in the sleeve the probes are secured in position.

The combination of a flexible sleeve with high-density filling liquid (1.6 g cm^{-3}) assures that the borehole is sealed all along its length as the sleeve expands to fill the entire void, including small cavities or diameter irregularities (Figure 12.4). Accordingly, the borehole itself becomes inert and potential preferential flow along its length is reduced. In addition, the high-density liquid used in the sleeve could potentially restore the tension-release cracks that might have been generated during drilling.

Although the method can be implemented at any drilling angle, including vertical boreholes, a slanted drilling method is preferred to assure monitoring of an undisturbed soil column. Assuming that the general flow direction in the vadose zone is vertical, every point on the upper side of a slanted borehole faces an undisturbed sediment column. Though the sediments on the borehole side wall could potentially be disturbed by the drilling activity, it is obvious that from the borehole walls to land surface the sediment column is undisturbed. Accordingly, it is assumed that the flow regime from land surface to any given point on the slanted borehole takes place in an undisturbed soil column. This assumption is not valid for vertical boreholes where the probes are distributed along the drilling axis, which is also the general flow direction. Throughout our experience with slanted-borehole drilling, the drilling action has been shown to have very little disturbance effect on the upper part of the borehole side wall, since the heavy drilling shafts lie on the lower side of the borehole.

Figure 12.1. Schematic illustration of the monitoring setup, demonstrating the simultaneous monitoring of the three hydrological domains: surface water, vadose zone and groundwater

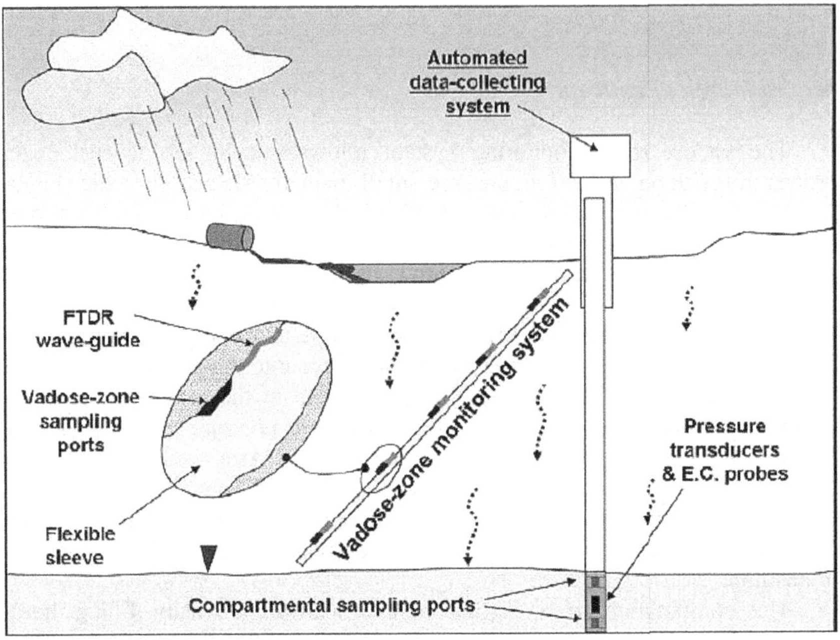

The drilling method depends on the site's geological characteristics and, of course, the available drilling facility. Nevertheless, we have drilled and successfully installed the monitoring system in a wide range of sedimentological setups, from fine sand dunes to coarse gravel alluvium, with a variety of drilling methods, such as simple flight-auger, odex and rock-bits.

To date, several probe types have been specially developed for this monitoring system and tested in the laboratory as well as in full-scale field experiments. These include flexible TDR (FTDR) or capacitance probes, which were previously presented by Dahan, McDonald and Young (2003), and vadose-zone sampling ports (VSP), which allow either sampling of the vadose-zone pore water or measurement of the water potential. In addition, Eh sensors are in the process of being developed to allow redox-potential measurement with the monitoring system.

Figure 12.2. Insertion of FTDR probes on a PVC sleeve into a slanted borehole

Figure 12.3. VSP and FTDR probes on flexible sleeves

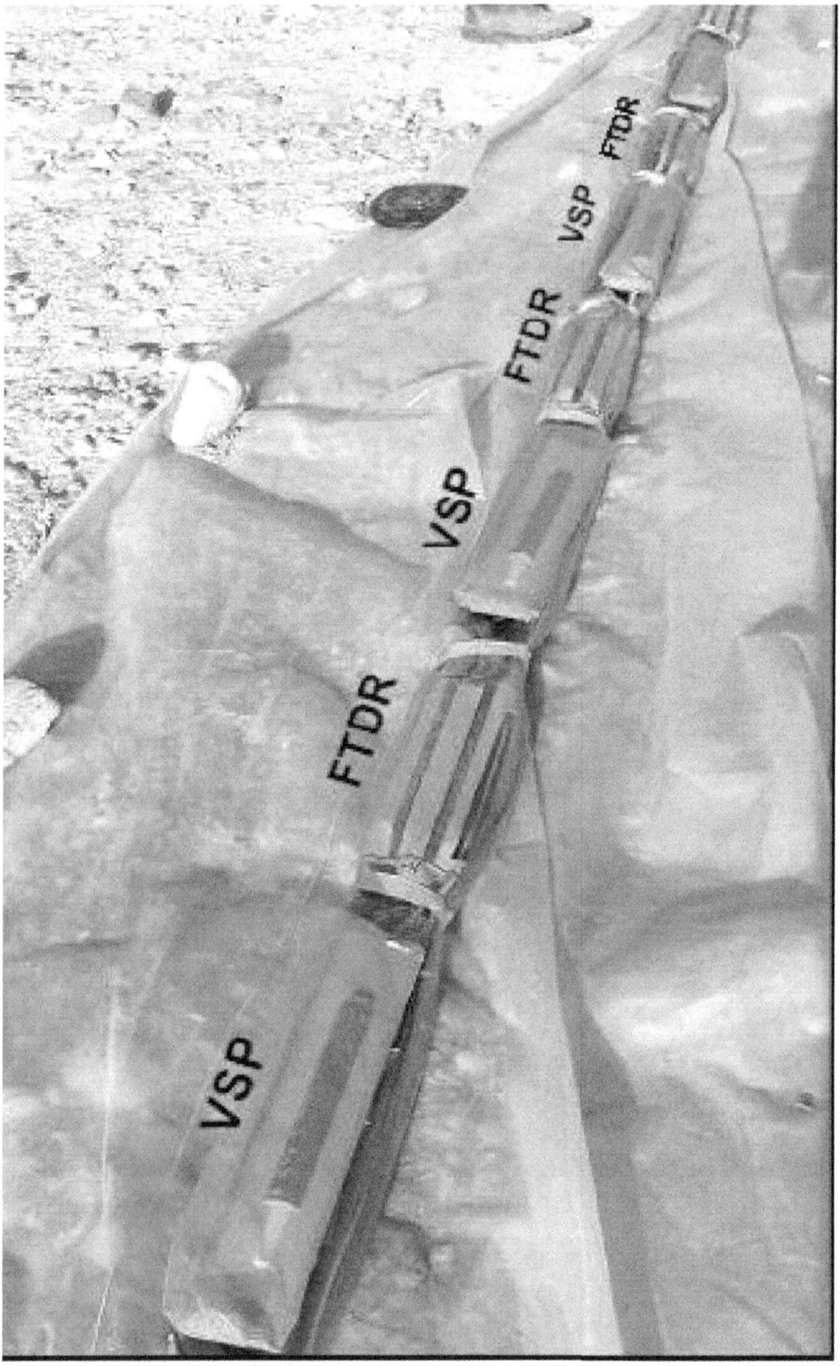

Figure 12.4. Flexible sleeve and FTDR probe exposed after the installation. The picture shows the attachment of the sleeve and probe to the borehole walls

Flexible TDR probes—FTDR

One of the most important hydrological parameters required for the investigation of water flow and pollutant transport in the vadose zone is the water-content variation with time and depth. A variety of methods and commercial tools exist for this purpose. Among them, the most common method for water-content measurement is based on time-domain reflectometry (TDR). In general, the method measures the bulk dielectric properties of the moist sediments and infers the volumetric water content from the ratio between the high dielectric constant of the water (~80) and the very low dielectric constant of most geological material (~2–7). The TDR method was first developed for measurement of soil moisture in the early 1980s (Topp et al., 1984). Since then it has been refined, and studies on its performance, potential applications and limitations have been carried out by several researchers (Ledieu et al., 1986; Herkelrath, Hamburg and Murphy, 1991; Dalton, 1992; Ferré et al., 1998). A thorough review on TDR was recently published by Robinson and others (2003).

Dahan, McDonald and Young (2003) presented flexible TDR (FTDR) probes that allow water-content measurement in the deep vadose zone. The method uses thin and flexible metal strips as waveguides (Figure 12.3). The flexible waveguides are attached to the flexible sleeve (Figure 12.2). As the sleeve is filled with resin, the waveguides are forced against the borehole walls and follow its roughness, undulation and diameter irregularities (Figure 12.4).

Accordingly, good contact of the probe with the borehole walls is achieved. The FTDR is usually operated with classic TDR operation systems (such as Tektronix or TDR 100). However, those systems are usually limited to a cable length of ~40 m, and beyond that length the measurement accuracy is reduced. To avoid cable-length problems and allow proper measurements at greater depths, water-content reflectometers (such as CS505, Campbell Scientific) may be connected directly to the flexible waveguides (Dahan et al., 2003).

Vadose-zone sampling ports—VSP

The VSP is designed to allow continuous monitoring of the water potential and frequent sampling of the sediment pore water all along the vadose zone. The VSP system was designed to be implemented together with the FTDR probes on a flexible sleeve system. Accordingly, the entire system provides a complete monitoring scheme that includes measurements of water content, water potential and the ability to sample the sediment pore water.

The VSP is based on a structural modification of standard tensiometers and suction cups. The probes were designed to allow their attachment to the sediments on the upper side wall of an uncased slanted borehole, similar to the FTDR probe. The sampling ports are mounted on the outer side of a flexible sleeve in the desired distribution, according to the lithological profile. After lowering the sleeve into the borehole, it is filled with two-component urethane. The hydrostatic pressure that is generated within the sleeve pushes the VSP against the borehole walls to achieve intimate contact between the probe and the sediments.

The general structure of the VSP is very similar to standard tensiometers or suction cups. It is constructed from a closed chamber with a porous section that needs to be in contact with the sediments and allows water continuity between the sediment pore water and the water within the chamber. Once hydraulic continuity between the sediment pore water and the closed chamber is achieved, the soil water pressure may be interpreted from the water pressure within the chamber. Alternatively, if sampling of the pore water is desired, then application of low pressure (vacuum) within the chamber will draw the sediment pore water into the chamber for later sampling.

Each cell has a set of small-diameter access pipes that extend from the cell to land surface. The access pipes allow water sampling and maintenance of the cell water solution. A set of valves and micro-pressure transducers in each cell allows measurement of the cell water pressure. A water-level mechanism monitors the water level in the cell.

Direct Contact VSP (DCVSP)

The VSP has two main models, designed for two environments. The first is called Direct Contact VSP (DCVSP). This model is designed for "soft" geological formations where the borehole drilling results in clean and smooth borehole walls (typical of sandy formations). For such boreholes, the VSP has a customized curved ceramic plate that is shaped to the exact borehole radial curvature (Figure 12.5). The curved ceramic plate is attached to the cell chamber to form the DCVSP body. Accordingly, the DCVSP body is designed to allow full contact of the curved ceramic plate with the borehole side walls. Once the sleeve with the DCVSP is inserted into the borehole and filled with urethane, the hydrostatic pressure in the sleeve pushes the DCVSP towards the borehole's upper side wall, allowing contact of the curved ceramic plate with the sediments. In this model, the curved ceramic plate is in direct contact with the soil. Accordingly, this model is less suitable for coarse gravelly sediments such as alluvial formations, since drilling in such material usually results in rough and undulating borehole walls, which may ultimately result in poor contact of the curved ceramic plate with the sediment.

Pillow VSP (PVSP)

The second model is specially designed for coarse material where drilling produces rough borehole walls with irregular diameter. In such boreholes, contact between the solid ceramic plate and the coarse sediment may not be achieved. In this case, the hydraulic continuity between the ceramic cup and the borehole side wall is achieved through an unconsolidated agent that bridges between the coarse material and the cell (Figure 12.6). The connecting agent is customized from a soft sand "pillow" that helps create hydraulic continuity between the cell's porous tip and coarse sediments on the borehole walls. The pillow is made of a flexible fiber mesh that covers the cell with very find sand. The installation technique is identical to that described for the FTDR probes and DCVSP. While pressed against the rough wall, the sand from the pillow is pushed out of the mesh into the sediment cavities to generate fine pore continuity between the cell and the coarse sediments.

Figure 12.5. Photograph and schematic illustration of DCVSP. Above: curved ceramic plate and cell body. Below: schematic demonstration of the DCVSP in a slanted borehole

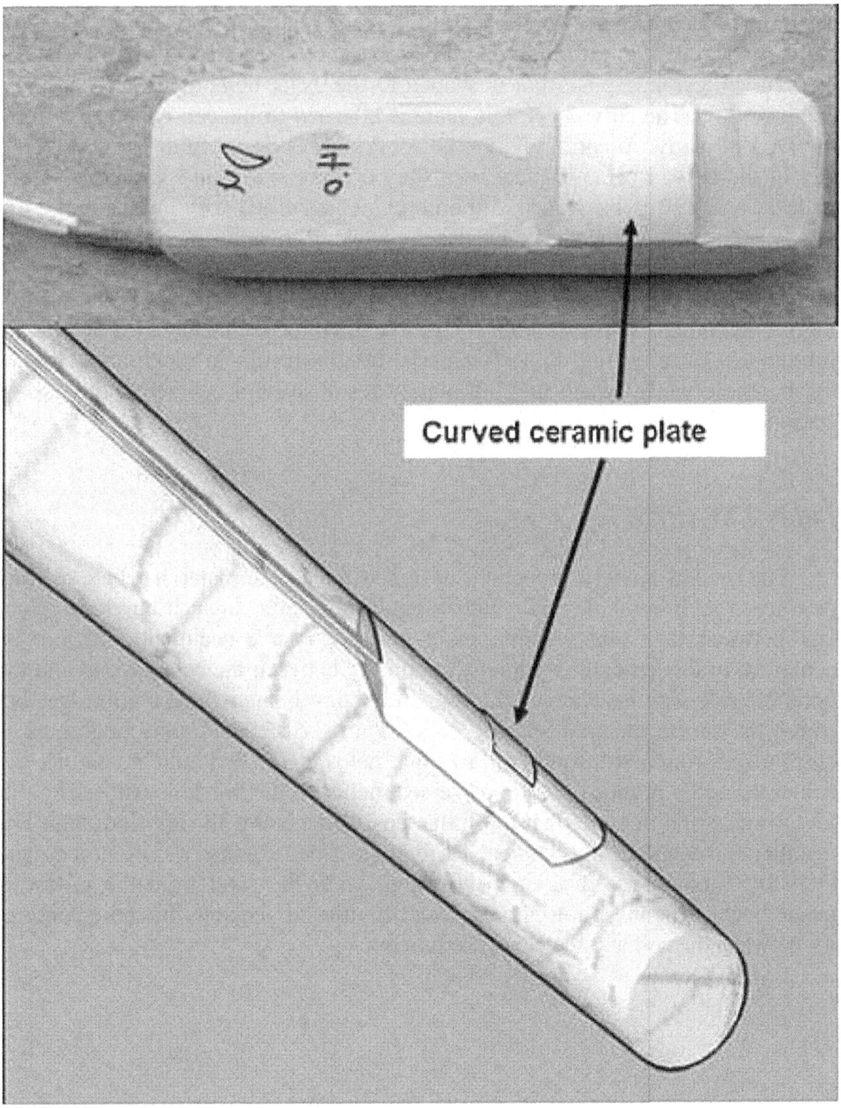

Figure 12.6. (a) Schematic illustration of the PVSP installed in a borehole drilled in a coarse alluvial formation. The illustration shows the sand agent creating the contact between the coarse sediments and the ceramic tip. (b) A photo demonstrating the sand pillow's flexibility and ability to follow the borehole wall undulations

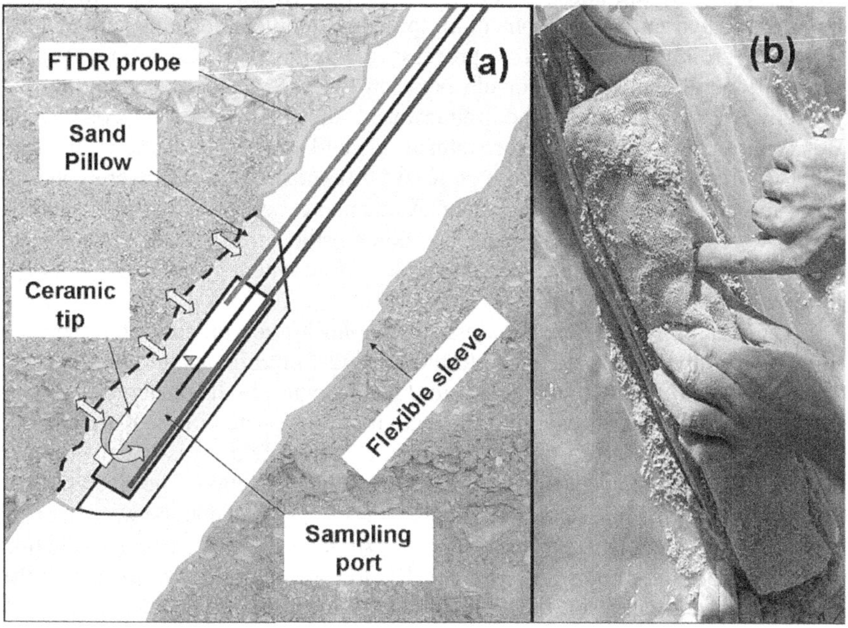

Results

To date, the method has been successfully installed in numerous studies on water infiltration and groundwater recharge in several countries, including Israel, Namibia, South Africa, Spain and the US (Arizona). The studies have included investigation of floodwater infiltration from ephemeral stream channels into alluvial aquifers as well as the impact of land use on groundwater recharge in a coastal aquifer with a thick vadose zone of ~20 m.

The results presented here are data samples from three different field experiments. The experiments were conducted under different climatic, geological and hydrological conditions. The data samples demonstrate the system's performance and the ability to monitor hydrological aspects of the infiltration process.

Monitoring of floodwater infiltration using FTDR

A study on floodwater infiltration and groundwater recharge was conducted in several ephemeral rivers in hyper-arid deserts around the world. The study sites were located along Wadi Arava (Israel), Kuiseb River (Namibia), Buffels River (South Africa), and Rio Andarax (Spain). Each site was instrumented to monitor the hydraulic conditions of all three hydrological domains participating in and controlling the infiltration process: (a) the flood on the surface, (b) the vadose zone as the linking domain between the surface water source and the groundwater below, and (c) the groundwater. The floodwater and groundwater were continuously monitored for water levels and electrical-conductivity variation with time, and in the vadose zone, water-content variations were continuously monitored by the FTDR system. At each site, several independent FTDR systems were installed in different boreholes drilled in different directions to allow an evaluation of the potential impact of sediment heterogeneity on the infiltration process.

The data presented here demonstrates the infiltration process from the relation between the floodwater stage to the wetting-front propagation, through the vadose zone and finally the groundwater level response to the recharge event. The data are taken from a monitoring station that is located in the Kuiseb River by the Gobabeb Research Center, Namibia. Figure 12.7a gives the river water stage during a flood event that was recorded in January 2006. The river streambed is composed of fine sand with small amounts of gravel and the water table is located ~5 m below land surface. Figure 12.7b gives the water content variation with depth as measured by one of the FTDR systems that was placed in the streambed. Each curve on the figure presents the water-content variation at a different depth and clearly shows the infiltration process along the entire vadose zone, from land surface to the groundwater. This data allows calculation of wetting-front propagation with depth as well as the rate of change in water content.

Accordingly, flow velocity and flux rate may be calculated directly from multiplication of the velocity by the change in water content. The floodwater infiltration through the vadose zone and groundwater recharge may be confirmed from the rise in the water table (Figure 12.7c).

Although the data presented here is only a sample from a single monitoring station during a single flood event, very similar results were observed from all monitoring stations. Though the wetting pattern is similar, the measured magnitude of each parameter varies from site to site according to the specific hydrological, geological, and geomorphological settings that characterize them.

Monitoring of deep vadose zone water potential—DCVSP

Land-use impact on groundwater recharge of the coastal aquifer is currently being studied using the deep vadose zone monitoring system. The study is being conducted in urban, agricultural, and open sand dune spaces around the coastal city of Ashdod, Israel. The sand-dune study site is instrumented with several independent monitoring FTDR systems and two additional DCVSP systems. Though the two DCVSP systems are identical, one system is dedicated to pore-water sampling and the other to water-pressure measurements. The systems were installed away from each other (6 to 15 m) to prevent any influence of the pore-water sampling on the water-pressure measurements. The vadose zone in this site is composed of an ~20-m thick unconsolidated sand dune and sandstone with a single, ~1.5-m thick clay layer ~7 m below surface level. The climate is characterized by rainy winters (~500 mm from October to April) followed by dry summers. Since the study site is located on bare sand dunes with no vegetation, the recharge process is controlled primarily by direct percolation of intense rain events.

Preliminary results from this site demonstrate a pore-water pressure response to an infiltration process generated by a rain event (Figure 12.8). The water pressure is measured by the DCVSP as absolute pressure (1000 mb represents atmospheric pressure) at three different depths of 4, 6.5 and 11.5 m below surface level. The water-content variation above and below each DCVSP is monitored with FTDR probes. The figure shows that each monitored wetting process is followed by a water-pressure increase. Alternatively, reduction in water content is followed by a water-pressure decrease. Deviation from this pattern can be seen on the set of probes that was installed at 11.5 m. Prior to the increase in water pressure that follows the water-content increase, a pressure rise is observed with no change in water content. Since the pressure transducers measure absolute pressure, it is possible that the pressure rise occurred due to a pressure increase in the vadose-zone gas phase and not directly as a result of a change in the water tension. The pressure increase may be generated by the infiltration event that, while propagating down the vadose zone, entraps and pressurizes air in the lower layers. The other DCVSP system that was installed at this site allowed frequent water sampling at various depths from land surface to the groundwater throughout the season. The water samples are still undergoing analysis and results from that part of the study are not available at this stage. This set of DCVSP was the first prototype to be installed in the field and a new, more efficient and accurate monitoring set will be installed in the near future.

Figure 12.7. Monitoring of floodwater infiltration and groundwater recharge from a streambed of an ephemeral river in a hyper-arid environment. (a) Floodwater stage. (b) Water-content variation in the vadose zone underneath the streambed during the flood. (c) Water-table response to the flood event

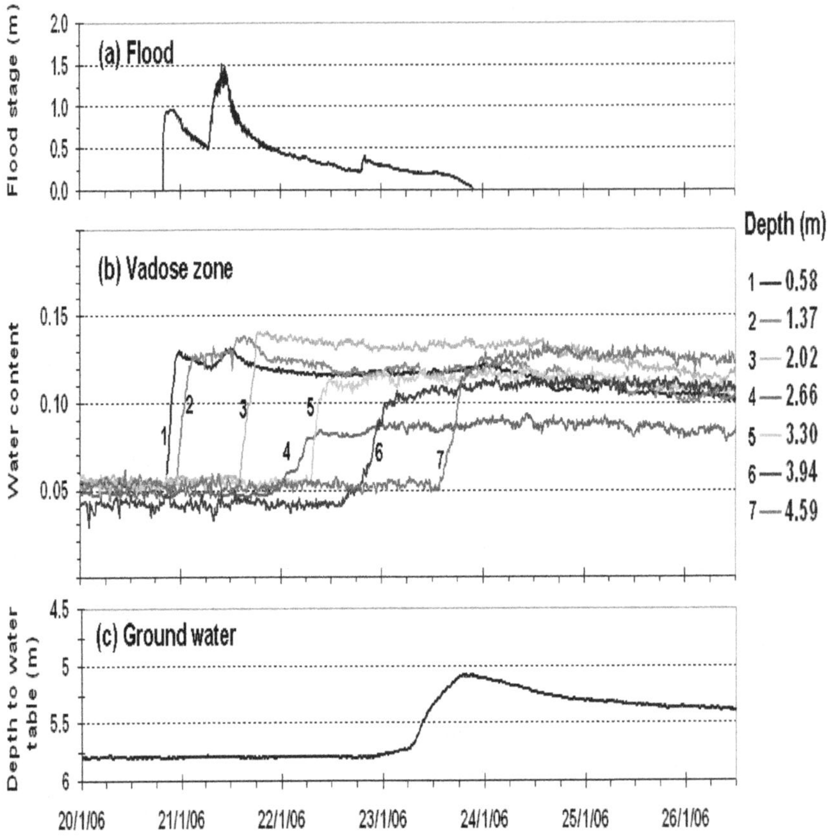

Figure 12. 8. Water-content and water-pressure variations at three different depths as measured by the vadose-zone monitoring system following a rain event

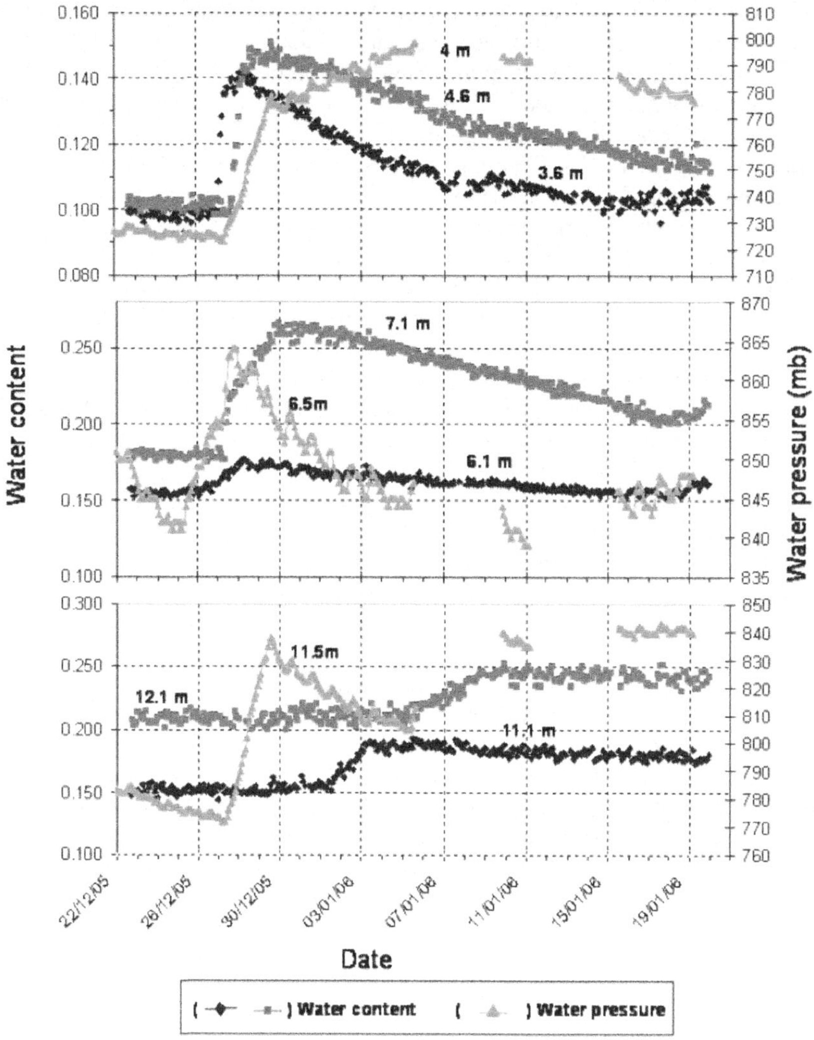

Pore-water sampling using PVSP

Water flow and contaminant transport from land surface to the groundwater was investigated in an alluvial formation in wadi Arava, Israel. To investigate the solute-transport pattern, a multi-tracer test was conducted on a section of the stream channel that was instrumented with a vadose-zone monitoring system. The system included five PVSP to sample the percolating water and five FTDR probes to monitor the vadose-zone wetting process. The sampling ports were equally distributed on a 5-m long flexible sleeve that was installed in a slanted borehole (Figure 12.6b). Figure 12.9 shows a schematic illustration of the experimental setup. The experiment was conducted in a pond that was constructed above the sampling system. Large ring infiltrometers were installed above each sampling cell to allow flux measurements in several sections of the stream channel and allow infiltration of different tracer solutions above the sampling cells.

The pond was flooded with water to achieve a constant water head. Equal water head between the ring's inner side and the pond was maintained throughout the experiment. The water-content profile in the vadose zone was continuously monitored by the FTDR system, and the sampling cells were frequently sampled. Two tracer tests were conducted in two different sessions, using 10 different tracers [five isomers of fluorobenzoic acid (FBA), Br, Li, eosin, Na-naph and uranine]. During each tracer test, a certain mass of traced solution at a constant concentration was allowed to percolate through each ring.

Monitoring the wetting/draining process with the FTDR system facilitated tracing of the wetting-front propagation. In addition, analysis of water samples obtained by the sampling cells facilitated breakthrough-curve determinations for all the tracers in all the samples.

Figure 12.10 presents the breakthrough curves of the FBA tracers that were injected into the different infiltration rings on the surface during the first tracer test, as they appeared in the sampling cell located 2.3 m below the land surface. The tracer breakthrough curves emphasize the complexity of the infiltration process through layered soil. The graph clearly shows the flow lines' mixing pattern, and demonstrates that the flow pattern is far from being vertical. Moreover, some of the tracers appear in the cells with a breakthrough curve characterized by two distinct concentration peaks. A multi-peak breakthrough curve is a typical indication of dominancy of a preferential flow mechanism with a dual-porosity system. Analysis of the results from all sampling cells in the two tracer tests revealed similar flow behavior.

Figure 12.9. Schematic illustration of the tracer test experimental setup and sampling port system (above), and a photo of the infiltration pond with the ring infiltrometers for the traced solution (below)

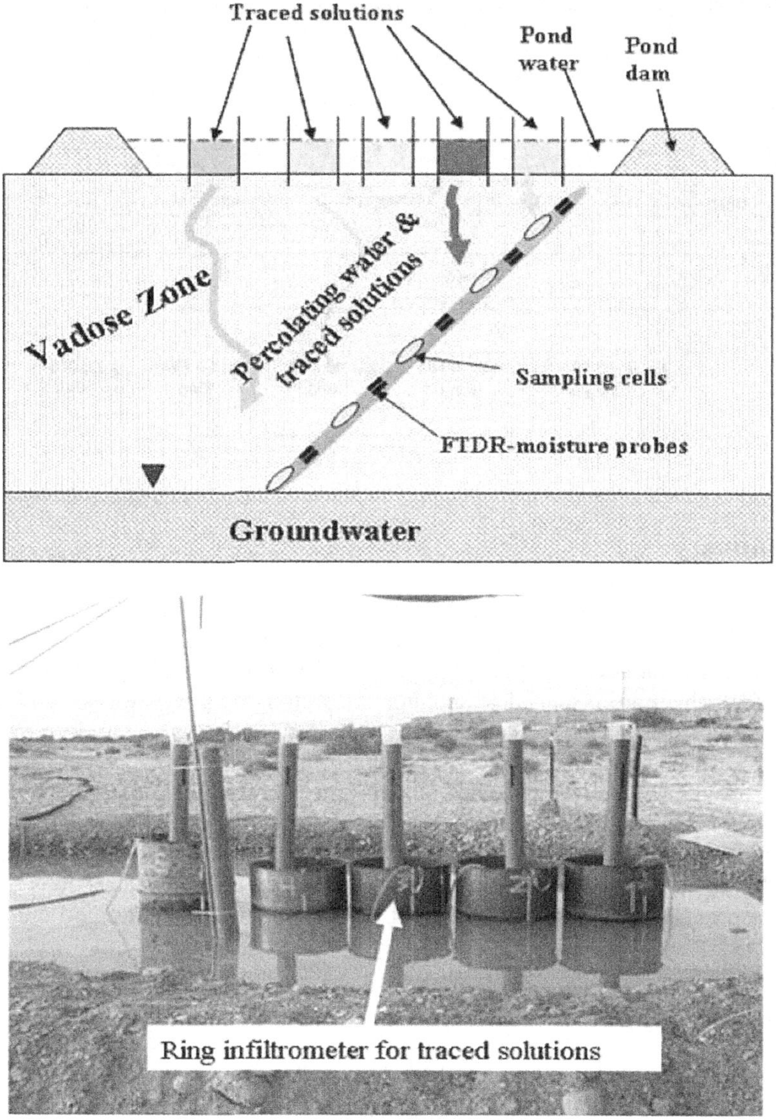

Figure 12.10. Breakthrough curves of the five FBA tracers that were injected through the ring infiltrometers on the surface as they appear in the sampling cell that is located 2.3 m below surface

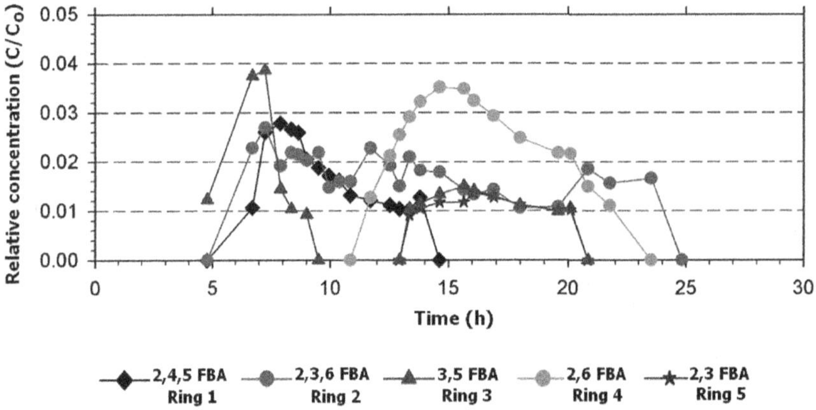

Summary

A vadose-zone monitoring system that provides real-time information on the infiltration process through the deep vadose zone was developed and tested in several laboratory and field experiments. The system is composed of FTDR probes, which are designed to monitor the wetting/draining process, and VSP sampling cells, which are designed to allow measurements of the deep vadose zone's water potential and allow sampling of the sediment pore water. Preliminary results from several studies demonstrate the ability to trace the wetting-front propagation velocities and thus allow direct calculation of the water fluxes. Continuous monitoring of water content and water pressure combined with the ability to sample the sediments' pore water in deep undisturbed sections of the vadose zone allows better evaluation of vadose-zone flow and transport models.

Acknowledgements

We thank Michael Kugel and Michael Fulton for their technical assistance and creative ideas. Funding for the studies was provided by the Israel Science Foundation (Grant #276/03) and the 6th framework program of the European Commission (WADE). We would like to express our appreciation to the Gobabeb Research Center and the Desert Research Foundation of Namibia.

Notes

1. A patent is pending on the system described in this chapter.

2. **Ofer Dahan** completed his Ph.D. at The Hebrew University of Jerusalem in 1999. Since 2002, he has been a staff member at the Zuckerberg Institute for Water Research and a lecturer at Ben-Gurion University of the Negev, Israel. His research interests focus mainly on water percolation and contaminant transport through the vadose zone.

Boaz Tatarsky obtained his B.Sc. degree in geological and environmental science from the Department of Earth Sciences at The Hebrew University of Jerusalem. He has completed his M.Sc. degree at the Zuckerberg Institute for Water Research, Ben-Gurion University of the Negev.

Yaara Rimon completed her B.Sc. degree in soil and water science at the Faculty of Agricultural, Food and Environmental Quality Sciences, The Hebrew University of Jerusalem in 2004. She is currently working towards a Ph.D. degree in Hydrology at The Hebrew University of Jerusalem.

Ruth Talby obtained her B.Sc. degree in geological and environmental sciences, from Ben-Gurion University of the Negev. She has completed her M.Sc. degree in the same department, in collaboration with the Department of Environmental Hydrology and Microbiology at Zuckerberg Institute for Water Research.

References

Dahan, O., McDonald, E., & Young, M. (2003) Development of a flexible TDR probe for deep vadose zone monitoring. *Vadose Zone Journal*, 2:270–275.

Dahan, O., Nativ, R., Adar, M. E., & Berkowitz, B. (1999) Field observation of flow in a fracture intersecting unsaturated chalk. *Water Resources Research*, 35(11):3315–3326.

Dalton, F. N. (1992) Development of time-domain reflectometry for measuring soil water content and bulk soil electrical conductivity. *Soil Science Society of America Journal*, 30:143–167.

Doughty, C., Salve, R., & Wang, S. Y. (2000) Liquid-release tests in unsaturated fractured welded tuffs: II. Numerical modeling. *Journal of Hydrology*, 256:80–105.

Ferré, P. A., Knight, H. J., Rudolph, L. D., & Kachanoski, G. R. (1998) The sample area of conventional and alternative time domain reflectometry probes. *Water Resources Research*, 34:2971–2979.

Flint, A., & Flint, N. (2000) Near surface infiltration monitoring using neutron moisture logging, Yucca Mountain, Nevada. In B. Looney, & R. Falta (Eds.), *Vadose zone science and technology solutions*. Columbus, OH: Battle Press.

Flint, A., Flint, L., Kwicklis, E., Fabryka-Martin, J., & Bodvarsson, G. (2002) Estimating recharge at Yucca Mountain, Nevada, USA: comparison of methods. *Hydrogeology Journal*, 10:180–204.

Glass, R. J., Nicholl, M. J., & Tidwell, V. C. (1995) Challenging models for flow in unsaturated, fractured rock through exploration of small scale process. *Geophysical Research Letters*, 22:1457–1460.

Haimerl, G. (2004) Groundwater Recharge in Wadi Channels Downstream of Dams. Unpublished doctoral thesis. Technische Universität München.

Heilig, A., Steenhuis, S. T., Walter, M. T., & Herbert, S. J. (2003) Funneled flow mechanisms in layered soil: field investigations. *Journal of Hydrology*, 27:210–223.

Herkelrath, W. N., Hamburg, S. P., & Murphy, F. (1991) Automatic, real-time monitoring of soil moisture in a remote field area with time domain reflectometry. *Water Resources Research*, 27(5):857–864.

Hillel, D. (1980) *Application of soil physics*. Orlando, FL.: Academic Press.

Kool, J. B., Parker, J. C., & van Genuchten, M. Th. (1987) Parameter estimation for unsaturated flow and transport models: A review. *Journal of Hydrology*, 91:255–293.

Kowalsky, M. B., Finsterle, S., & Rubin, Y. (2004) Estimating flow parameter distributions using ground-penetrating radar and hydrological measurements during transient flow in the vadose zone. *Advances in Water Resources*, 27:583–599.

Ledieu, J., de Ridder, P., de Clerck, P., & Dautrebande, S. (1986) A method of measuring soil moisture by time-domain reflectometry. *Journal of Hydrology*, 88:319–328.

Liu, H. H., Salve, R., Wang, J. S., Bodvarsson, G. S., & Hudson, D. (2004) Field investigation into unsaturated flow and transport in a fault: model analyses. *Journal of Contaminant Hydrology*, 74:39–59.

Looney, B., & Falta, R. (Eds.). (2000) *Vadose zone science and technology solutions*. Columbus, OH: Battle Press.

Murdoch, L. C., Slack, W. W., Harrar, W., & Siegrist, R. (2000) Embedded sidewall sampler and sensors to monitor the subsurface. *Ground Water*, 38:657–667.

Nativ, R., Adar, E., & Becker, A. (1999) Designing a monitoring network for groundwater in fractured media. *Ground Water*, 37:38–47.

Philip, J. R. (1957). The theory of infiltration: 4. Sorptivity and algebraic infiltration equations. *Soil Science*, 84:257–264.

Robinson, D. A., Jones, S. B., Wraith, J. M., Or, D., & Friedman S. P. (2003) A review of advances in dielectric and electrical conductivity measurement in soils using time domain reflectometry. *Vadose Zone Journal*, 2:444–475.

Sidle, R. C., Nilsson, B., Hansen, M., & Fredericia, J. (1998) Spatially varying hydraulic and solute transport characteristics of a fractured till determined by field tracer tests, Funen, Denmark. *Water Resources Research*, 34:2515–2527.

Šimůnek, J., Šejna, M., & van Genuchten, M. Th. (1998) The Hydrus-1D software package for simulating the one-dimensional movement of water, heat, and multiple solutes in variably-saturated media: Version 2.0. *IGWMC-TPS-70*. International Groundwater Modeling Center. Colorado School of Mines, Golden, CO.

Topp, G. C., Davis, J. L., Bailey, W. G., & Zebchuk, W. D. (1984) The measurement of soil water content using a portable TDR hand probe. *Canadian Journal of Soil Science*, 64(3):313–321.

Tseng, P. H., Soll, W. E., Gable, C. W., Turin, H. J., & Bussod, G. Y. (2003) Modeling unsaturated flow and transport processes at the Busted Butte Field Test Site, Nevada. *Journal of Contaminant Hydrology*, 62-63:303–318.

Van Geel, J. P., & Sykes, J. F. (1994) Laboratory and model simulations of a LNAPL spill in a variably-saturated sand: 1. Laboratory experiment and image analysis techniques. *Journal of Contaminant Hydrology*, 17:1–25.

Verburg, K., Ross, P. J., & Bristow, K. L. (1996) SWIMv2.1 User Manual. *Divisional Report 130*. CSIRO, Australia.

Walser, G. S., Illangasekare, T. S., & Corey, A. T. (1999) Retention of liquid contaminants in layered soils. *Journal of Contaminant Hydrology*, 39:91–108.

Wolfsberg, A. V., Fabryka-Martin, J. T., Campbell, K. S., Levy, S. S., & Tseng, P. H. (1999) Use of chlorine-36 and chloride data to evaluate fracture flow and transport models at Yucca Mountain. *Proceedings, International Symposium on Dynamics of Fluids in Fractured Rocks*, LBNL, Berkeley, CA.

Yokuda, E., & Smith, R. (1993) A new probe for in situ TDR moisture measurement. *Proceedings, 39th International Instrumentation Symposium*, Albuquerque, NM.

Zilberbrand, M., & Gvirtzman, H. (1996) Monitoring of water flow and solute transport through the unsaturated zone using a large-diameter borehole. *Groundwater*, 34:57–65.

Chapter 13

Ectoparasite Population Dynamics: Determinants of Ectoparasite Species Composition[1]

Boris R. Krasnov

Introduction

Carrying out similar research in different geographic locations has two important implications. First, the technology and approaches developed in one region can thus be transferred to other regions. Second, the results obtained in one region can thus be validated via results obtained in other geographic locations. Indeed, any scientific study reveals some patterns and processes. However, a question that remains is how general are these patterns and processes and to what extent do they apply to objects, settings, or times other than those that were the subject of a study? In this chapter, I describe one of the examples from two series of similar studies carried out in the Negev Desert and Slovakia on fleas and their small mammalian hosts.

Composition of parasite communities can vary across host individuals, populations, species and communities (Carney and Dick, 2000; Poulin and Valtonen, 2002; Calvete et al., 2004). The source of this variation is diversity of the host's biotic and abiotic environment. In other words, some part of a parasite community encountered in a host is due to its specific location, another part due to host identity and yet another part due to the host's environment (Kennedy and Bush, 1994). However, the relative importance of spatially variable factors in variation of community composition is poorly known for most parasite and host taxa. This is especially true for ectoparasites, which, in contrast to endoparasites, are strongly influenced by the off-host environment.

Starting this research in the Negev Desert in 1992, I addressed the following questions: (a) how variable is the composition of flea assemblages among different populations of the same host occurring in different habitats; and (b) whether the composition of flea assemblages in a habitat is affected either by species composition of hosts or by the environment of the habitat. The results of this study were published in 1997 and 1998 (Krasnov et al., 1997; Krasnov et al., 1998). About 10 years later, I decided to validate these results using the data collected by my colleagues in Slovakia during 18 years of continuous field surveys (Stanko, 1987, 1988, 1994; Stanko et al., 2002). The results of the Slovakian study were published in 2006 (Krasnov et al., 2006). These two studies carried out in a hyper-arid desert of the Middle East and forests, meadows and brook valleys of central Europe contribute to understanding the relative importance of "environmental" and "host" parameters of a habitat as factors affecting composition of flea assemblages.

Negev Desert

Habitats, hosts and fleas

The Ramon erosion cirque is an area of about 200 km^2 situated at the southern boundary of the Negev Highlands. The cirque is incised into the crest of a northeast-southwest trending asymmetrical anticline and is a valley surrounded by steep walls of hard rocks (limestone and dolomite) at the top and friable rocks (sandstone) at the bottom. The altitude of the north rim of the cirque ranges between 900 and 1200 m above sea level, while the south rim is about 510 m above sea level. The level of the cirque bottom decreases gradually from the southwest (900 m above sea level) to the northeast and in the deepest part reaches 420 m above sea level.

The climate is characterized by hot, dry summers (mean daily air temperature of 34 °C in July) and relatively cool winters (mean daily temperature of 12.5 °C in January). There is a sharp decrease in annual rainfall from 100 mm on the north rim to 56 mm in the bottom of the cirque. Rainfall also decreases from southwest of the cirque bottom to its northeast.

Habitat classification was based on 13 environmental variables including soil, vegetation and relief parameters and resulted in six main habitat types as follows:

1. sand dunes with cover of *Calligonum comosum* or *Echiochilon fruticosum*;
2. dry river beds (*wadis*) of the eastern and central parts of the cirque with sandy-gravel soils and with cover of *Retama raetam*, *Moricandia nitens*, *Tamarix nilotica* and *Artemisia monosperma* among gravel plains (which will be referred to as east *wadis*);
3. flat gravel plains with sparse vegetation of *Hammada salicornica*, *Anabasis articulata* and *Gymnocarpos decandrum*;
4. limestone cliffs and rocks with sparse cover of *Zygophyllum dumosum*, *Helianthemum kahiricum* and *Reaumuria hirtella*;

5. complex of narrow *wadis* and hills of the western part of the cirque covered with thin loess cover and vegetation of *Anabasis articulata, Atriplex halimus* and *Artemisia herba-alba* (which will be referred to as west *wadis*); and
6. complex of loess hills and wide *wadis* with vegetation of *Anabasis articulata, Atriplex halimus, Artemisia herba-alba, Salsola schweinfurthii* and *Noaea mucronata.*

The region is inhabited by 11 common rodent species parasitized by nine common flea species. However, the house mouse *Mus musculus* and the lesser Egyptian jerboa *Jaculus jaculus* have extremely low levels of flea infestation.

Habitat variation in flea-species composition

To visualize habitat variation in flea-species composition, flea assemblages from each individual host were ordinated using principal component analysis. Then these assemblages were plotted according to their habitat occurrences. This procedure facilitated construction of ordination-space axes that were linear combinations of abundances of each flea species and, thus, presented the main directions of change in flea-species composition among individual hosts. Importantly, these axes were obtained independently of the affinity of either host species or habitat type.

The resulting four ordination axes explained 86% of the total variance, with each of them reflecting a change in flea-species composition (Figure 13.1). The first axis represented the change from *Xenopsylla dipodilli, Nosopsyllus iranus, Stenoponia tripectinata* and *Rhadinopsylla masculana* on gerbils (positive zone of the first axis) to *Parapulex chephrenis* on the Egyptian spiny mouse *Acomys cahirinus*. This change is illustrated by segregation of Wagner's gerbil, *Gerbillus dasyurus*, bushy-tailed jird *Sekeetamys calurus* and partly the pygmy gerbil *Gerbillus henleyi* from *A. cahirinus* and *Acomys russatus* along the first axis. The second axis responded mainly to fleas on Sundevall's jird *Meriones crassus* (*Xenopsylla conformis* and *Coptopsylla africana*). So, *M. crassus* was distanced from all other host species to the positive zone of the second axis. The third and fourth axes were correlated with the relative abundances of *Xenopsylla ramesis* and *Myoxopsylla laverani*, respectively, and thus, they determined positions of the fat sand rat *Psammomys obesus* and the Asian garden dormouse *Eliomys melanurus*, respectively.

When flea assemblages were plotted in the resulting ordination space according to their habitat occurrences, five among- and within-habitat directions of shift in flea-species composition could be envisaged (Figure 13.2). Among-habitat shifts were (a) from rocky habitats to east *wadis* and sand dunes along the first axis and (b) from sand dunes and east *wadis* to west *wadis* and loess hills along the second axis. These shifts can be interpreted as responses of flea assemblages to a gradient of rodent burrow availability and a gradient of soil humidity and temperature, respectively. Within-habitat shifts were (a) within east *wadis* along the first axis; (b) within rocky habitats along the first axis; and

(c) within loess habitats along the third axis. Obviously, these shifts in flea-species composition were related to among-host differences as follows: (a) between *G. dasyurus* and *M. crassus*; (b) between gerbillids (*D. dasyurus* and *S. calurus*) and murids (*A. cahirinus* and *A. russatus*); and (c) between gerbillids (*G. dasyurus* and *M. crassus*) and *Acomys* or *E. melanurus*, respectively. The composition of a flea community is therefore determined by two directions of change in flea-species composition, namely among hosts within a habitat and among habitats within a host.

Figure 13.1. 75% confidence ellipses for flea assemblages from each examined animal in the Negev Desert in the space of two first principal component axes according to the host species. Data for the fat sand rat *Psammomys obesus*, lesser Egyptian gerbil *Gerbillus gerbillus*, and Asian garden dormouse *Eliomys melanurus* are squares, circles and diamonds, respectively. Recalculated and redrawn after Krasnov, et al. (1997) (reprinted with permission from Cambridge University Press)

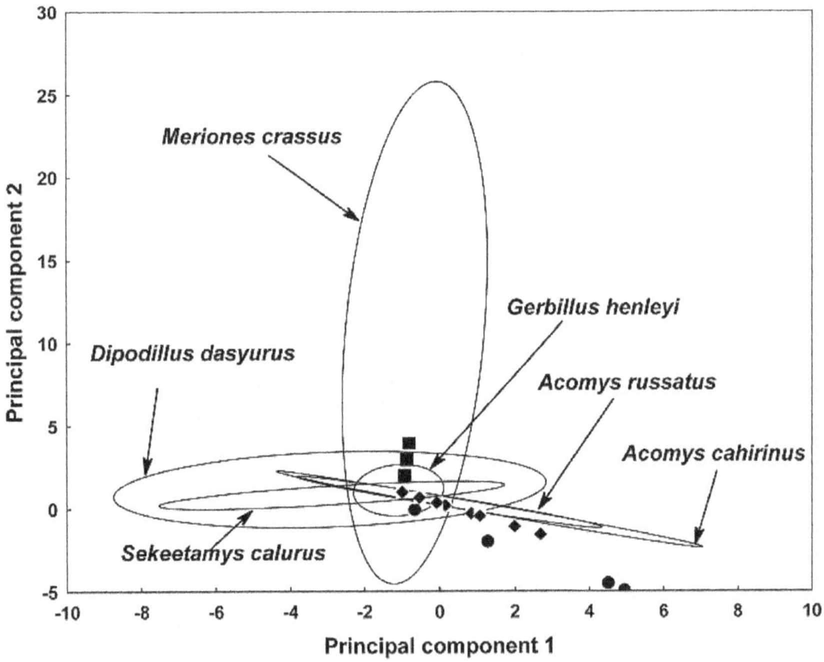

Figure 13.2. Scatterplots for flea assemblages from each examined animal in the Negev Desert in the space of (a) the first and the second and (b) the first and the third principal component axes according to habitat type. Habitat types are sand dunes (closed squares), east *wadis* (opened squares), gravel plains (closed triangles), rocks (opened triangles), west *wadis* (circles) and loess hills (diamonds). Recalculated and redrawn after Krasnov, et al. (1997) (reprinted with permission from Cambridge University Press)

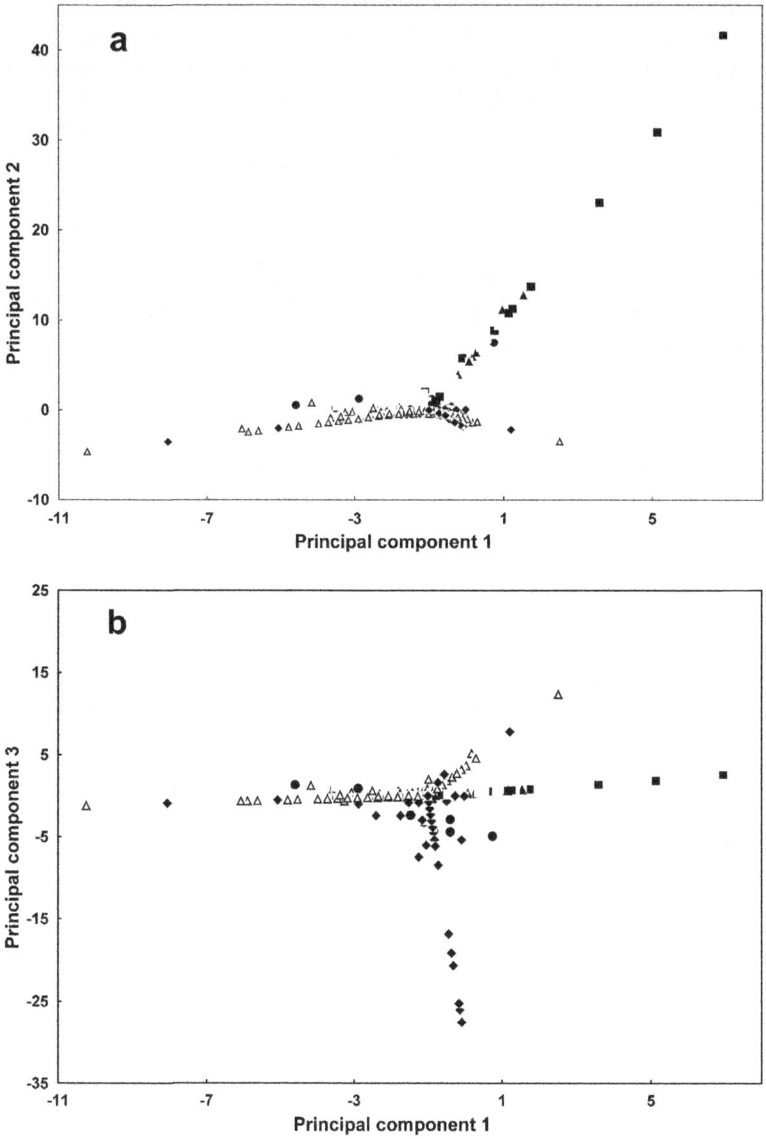

The main conclusion from this study is that host-habitat relations affect species composition of flea assemblages and, therefore, species composition of flea infra- and xenocommunities is determined not only by host-flea relations, but also by host-habitat relations. In turn, flea-species composition in a habitat is determined not only by host-species composition there, but also by environmental parameters of that habitat. These parameters determine the conditions of the burrow or the nest of the host (temperature, humidity, material) and, thus, affect the flea assemblage.

Among fleas of the studied region, host-dependent and host-habitat-dependent species can be clearly distinguished. Furthermore, distribution of even host-dependent fleas suggests that the relatively strong host-specificity is also restricted environmentally. For example, *P. chephrenis* is obviously host-dependent and associated with *A. cahirinus*. Nevertheless, this flea does not occur on *A. cahirinus* throughout its entire distribution within Israel; it occurs only in the southern region and in the Jordan Valley north to Lake Tiberias, where climatic conditions are similar to those of southern Israel (Theodor and Costa, 1967). *Acomys cahirinus* does not build deep burrows but lives in rock crevices and between stones. This results in the strong dependence of a flea on an individual host, on the one hand, and its relatively high susceptibility to climatic conditions, on the other hand.

Xenopsylla conformis and *X. ramesis* on *M. crassus* exemplify host-habitat-dependent fleas. The former parasitized *M. crassus* in sand dunes, east *wadis* and gravel plains, whereas it was replaced by the latter in loess hills. Both species occurred in habitats of intermediate environment (west *wadis*; Krasnov, et al., 1996), although their abundances were relatively low. It seems that the distribution of these fleas conforms to the distinct-preference model of community organization (Pimm and Rosenzweig, 1981). The pattern of among-habitat distribution of these two fleas on *D. dasyurus* was the same, except that, in contrast to *M. crassus, X. conformis* and *X. ramesis* were subdominants whereas *X. dipodilli* was dominant.

In addition, there were also fleas in which host- or host-habitat dependence was weakly expressed. For example, *N. iranus* occurred on seven host species in all habitat types.

Central Europe

Description of habitats, hosts and fleas

The study area in Slovakia is characterized by a typical temperate climate with a mean air temperature of 29.3 °C in July and -3.8 °C in January. Two habitat groups (lowland and mountain) with nine habitat types, based on physiognomy, may be distinguished.

Lowland habitats are those situated at elevations between 100 and 200 m above sea level and include:

1. lowland river valleys with willow-poplar and ash-alder floodplain forests dominated by *Salix alba, Salix fragilis* and *Populus alba*;
2. woodland belts represented by 3–8 rows of *Populus canadensis* and various shrubs (*Prunus* sp., *Rosa* sp., *Sambuccus nigra*) with herbal floor composed mainly of *Urtica dioica*;
3. agricultural fields of wheat, maize and stubble;
4. floodplain lowland forests dominated by *Fraxinus angustifolia, Quercus robur, Carpinus betulus, Salix alba* and *Salix fragilis*; and
5. shrubbery dominated by *Prunus spinosus, Rosa canina* and *Crataegus* sp. with sporadic occurrence of poplar and willow trees.

Mountain habitats are situated at elevations from 300 to 1100 m above sea level and include:

6. narrow submontane and montane brook valleys (later referred to as mountain river valleys) with the main vegetation represented by *Alnus glutinosa, Alnus incana, Fagus sylvaticus* and *Carpinus betulus*;
7. submontane and montane forests dominated by *Fagus sylvaticus, Carpinus betulus, Quercus robur* and *Acer platanoides*; and
8. shrubbery patches on pastures dominated by *Prunus spinosus, Coryllus avellana* and *Rosa canina.*
9. urban habitats represented by gardens and orchards in public green spaces within cities (at elevation 650–750 m above sea level).

In total, a survey in this area resulted in 24 small mammal species (rodents and soricomorphs) and 30 flea species (see complete species lists in Stanko, 1987, 1988, 1994; Stanko, et al., 2002). Of these, 14 common host species and 25 common flea species were used in the analyses to answer whether the composition of flea assemblages in a habitat is affected by species composition of hosts or environment of the habitat.

Habitat variation in flea-species composition

The first step in the test for matching between host-species composition and flea-species composition in a set of habitats was evaluation of between-habitat similarity in either flea- or host-species composition, based on abundances of either fleas or hosts, respectively. This resulted in two between-habitat similarity matrices. Among-habitat similarity in host-species composition averaged 52.6%, being maximal between lowland river valleys and lowland woodland belts and minimal between lowland fields and mountain shrubbery. Similarity in flea-species composition was slightly higher (average 61.5%), being maximal also between lowland river valleys and lowland woodland belts, but minimal between lowland shrubbery and mountain shrubbery. Ordinations of habitats, based on similarity in either flea- or host-species composition, distinguished groups of habitats that were more similar to each other than to other habitats in both flea- and host-species composition (Figure 13.3). These were three mountain habitats (river valleys, forests and shrubbery) and three lowland habitats (river valleys, forests and woodland belts). Field and urban habitats differed sharply from other habitat types by their flea and host composition. However, lowland shrubbery was similar in its host-species composition to other lowland habitats, but flea-species composition in this habitat differed substantially from any other habitat.

Then, the questions that should be answered are (a) whether and (b) how closely the two sets (host-similarity and flea-similarity) were related. This was done using the ρ statistic that correlates the elements of two similarity matrices and, thus, indicates significant disagreement or agreement between these matrices ranging from a minimum of 0 (no relation between two matrices) to a maximum of 1 (perfect match between two matrices). The resulting ρ equaled 0.36, demonstrating that among-habitat similarity in flea-species composition was reflected by among-habitat similarity in host-species composition. Furthermore, between-habitat flea-species composition similarity was positively correlated with between-habitat host-species composition similarity.

When flea assemblages from each host individual were ordinated using principal component analysis, as was done for the Negev data (see above), the first five principal components explained 64% of the total variance in flea-species composition. When flea assemblages were plotted according to their hosts, the segregation in flea-species composition between rodents and shrews can be discerned easily (Figure 13.4a). However, when flea assemblages were plotted according to their habitat affinities, the distinction of habitats based on variation in flea composition was not as clear (Figure 13.4b). Moreover, the variation in each principal component (analyzed using ANOVAs) was explained better by the factor of host species than by the factor of habitat type.

Nevertheless, habitat type, in general, affected flea-species composition within a host species, although this effect was manifested differently in different hosts (Figure 13.5). For example, flea assemblages on the striped field mouse *Apodemus agrarius* and the pygmy wood mouse *Apodemus uralensis* were similar across most habitats except for lowland shrubbery and mountain river valleys, respectively. Flea assemblages on *A. agrarius* in lowland shrubbery differed from all other habitats due to the high abundance of *Ctenophthalmus solutus*, whereas *Megabothris turbidus* was mainly responsible for the difference in flea composition on *A. uralensis* between mountain river valleys and the rest of habitats. Flea assemblages on the common vole *Microtus arvalis* and the common shrew *Sorex araneus* within lowland habitats and within mountain habitats tended to cluster along the first dimension of the ordination space, but differed sharply between these two habitat groups. The reason for these differences in *M. arvalis* was the relatively high abundance of *Ctenophthalmus assimilis* in lowland habitats, whereas lowland and mountain flea assemblages on *S. araneus* differed due to relatively high abundance of *Palaeopsylla soricis* and relatively low abundance of *Palaeopsylla similis* in lowland as compared to mountain habitats. Apparent clusters of habitats (= similar flea assemblages) were also evident for the wood mouse *Apodemus sylvaticus*, bank vole *Myodes glareolus* and the European pine vole *Microtus subterraneus*. In most cases, these clusters were represented by habitats belonging to either lowland or mountain areas. *Nosopsyllus fasciatus* was the main reason for among-habitat differences in *A. sylvaticus*, whereas flea assemblages of both voles differed among habitats mainly due to *M. turbidus* (28–50% of contribution), *Amalaraeus penicilliger* (24–42% of contribution) and *C. assimilis* (25–43% of contribution).

These results suggested that species composition of flea infra- and xeno-communities in a given habitat of Slovakia was determined mainly by host identity and, to a lesser extent, by habitat identity. To confirm this, repeatability analysis (Arneberg, Skorping and Read, 1997) based either on host or on habitat identity was carried out. It appeared that flea-species composition varied less (a) among populations of the same host species than among host species and (b) among habitats of the same type than among different habitat types (Krasnov, et al., 2006). However, the proportion of the total variance in flea-species composition originating from differences among host species, as opposed to within species, was 18.8%, whereas the proportion of the total variance originating from differences among habitat types, as opposed to within habitat type, was only 8.7%.

Nevertheless, the between-habitat similarity in host-species composition explained only about 20% of variance of the between-habitat similarity in flea-species composition. This means that some habitat pairs were characterized by similar flea assemblages but different host composition, and other habitat pairs by similar host communities harboring different flea assemblages. For example, the difference in dominating host species between mountain shrubbery (*Apodemus flavicollis* and *M. glareolus*) and agricultural fields (*A. agrarius*, *A. uralensis* and *M. arvalis*) led to a coefficient of similarity as low as 8%. However, flea assemblages on these hosts in both habitats were dominated by *Ctenophthalmus agyrtes*, *C. assimilis*, *Ctenophthalmus solutus* and *M. turbidus*, leading to a coefficient of similarity as high as 47%. On the other hand, mountain and lowland shrubbery were occupied by host communities with 51% similarity, although the presence of *A. penicilliger*, *Amphipsylla rossica*, *Ctenophthalmus bisoctodentatus*, *Doratopsylla dasycnema*, *Hystrichopsylla talpae*, *Monopsyllus sciurorum*, *Peromyscopsylla bidentata*, *P. similis* and two *Rhadinopsylla* species in mountains and their absence in lowland resulted in only 30% similarity in flea assemblages. This difference can be due to the difference in elevation and, consequently, in air temperature between mountain and lowland area.

Figure 13.3. Multidimensional scaling distribution of habitats based on Bray-Curtis similarity in (a) flea and (b) host composition. Abbreviations for habitat names are LRV – lowland river valleys, Blt – lowland woodland belts, Fld – lowland agricultural fields, LFrst – lowland forests, LShrb – lowland shrubbery, MRV – mountain river valleys, MFrst – mountain forests, MShrb – mountain shrubbery, and Urban – urban habitats. Redrawn after Krasnov, et al. (2006) (reprinted with permission from Springer Science and Business Media)

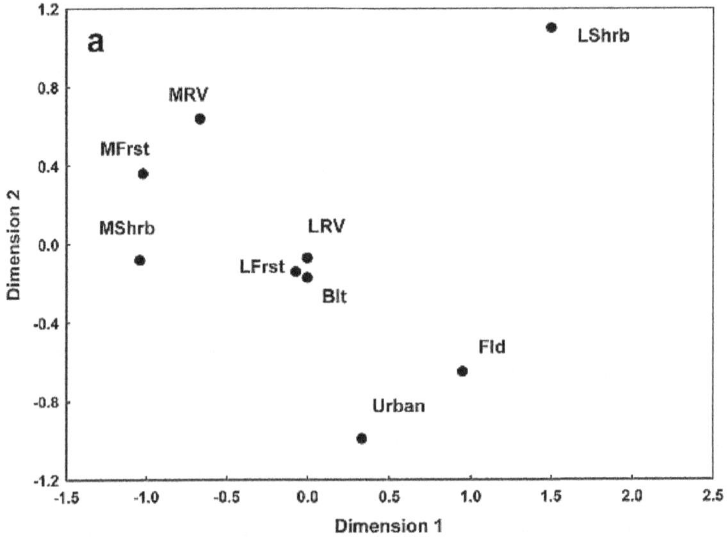

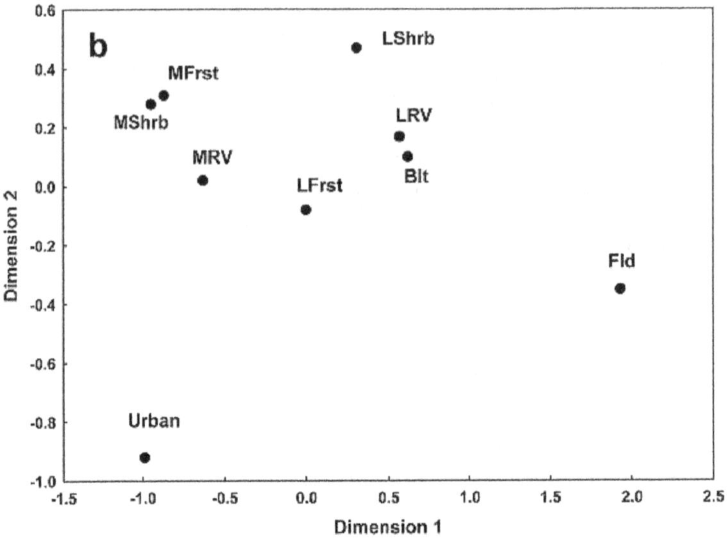

Figure 13.4. 95%-confidence ellipses for flea assemblages from each individual small mammal from Slovakia in the space of two first principal component axes according to (a) the host species and (b) habitat type. See Figure 13.3 for the abbreviations of habitat names. Redrawn after Krasnov, et al. (2006) (reprinted with permission from Springer Science and Business Media)

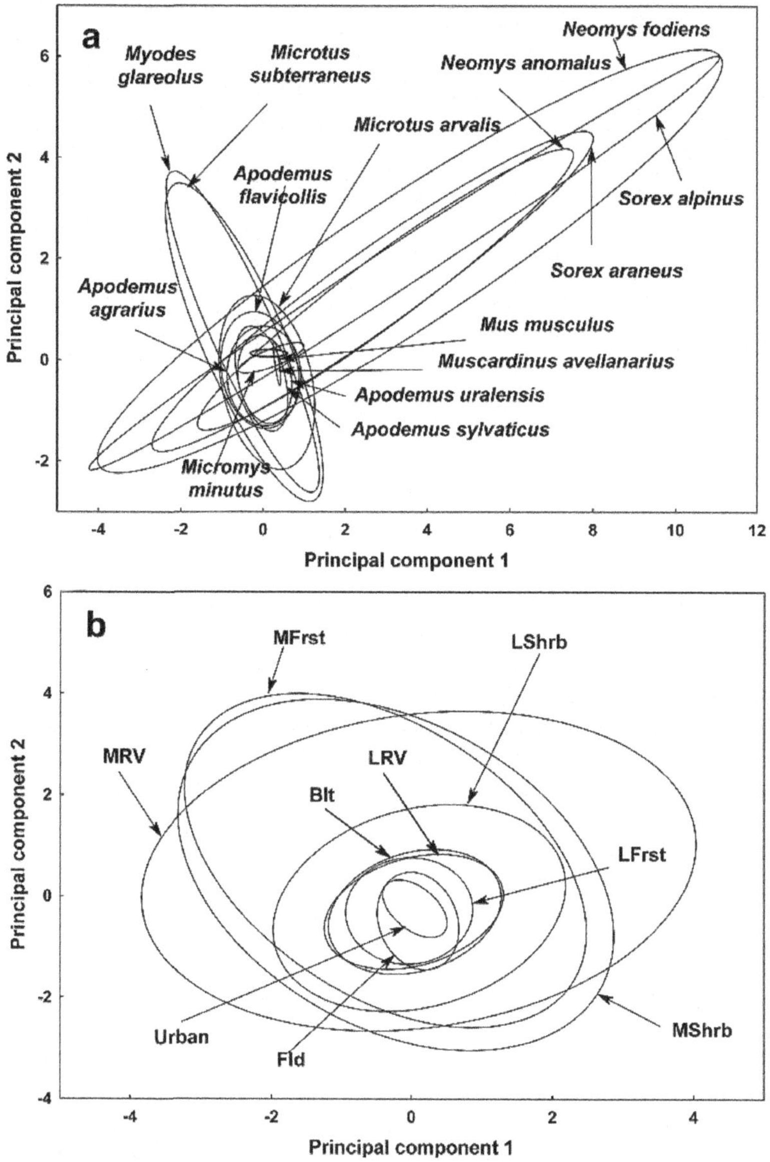

Figure 13.5. Multidimensional scaling distribution of habitats of Slovakia based on Bray-Curtis similarity in flea composition. See Figure 13.3 for the abbreviations of habitat names. Redrawn after Krasnov, et al. (2006) (reprinted with permission from Springer Science and Business Media)

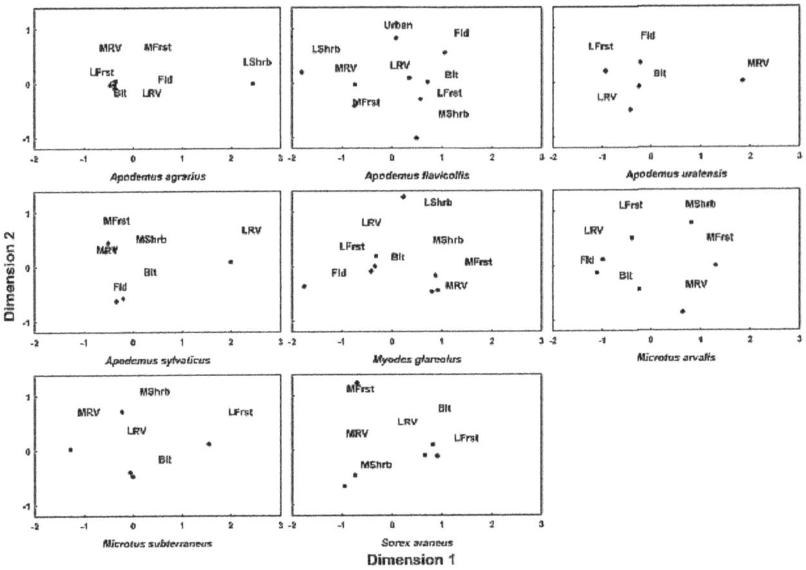

In other words, although the effect of habitat on flea-species composition was lower than that of host-species composition, it cannot be ruled out. This effect can be associated with both biotic and abiotic components of a habitat. Among-habitat variation in the biotic component can be related to the number of co-habitating hosts that facilitate between-host flea transfer. High probability of this transfer might, therefore, mask the effect of host identity on the within-host–among-habitat composition of flea assemblages. Another biotic component of a habitat is represented by the fleas. Different combinations of flea-species composition might be a result of competition between some flea species, at least, in their larval stages (Krasnov et al., 2005).

Abiotic components of a habitat are represented by environmental factors such as air temperature, relative humidity and substrate structure. All these components have a strong effect on survival and development of both preimaginal and imago fleas (e.g., Krasnov et al., 2001, 2002). Comparison of among-habitat variation in composition of flea assemblages within host species supports, albeit indirectly, the important role of environmental factors in determining flea community structure. Apparent clusters of flea assemblages corresponding to lowland and mountain habitats can be distinguished in six of eight species. Environmental variation between these two areas is likely more pronounced than that among habitats within each of these areas.

Concluding remarks

The study based on Slovakian data thus confirmed the conclusions of the earlier study carried out in the Negev Desert: that species composition of fleas on a host species is determined not only by host-flea relations, but also by host-habitat relations. Consequently, a habitat for a flea is not a particular host or a group of hosts but rather a particular host or a group of hosts in a particular habitat. Nevertheless, among-habitat differences in flea assemblages within a host species in Slovakia appeared to be less pronounced than those in the Negev desert. Indeed, flea assemblages of some desert hosts were composed, at least seasonally, of completely different species (Krasnov et al., 1998). This was not the case for any host species in the Slovakian study. Flea assemblages of the same host in different habitats differed by relative abundances of fleas rather than by their species assortment. The explanation of this difference between temperate and arid environments can be related to differences in the sheltering pattern of host species. Imago fleas spend a considerable part of their lives in a host shelter, whereas this shelter is an ultimate habitat for pre-imagoes. Most desert hosts construct deep belowground burrows, whereas most temperate rodents and insectivores either construct aboveground nests or dig shallow burrows (Kucheruk, 1983). Between-habitat differences in the environmental conditions of the burrows are likely more pronounced than that of the aboveground nests. An alternative explanation might be related to the frequency of interspecific visits of each other's burrows in small mammals, which presumably is higher in temperate regions (Kucheruk, 1983) because of relatively higher small mammal density than in deserts. This could increase host switching and, therefore, lead to more scattered presences of flea species across the populations of a host species occupying different habitats.

Note

1 This chapter contains previously published material from: "Immune response to fleas in a wild desert rodent: Effect of parasite species, parasite burden, sex of host and host parasitological experience," by Irina S. Khokhlova, Marina Spinu, Boris R. Krasnov and A. Allan Degen (2004), *Journal of Experimental Biology* 207: 2725–2731, published by The Company of Biologists; from: "Distribution of fleas (*Siphonaptera*) among small mammals: Mean abundance predicts prevalence via simple epidemiological model," by Boris. R. Krasnov, Michal Stanko, Dana Miklisova and Serge Morand (2005), *International Journal for Parasitology* 35: 1097–101, published by Elsevier Science; and from "Relationship between host abundance and parasite distribution: Inferring regulating mechanisms from census data," by Michal Stanko, Boris R. Krasnov and Serge Morand (2006), *Journal of Animal Ecology* 75: 575–583, published by Blackwell. It is reprinted here with permission from Boris R. Krasnov. The figures are reprinted with permission (as noted) from Cambridge University Press, and from Springer Science and Business Media.

References

Arneberg, P., Skorping, A., & Read, A. F. (1997) Is population density a species character? Comparative analyses of the nematode parasites of mammals. *Oikos*, 80:289–300.

Calvete, C., Blanco-Aguiar, J. A., Virgós, E., Cabezas-Díaz, S., & Villafuerte, R. (2004) Spatial variation in helminth community structure in the red-legged partridge (*Alectoris rufa* L.): Effects of definitive host density. *Parasitology*, 129:101–113.

Carney, J. P., & Dick, T. A. (2000) Helminth communities of yellow perch (*Perca flavescens* (Mitchill)): Determinants of pattern. *Canadian Journal of Zoology*, 78:538–555.

Kennedy, C. R., & Bush, A. O. (1994) The relationship between pattern and scale in parasite communities: A stranger in strange land. *Parasitology*, 109:187-196.

Krasnov, B. R., Burdelova, N. V., Khokhlova, I. S., Shenbrot, G. I., & Degen, A. A. (2005) Pre-imaginal interspecific competition in two flea species parasitic on the same rodent host. *Ecological Entomology*, 30:146–155.

Krasnov, B. R., Khokhlova, I. S., Fielden, L. J., & Burdelova, N. V. (2001) The effect of temperature and humidity on the survival of pre-imaginal stages of two flea species (Siphonaptera: Pulicidae) *Journal of Medical Entomology*, 38:629–637.

Krasnov, B. R., Khokhlova, I. S., Fielden, L. J., & Burdelova, N. V. (2002) Time to survival under starvation in two flea species (Siphonaptera: Pulicidae) at different air temperatures and relative humidities. *Journal of Vector Ecology*, 27:70-81.

Krasnov, B. R., Shenbrot, G. I., Khokhlova, I. S., & Ivanitskaya, E. Y. (1996) Spatial structure of rodent community in the Ramon erosion cirque, Negev highlands (Israel). *Journal of Arid Environments*, 32:319–327.

Krasnov, B. R., Shenbrot, G. I., Medvedev, S. G., Khokhlova, I. S., & Vashchenok, V. S. (1998) Habitat-dependence of a parasite-host relationship: flea assemblages in two gerbil species of the Negev Desert. *Journal of Medical Entomology*, 35:303–313.

Krasnov, B. R., Shenbrot, G. I., Medvedev, S. G., Vashchenok, V. S., & Khokhlova, I. S. (1997) Host-habitat relations as an important determinant of spatial distribution of flea assemblages (Siphonaptera) on rodents in the Negev Desert. *Parasitology*, 114:159–173.

Krasnov, B. R., Stanko, M., Miklisova, D., & Morand, S. (2006) Habitat variation in species composition of flea assemblages on small mammals in central Europe. *Ecological Research*, 21:460-469.

Kucheruk, V. V. (1983) Mammal burrows: Their structure, topology and use. *Fauna and Ecology of Rodents*, 15:5–54 (in Russian).

Pimm, S. L., & Rosenzweig, M. L. (1981) Competitors and habitat use. *Oikos*, *37*, 1–6.

Poulin, R., & Valtonen, E. T. (2002) The predictability of helminth community structure in space: A comparison of fish populations from adjacent lakes. *International Journal for Parasitology*, 30:1235–1243.

Stanko, M. (1987) Siphonaptera of small mammals in the northern part of the Krupina plain. *Stredné Slovensko, Zborník Stredoslovenského Múzea, Banská Bystrica,* 6:108–117 (in Slovak).

Stanko, M. (1988) Fleas (Siphonaptera) of small mammals in eastern part of Volovské vrchy mountains. *Acta Rerum Naturalium Musei Nationalis Slovaci Bratislava,* 34:29–40 (in Slovak).

Stanko, M. (1994) Fleas synusy (Siphonaptera) of small mammals from the central part of the East-Slovakian lowlands. *Biologia, Bratislava,* 49:239–246.

Stanko, M., Miklisova, D., Goüy de Bellocq, J., & Morand, S. (2002) Mammal density and patterns of ectoparasite species richness and abundance. *Oecologia*, 131:289–295.

Theodor, O., & Costa, M. (1967) *A survey of the parasites of wild mammals and birds in Israel. (I) Ectoparasites.* Jerusalem: Israel Academy of Science and Humanities.

Chapter 14

Crust and Dust

Eli Zaady

This chapter has two parts: The first part, entitled "Biogenic Soil Crusts Enhance Runoff Production in the Northern Negev Desert" (by Eli Zaady, David J. Eldridge and Moshe Shachak), deals with the soil surface in desert ecosystems and relates to the effect of soil surface crust on desert hydrology. The second part, entitled "Applicative Methods for Aeolian Particle Study in Desert Environments" (by Zvi Y. Offer and Eli Zaady), deals with aeolian dust dynamics and relates to field research methods, used in the Negev desert, to analyze particle size distribution, micromorphology and chemistry. These subjects are applicable to similar environments worldwide.

Biogenic Soil Crusts Enhance Runoff Production in the Northern Negev Desert

Eli Zaady, David J. Eldridge and Moshe Shachak[1]

Introduction

The landscape of the northern Negev desert of Israel is predominantly natural shrubland comprising two biotic patch types: shrubs and biogenic soil crust (BSC) communities (Evenari, 1985; West, 1990; Shachak, Sachs and Moshe, 1998). In the Negev region, the crust could be of biogenic origin (BSC; West, 1990), or of physical-chemical origin (e.g. depositional crusts; Eldridge, Zaady and Shachak, 2000). A BSC is characterized by a tightly structured surface (Fletcher and Martin, 1948), colonized by a complex community of mosses, lichens, soil algae, fungi, cyanobacteria and soil bacteria (Friedmann and Galun, 1974; Skujins, 1984; West, 1990; Johansen, 1993.). This "cryptogamic" or "microphytic" crust community varies markedly from relatively homogeneous 2–3-mm thick cyanobacterial crusts to very complex crusts 15 mm thick. Homogeneous crusts occur in the central Negev in areas that receive 75–125 mm average annual rainfall, while complex crusts occur in the northern Negev in areas that receive 200–300 mm average annual rainfall (Zaady, Gutterman and Boeken, 1997).

There is some confusion about the role of BSC in infiltration processes (Yair, 1990; Belnap, 2006). Well-developed BSCs found in dry regions around the world are hydrophobic, due to the gels and polysaccharides secreted by their constituent microphytes—cyanobacteria and soil algae (Mehta and Vaidya, 1978; Bertocchi, Navarini and Cesaro, 1990; De-Philipis et al., 1993)—and the mucilaginous sheaths these exude directly onto the soil surface. This mucilage binds the soil surface particles (Bailey, Mazurak and Rosowski, 1973; Metting and Rayburn, 1983), making them effective at preventing soil erosion by wind and water (West, 1990; Gillette and Dobrowlski, 1993; Belnap, 1995; Belnap and Gillette, 1998; Eldridge, 1993). The cyanobacteria track water and light in desert soils and move upward within the soil surface (Garcia-Pichel and Pringault, 2001). Brotherson and Rushforth (1983) and Greene, Chartres and Hodgkinson, (1990) reported that BSC enhances water infiltration. Studies in the northern Negev desert of Israel on loess soils have suggested that BSC often reduces infiltration (Zaady and Shachak, 1994; Eldridge et al., 2000) and thus might enhance runoff generation (Shachak et al., 1998).

In this article we report the results of a two-part study (two different spatial scales) that assessed runoff generation by and infiltration through BSC. In the first small-spatial-scale study we examined infiltration through crusts in Petrie dishes (64 cm^2). We compared these results with observations from the second large-spatial-scale study of natural rainfall (0.5 m^2) and rainfall simulation (1 m^2) *in situ* on field plots. Our **objective** was to explore the relationship between BSC and water flow through (infiltration) and across (runoff generation) crusted surfaces in the northern Negev. Our **hypothesis** was that crusts retard infiltration and enhance runoff of water, and that this phenomenon is consistent across a range of spatial scales. We tested this hypothesis by measuring 1) infiltration through small cores of intact crust under laboratory-induced rainfall, and 2) runoff from *in-situ* crusts using both natural rainfall and artificial rainfall.

Study site

The field-based studies were conducted at the Long Term Ecological Research Station (LTERS) (Gosz, French and Sprott, 2000) at Sayeret Shaked (31°17' N, 34°37' E), located 150 m above sea level in the northern Negev near Beer Sheva, Israel. The landscape of this 2-km^2 area, closed off from grazing livestock since 1987, consists of low hills. The average daily minimum temperature in winter is 6–8 °C; the average daily maximum temperature in summer is 32–34 °C (Stern et al., 1986). Most rainfall events in this semiarid region occur between November and March. Long-term average annual rainfall is approximately 200 mm (the 200-mm isohyet forms the transition zone between arid and semiarid deserts in Israel; Bruins, 1990) (see Table 14.1).

The loess soil, composed of 14% clay, 27% silt and 59% sand (USA classification: loess soil with sandy loam texture—Calcixerollic, Xerochrepts) is up to 1 m thick and sits atop Eocene bedrock. The salt content of the upper 25 cm of soil is low, with an electrical conductivity of 0.4 dS m^{-1} (Teomim, 1990).

The soil surface is covered with BSC consisting of bacteria, cyanobacteria, algae, mosses and lichens (Zaady, Groffman and Shachak, 1996). Interspersed with the BSC are scattered patches of perennial shrubs. The dominant perennials in the research area are: *Asphodelus ramosus* (Liliaceae), *Thymelaea hirsuta* (Thymelaeaceae), *Noaea mucronata* (Chenopodiaceae), *Atractylis serratuloides* (Asteraceae) and *Pituranthos tortuosus* (Apiaceae) (Feinbrun-Dothan and Danin, 1991).

Soil covered with a well-developed BSC has a tightly structured surface (Fletcher and Martin, 1948). In contrast, the soil under patches of shrubs lacks a well-developed BSC, and its surface is covered with loose soil particles.

Table 14.1. Average annual rainfall (mm) and runoff production (liters m^{-2}) in plots with biogenic soil crust cover and calculated mean overland water runoff at the Sayeret Shaked Long-Term Ecological Research station, between 1993 and 2006

Rainy season	Average annual rainfall (mm)	Mean runoff (25.5%) (liters m^{-2})
1993-1994	97.5	24.9
1994-1995	285.0	72.7
1995-1996	108.0	27.5
1996-1997	200.0	51.0
1997-1998	177.0	45.1
1998-1999	68.6	17.5
1999-2000	89.6	22.8
2000-2001	189.1	48.2
2001-2002	150.8	38.5
2002-2003	176.7	45.1
2003-2004	158.1	40.3
2004-2005	282.8	72.1
2005-2006	148.0	37.7

Methods

Undisturbed samples of surface soil dominated by BSC were collected in 90-mm diameter plastic Petrie dishes and transported intact to the laboratory. The experiment consisted of a fully orthogonal design of three surface types: natural BSC, physico-chemical (compacted) crust, and loose (plowed) soil as control, each type was replicated 20 times. The BSC samples were oven-dried for 24 hours at 48 °C before commencement of the study. The physico-chemical soil crust samples were prepared by watering loose, crust-free soil, compacting it, and then rapidly oven drying it at 48 °C for 24 hours. Five 0.5-mm holes were drilled into the base of each Petrie dish and Whatman No. 1 filter paper was placed over the holes to prevent any loss of soil. Each of the samples weighed 80 g at the beginning of the experiment. The Petrie dishes were placed inside plastic tubes of the same circumference and sealed there with silicon rubber sealant to prevent water leakage. A 100-ml column of distilled water was added to each tube above the Petri dish (Maestre et al., 2002). Infiltration rates for each treatment were measured following application of the distilled water.

In the field studies, we tested whether runoff water was generated by BSC cover on the slopes of the watershed. Fifteen small runoff plots were established with internal areas of 0.5 m^2 on a BSC-dominated north-facing slope (7° inclination; Zaady, Levacov and Shachak, 2004). The plots were bound with sheets of

PVC hammered into the soil to a depth of about 5 cm. A flume on the downhill side of each plot led to an 18-liter capacity runoff-water collecting device. The flume and the collector were covered to prevent raindrops from entering them from above (Zaady et al., 2004). After each runoff-producing rainfall event, the runoff found in the containers was collected and measured.

Ten additional sites were chosen to measure runoff from 1-m^2 plots subjected to simulated rainfall from a rotating-disc rainfall simulator (Morin and Cluff, 1980). Five of the plots had undisturbed BSC covers; the other five plots had been plowed by LTERS foresters to control weeds and thus had no crust. Each runoff plot was constructed from four sheets of steel hammered into the soil to a depth of about 5 cm. A sloping flume was connected to each plot at the downhill edge to collect runoff water. The flumes and collecting device were covered from above to prevent raindrops from entering as they collected runoff. A flexible hose attached to a vacuum pump was connected to the lower end of the flume, by which runoff from the flume was pumped into a graduated cylinder to measure the volume of water collected at 1 min intervals from the moment water ponding commenced until a steady-state runoff rate was achieved. The rainfall simulator nozzle delivered simulated raindrops from a standard height of 2.05 m, producing rainfall of 2.5-mm diameter mean drop size with energy of approximately 30 kJ m^{-2} min^{-1} at a pressure of 60 k Pa. Although rainfall intensity was set at 37.5 mm h^{-1}, measured intensity (mean = 47.2 mm h^{-1}) varied slightly among plots according to height above the ground (Eldridge, Zaady and Shachak, 2002). Distilled water was used for rainfall simulations to avoid any influence of cations. Intensity was calibrated by collecting all of the rainfall at the commencement and cessation of simulations each day. Simulated rainfall was applied at a constant rate for 30 minutes, which was sufficient to achieve steady-state runoff.

The source of the biogenic soil crusts and soil samples, and the location of the field experiments, was—as stated above—the north-facing slopes of the watershed.

Statistical Analyses

One-way ANOVA, with Duncan New Multiple Range (Sokal and Rohlf, 1995), were used to test differences in infiltration and runoff among crust types for all three studies. Results are expressed in the text as mean ± standard error of the mean (S. E. M.).

Results

Small-scale laboratory studies

With the small-scale study, infiltration was lowest in the Petrie dishes containing BSC and highest in loose soil ($P<0.001$; Figure 14.1). The infiltration rate in the loose soil was about 30 times greater than that in the controls (Figure 14.1). Water began seeping through the base of the Petrie dishes with loose soil about 2–3 minutes after addition of water compared to 28 minutes on the dishes with the BSC. Most of the water that did infiltrate though the crust did so along the edges of the crust samples or via the narrow seams between individual crust platelets.

Figure 14.1. Infiltration rates for three different crust types in a laboratory experiment; biogenic soil crust as compared to physico-chemical (compact) crusts and loose soil as controls. Different letters indicate a significant difference in that variable at P<0.05. Bar represents S.E.M.

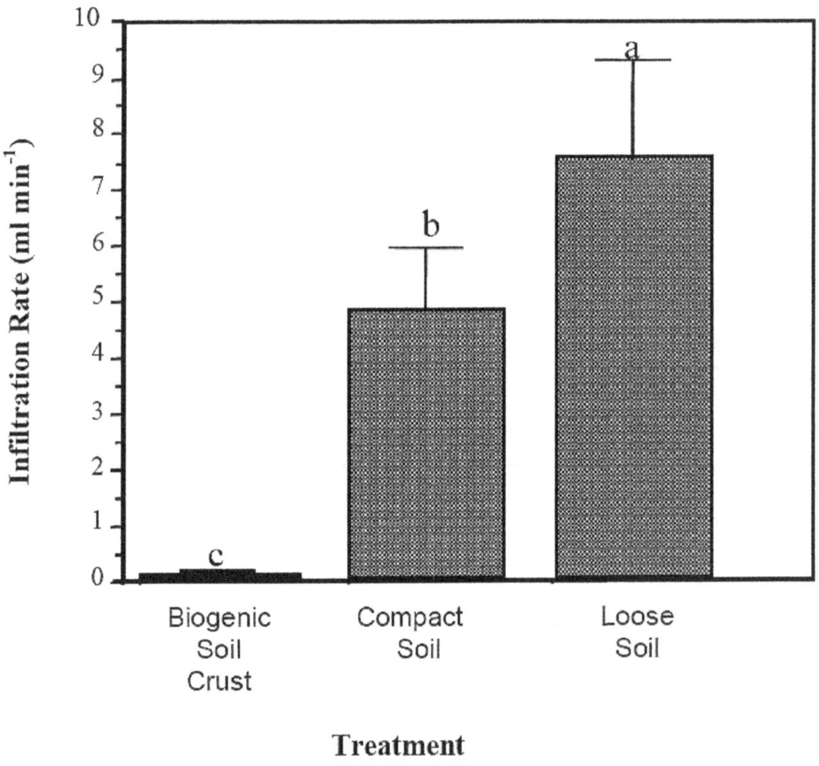

Field and simulator studies

Although runoff was produced in each of the five years, not all rainfall events result in runoff. Small, scattered storms of intensities less than 5 mm h^{-1} produced no runoff due to the low moisture content of the soils and the fact that the interception capacity of the soil surface was not exceeded. Over the five years of measurements, 19–32% of the rainfall was lost as runoff. For rainfall events that produced runoff, increasing annual rainfall was associated with increasing runoff ($F_{1,4}$ = 38.44, P=0.003, R^2=0.882; Eldridge et al., 2002).

With the rainfall simulation study, BSC disturbance significantly affected runoff rates (P=0.027). Runoff from the five biogenically crusted plots was significantly greater than from the plowed plots (Table 14.2), and the increase in runoff rate from the control plots (mean = 1.5 mm h^{-1} min^{-1}) was three times greater than from the other treatments (0.5 mm h^{-1} min^{-1}). Approximately one-third of all applied water was lost as runoff on the natural BSC plots, whereas the plowed plots lost less than 15% of applied rainfall as runoff (Eldridge et al., 2002).

Discussion

These studies, at spatial scales varying from centimeters to meters, provide supporting evidence that BSC communities in the northern Negev Desert enhance runoff production (Table 14.2). At both scales, infiltration was retarded by the presence of BSC, which, when disturbed in the field experiments, resulted in enhanced infiltration.

Redistribution of water is crucial for the sustained functioning of landscapes in the Negev Desert. Our laboratory study showed that despite low hydraulic conductivity (Eldridge et al., 2000), BSC did allow some infiltration of rainfall into the soil crust. This is supported by our studies at the broader sub-landscape scales, where 19–32% of the rainfall (or applied simulated water) was shed as runoff (Eldridge et al., 2002).

Table 14.2. Mean runoff production after rainfall simulation on plots with biogenic crusts and on plots where crusts have been removed by plowing

Soil surface type	Runoff water (mm hr^{-1})[a]
Biogenic soil crust (n=5)	13.7 a
Plowed (n=5)	4.1 b

[a]Means with the different letters within a column indicate a significant difference in that variable at P<0.05.

The runoff coefficients recorded in the rainfall simulation study were almost double those recorded from the small plots at the LTERS under natural rainfall conditions (Eldridge et al., 2002; Zaady et al., 2004). The reason for this lies partly in the characteristics of natural rainfall (rainfall discontinuity), which are often poorly reproduced by rainfall simulators, and partly in the discontinuous spatial distribution of water ponding. Natural rainfall is highly intermittent (Kidron and Yair, 1997), with short-duration high-intensity storms often less than a few minutes in duration. Typically, these are separated by rain-free periods of many minutes (Kidron and Yair, 1997), or by longer intervals where the intensity is lower than the hydraulic conductivity of the substrate. This rain-free period may even be sufficient to allow the sediments to drain and dry. The result is that the pattern of water ponding is unlikely to be contiguous across a plot. Under natural rainfall, zones of water ponding will develop (Dunne, Zhang and Aubry, 1991), separated from other zones of high hydraulic conductivity that often do not normally pond because of abundant plant roots or faunal macropores (Beven and Germann, 1982). Consequently, during natural rainfall, only small, spatially noncontiguous areas of the surface may pond. Depending on the arrangement of these ponds, runoff will often be short lasting and intermittent. The duration of water ponding and runoff is often less than the concentration time required for continuous flow along an entire slope—or from the top of a mini-catchment to the flume. In contrast, rainfall simulators deliver constant rainfall intensity with little or no temporal variability. Under simulation, water ponds rapidly, often spreading out onto areas which would not normally pond water under natural conditions. The result is greater runoff and therefore a higher runoff coefficient than would otherwise occur under natural conditions.

Reduced infiltration through the crust at the LTERS resulted from a combination of the high water-shedding capability of the crusts, surface sealing by biogenic elements at the surface, little or no surface obstruction by plants, an enhanced surface roughness created by the crusts themselves, and relatively long slopes. Our results showed a significant decrease in runoff water generation from heavily disturbed (plowed) plots as compared to natural undisturbed BSC plots (Table 14.2). This is supported by other studies in the Negev Desert using disc permeameters (infiltrometers) which showed that removal of the crust enhanced infiltration under water-ponding conditions but not necessarily under tension (Eldridge et al., 2000).

What then is the mechanism whereby BSC reduce infiltration and increase runoff? Cyanobacteria are widespread in the surface soils of the Negev Desert and at the LTERS they comprise a tightly structured surface community (Zaady et al., 2000). The two dominant cyanobacteria at this station are *Microcoleus vaginatus* (Chroococcales) a filamentous cyanobacterium that grows 1–2 mm below the soil surface and moves to maintain this depth, and *Nostoc punctiforme* (Oscillatoriales), a species that grows on the soil surface (Lange et al., 1992). These two species are known for their ability to secrete mucilaginous polysaccharide sheathes, which bind micro-aggregates (<0.25 mm in diameter) at the surface to produce macro-aggregates (Bailey et al., 1973; Belnap and Gardner, 1993; Moore, 1998). Matrix pores close to the soil surface often become colonized by cyanobacteria, typically creating a hydrophobic surface and reducing further water absorption (Rostango, 1989). Other less common cyanobacteria in the crust include *Chroococcus tugidus*, *Calothrix* sp. (Oscillatoriales) and the green algae *Palmella* sp. (Tetrasporales) (Metting, 1981; Zaady et al., 2000).

The development of a stable, hydrophobic crust is often dependent on other biogenic elements, such as mosses, to provide a structure upon which the cyanobacterial filaments can attach (West, 1990). The two most common mosses in the area are *Aloina bifrons* (Pottiaceae) and *Crossidium crossinerve* var. *laevipillum* (Pottiaceae), both of which are adapted to arid climates (Scott, 1982). Other mosses, such as *Bryum* spp. (Bryaceae), which colonize sheltered sites in the lower parts of the watershed, represent a higher successional stage in the development of a BSC (Zaady et al., 2000).

By generating runoff, BSC in a fully functional state controls the flow of resources (e.g., sediments, nutrients and seeds) across the landscape (Shachak et al., 1998). The northern Negev landscape is a mosaic of patches of BSC and shrubs. The arrangement of the landscape mosaic in a natural state is such that resources generated by the BSC patch are captured by and accumulate in the shrub patch. While the crust patches are sources for soil, nutrients and runoff water (Rostango, 1989; Eldridge, 1993; Abrahams, Parsons and Wainwright, 1994), shrub patches are resource sinks and the major loci of productivity and diversity. This is mainly due the accumulation of sediments, nutrients and water under the shrubs, which promotes a rich growth of associated herbaceous vegetation (Noy-Meir, 1985; Garner and Steinberger, 1989; West, 1989; Schlesinger et al., 1990; Zaady et al., 1996). The natural source-sink relationship between the two patch types is critical to the functioning of the ecosystem. Human activities (e.g., clear-cutting of shrubs for firewood, livestock grazing on shrubs) reduce the abundance of shrubs and disturb this relationship (Bruins, 1990; Ludwig and Tongway, 1995; Ludwig et al., 1997). Location of the ecological system along the precipitation gradient determines the relationship between the crust and shrub patches; therefore, there is an inverse relationship between annual precipitation and crust-to-shrub cover ratio (Zaady et al., 1997). Reduction in shrub cover leads to an increase in crust cover and vice versa. When resources shed from the crust patches are not sufficient to support the shrub patches, landscape diversity and productivity decrease. Heavy grazing or clear-cutting of shrubs increases crust cover, leading to leakage of resources from the system, followed by reduced productivity and diversity (Graetz, 1991; Thomas, 1995). To maintain the highest possible productivity and diversity in this system a balanced ratio of the two patch types must be maintained. Any change in the landscape mosaic, in either direction, promotes desertification processes.

Applications for increasing productivity in dry landscapes

Based on long-term ecological research, the Forest Department of the Jewish National Fund (JNF) in Israel uses contour catchments as a management practice in the forest-parks of the northern Negev to create sink patches on the slopes (Shachak et al., 1998). The contour catchments are strips of soil mounds 1 m high, constructed parallel to each other along the contour lines of the slopes. The distances between the contour catchments are 15–25 m. Their role is to trap and collect surface runoff water that leaks from the undisturbed natural areas of BSC and dwarf shrubs between the catchments. Soil for construction of these catchment mounds is scraped from a 1-m wide strip of surface soil immediately below the catchments and then piled up to form a mound. The crusted soil surface 1–2 m immediately above the mound is plowed to increase water infiltration for the trees, which are planted upslope of the mound. The trees are planted 7 m apart. Each tree exploits the rainfall it receives directly along with the runoff water it receives from the crusted area above it (Zaady, Shachak and Moshe, 2001).

In our study area, the average rainfall per year is 200 mm. Runoff water 1 mm in depth from a 1-m^2 catchment area amounts to 1 liter. The average shading area under the tree canopy, in this area, is 3×3 m. The amount of rain that falls directly on this tree is thus 200 liters $m^{-2} \times 9\ m^2 = 1{,}800$ liters. On average, 19–32% of the natural annual rainfall (and 25.5% of the applied simulated rainfall) was shed as runoff (Eldridge et al., 2002). The trees are planted 7 m apart, so a BSC-contributing area between two contour catchments 20 m apart and two trees 7 m apart is $7\ m \times 20\ m = 140\ m^2$. Therefore, $0.255 \times 200\ mm \times 140\ m^2 =$ 7,140 liters of water, which can be collected by this arrangement (Zaady et al., 2001). This amounts to 5,340 liters of additional runoff water for each tree.

Our results from this study showed that with proper ecological management of the landscape and its resources, afforestation is possible in dryland areas. The presence and the attributes of BSC cover can be utilized to enhance productivity, diversity and sustainability of desertified landscapes the world over.

Acknowledgements

We thank the Jewish National Fund; the Blaustein International Center for Scientific Cooperation for supporting the visit of one of the authors (DJE) to Sde Boker to undertake the research; the New South Wales, Australia Department of Land and Water Conservation for supporting the research; R. Garti and Y. Benyamini from the Soil Erosion Research Station, Israel Ministry of Agriculture for use of their portable rainfall simulator; E. Fizik and R. Levacov for assisting with the field experiment, and S. Brand and A. Leeper for commenting on and editing the manuscript.

Applicative methods for aeolian particle study in desert environments[2]

Wind may be considered the most dynamic of the agents that affect the surface of semi-arid and arid areas of the earth. Wind acts with variable intensities but without interruption all year long. It transports material resulting from soil erosion, weathering and disintegration, over short, medium or long distances, causing significant loss of soil in many areas, often causing considerable discomfort to local populations, finally settling down to form either uniform deposits (sand sheets, dust sheets) or very distinctive topographic relief forms (dunes, dust knolls). Wind may also sort the transported materials and deposit them accordingly.

Aeolian particles are a major constituent of the earth's atmosphere, especially in arid and semi-arid regions. They affect many aspects of daily life, including the economic development of these regions. Most aspects of human activity in desert regions and neighboring areas are affected by such particles, including agriculture, transport (by road and by air), solar energy and construction (in rural, urban or industrial settlements). Therefore, during the past 20 years, aeolian dust has become a major topic in environmental studies. The sources, entrainment, aerial transport and deposition of dust are of great importance to human activity and health (Rheis and Kihl, 1995) and to environmental evolution and development (Goudie and Middleton, 1992) in arid and semi-arid areas. Aeolian dust also plays a role in the development of soil, both as a controller of chemical pedogenic processes and as a supplier of mineral and biotic particles; it also improves the soil moisture regime.

Several studies focusing on the problems and implications of atmospheric dust have been published. They include a wide variety of subjects, e.g., the relationship between weather type and dust storm generation, synoptic conditions during dust events, local and intercontinental transport of dust, dust occurrence and climatic variation, reports of dust events in several parts of the world, current dust deposition compared with the past, the mechanisms of dust erosion, dust transport and dust deposition.

Large amounts of dust are usually found in the atmosphere above or near deserts. The quantity depends on many factors, including location, size and lithology of the source, regional and local meteorological circumstances, distance from the source, and general and local topographical location. These large amounts of dust may have an important impact upon the local weather or atmospheric circulation in the desert boundary layer. Since the fringe areas of many deserts and semi-deserts are expanding rapidly, atmospheric particles are one of the major natural components that should be studied to improve the economic potential of these areas. In deserts and adjacent regions affected by low rainfall, silt and clay-sized particles are propelled from different sources. These include alluvial deposits on dry flood plains, wadis, playas, loess-rich hamadas, alluvial fans, etc. Reduced fertility of systems is one of the most definitive and problematic aspects of desertification (Dregne, 1986; UNEP – United Nations Environment Programme, 1991).

In the Negev Desert, several aspects of aeolian dust have been studied by Offer and Goossens (2001a), particularly, the relationships between aeolian dust and meteorological parameters. Offer and Goossens (2001b) investigated the effects of desert topography on aeolian dust erosion, deposition and accumulation. Esser (1989) studied the role of aeolian dust deposition in a desert ecosystem. A general study of the relationships between dust, soils and landforms in the deserts of southern Israel and the Sinai was published by Gerson, Amit and Grossman (1985).

The following deals with field methods we used to study aeolian dust dynamics and to analyze particle size distribution, micromorphology and chemistry of dust in the Negev desert at the Avdat Experimental Station, near Sde-Boker. These methods are applicable to similar environments worldwide.

Comparisons among four aeolian dust collectors

Four different dust-collection methods were tested at the Avdat Experimental Station (Goossens and Offer, 1994). Each method in the Avdat experiment employed four plastic sedimentation trays installed in the field. These trays were circular (to eliminate the effects of wind direction change), 10 cm high, and 30 cm in diameter. Each collector contained a different sedimentation surface. Surface No. 1 was dry and flat (an empty collector tray). Surface No. 2 was dry and contoured, consisting of two superimposed layers of marbles (marble diameter: 1.6 cm). Surface No. 3 was flat and moist, consisting of filter paper fastened onto a wet sponge. Surface No. 4 was a free water surface (a collector filled with a 1 g liter^{-1} $CuCl_2$ solution to prevent algal growth). Surfaces No. 2, 3 and 4 were about 5 cm below the rim of the tray, whereas surface No. 1 was 10 cm below the rim at the bottom of the tray itself.

Collectors were in operation at the respective locations for a period of 24 months. Each month, the total amount of dust accumulated on the surfaces was measured, and new, fresh collectors were installed. Dust that had accumulated on surfaces No. 1 and 2 was collected with a brush, whereas dust that had accumulated on surfaces No. 3 and 4 was directly determined by weighing the sediment after the water had been evaporated in an oven.

Table 14.3 shows the total amount of dust accumulated on each collector during the Avdat experiment. The results were parallel for all months: the water surface always collected the largest quantity of dust, followed in order by the marble surface, the moist filter paper surface and the dry flat surface. The first year, dust accumulation on the dry flat surface was only 40% of the amount collected on the water-filled collector, while the moist filter paper collector and the

Table 14.3. Dust accumulation (g m^{-2}) on dust collector types at Avdat, and collector efficiency (percent, in italics) relative to the water collector, averaged over two years

	Dust accumulation each month												
	Jan	Feb	Mar	Apr	May	Jun	Jul	Aug	Sep	Oct	Nov	Dec	Whole Year
Dry flat surface	4.4	7.2	7.5	6.5	8.5	9.4	5.5	4.7	6.2	8.3	7.8	6.9	82.7
	41%	*50%*	*44%*	*40%*	*47%*	*52%*	*39%*	*37%*	*44%*	*52%*	*45%*	*45%*	*45%*
Flat moist filter	5.3	8.1	8.6	8.6	11.3	12.2	8.7	6.8	8.4	10.0	10.0	8.2	100.9
	50%	*57%*	*51%*	*52%*	*62%*	*67%*	*61%*	*53%*	*59%*	*63%*	*58%*	*53%*	*55%*
Glass marbles	7.9	12.0	11.7	10.9	13.9	14.2	9.5	8.0	10.4	13.3	14.6	12.2	138.3
	74%	*84%*	*69%*	*67%*	*76%*	*78%*	*70%*	*63%*	*73%*	*83%*	*84%*	*79%*	*75%*
Water	10.7	14.3	16.9	16.4	18.2	18.2	14.2	12.8	14.2	16.0	17.3	15.4	184.3

marble collector collected 49% and 71%, respectively. The second year, the relative order remained the same, but the relative percentages were somewhat different: 49%, 60% and 79%, respectively. Meteorological conditions, especially wind velocity, show an important variation with time (Goossens and Offer 1994). Furthermore, accumulation is largely determined by aeolian erosion, which is greatly dependent upon the frequency and intensity of the strongest winds.

Temporal variations in relative accumulation efficiency of the collectors were small over the 24-month period. No systematic trend was observed for any of the four collectors, and the differences within each collector remained small (a few % at maximum), except for the water collector, where some variation was observed during the first year.

Optimal measurement for aeolian dust deposition

Various experimental setups to measure dust deposition have been reported in the literature, including simple setups (Skärby, 1977) and complex setups: e.g., buckets (Köhler and Fleck, 1963), vertical gauges (Ralph and Barrett, 1976), filters under cover (N. J. Pattenden, 1977, cited in Steen, 1979) and moss-bags (Goodman et al., 1979). The efficiency of these methods, however, remains questionable. An optimal gross-deposition measuring method should both include free, undisturbed sedimentation on a surface, and a total prevention of redeflation of the settled sediment.

One of the four methods used in the Avdat experiment described above may be characterized as a "permanently moist surface method." The flat plastic tray was partly filled with clean water and a thin, fully wetted sponge was placed on the water surface. Due to its weight, the sponge sank into the water until its upper surface was nearly at the same level as the free water surface. The tray was then further filled with water until the sponge surface reached its maximum level. A filter paper (stabilized with inoxidable pins) was then placed on the sponge surface. It was immediately wetted by the water in the sponge. As dust particles settled on the filter paper, they were retained, due to the strong capillary forces of the water (Goossens and Offer, 1990).

The great advantage of this method was that the dust-collecting surface remained permanently moist, even if a very thick layer of dust formed on the filter paper. Due to capillary forces (which can reach very high values in fine-grained sediments such as desert dust) the top layer of the settled dust remained moist as long as water was available in the tray. Even under extremely hot desert conditions (with enormous evaporation rates) the collecting surface remained moist during prolonged periods of time, especially if the shape of the sponge matches the shape of the top of the tray (effectively preventing exposure of free water to the sun). The collecting surfaces remained moist the during the entire Avdat experiment (24 h), despite temperatures of up to 29°C and a water storage content of only 10 liters (Goossens and Offer, 1990). The filter paper used at Avdat was 28 × 32 cm, corresponding to an area of 0·0896 m^2. A plastic sheet on the ground prevented erosion of the soil in the immediate vicinity of the measuring site.

The efficiency of the method is dependent upon the position of the tray on the ground. If the tray is not dug into the ground, airflow patterns will be disturbed by the tray itself, resulting in a lower catching efficiency. A dug-in tray may, however, also catch material that has been supplied by saltation or surface rolling, and not only by suspension. To avoid saltation, it is advisable that a tray be used that is high enough to exclude saltated particles. The efficiency of the set-up should also be determined experimentally (i.e., in a wind tunnel). Tray efficiency is influenced by the size and shape of the tray and by its orientation with respect to wind direction; circular trays would be ideal. Efficiency is also dependent upon air velocity itself: high air velocity lowers efficiency, relative to low air velocity. Fortunately, most of these influences disappear when the tray is completely (but carefully) dug into the ground.

Aeolian dust erosion

Various techniques are used to measure aeolian dust erosion in the field. Most of these methods refer to erosion of sandy surfaces or agricultural soil. Examples are the use of erosion pins (Jungerious and Van der Meulen, 1989; Pluis, 1992) or tracers (Plasschaert, 1985). Other techniques are based upon topographic measurements (to determine the volume of the sediment eroded) or upon mass analysis of standard plots (Offer and Goossens, 1994).

At the Avdat Experimental Station a cardboard plate, 0.8 × 0.5 m in size and 4 mm thick, was installed on thirty 1.0 × 1.0 m plots from which the largest rock fragments had been removed (their removal facilitated placement of the cardboard plates flush with the surface to keep the wind from blowing underneath them, which would have seriously disturbed the erosion measurements). An erosion area 0.5 m × 0.5 m was marked off in the middle of each cardboard plate. These erosion squares were covered with exactly 1 kg of dust, prepared from loess that had been collected just north of Avdat and sieved in the laboratory to exclude all particles >60 μm. All the squares received the same quantity of dust, regardless of their topographic position. The inclination of all the cardboard plates was equal to that of their underlying slopes. Cardboard plates were affixed to the natural surface with metal pins. Before and after the experiment, they were covered with a 1 × 1 m plastic sheet to prevent unwanted deflation (Offer and Goossens, 1994). Erosion was calculated by determining the quantity of dust that remained on the erosion surface after the experiment. The measurements were made with a precision of 0.0001 g.

From the description above, it follows that all the erosion surfaces were smooth. This was done for two reasons. Firstly, and most importantly, it allowed us to install erosion surfaces of equal surface roughness. The effect of micro-roughness upon aeolian dust erosion is therefore eliminated; thus, the field experiment reflected only the effect of macro-roughness (i.e., topography). (This was also the aim of the wind tunnel simulations, as mentioned earlier). One should also realize that the effect of micro-roughness upon aeolian dust erosion is almost impossible to simulate on topographic scale models, simply because of the large reduction of scale. Moreover, micro-roughness in the wide surroundings of each plot at Avdat was almost identical: all the plots were located on stony or rocky surfaces of equal roughness. A small square 0.8 × 0.5 m within this rocky area would not significantly influence the general roughness of the field.

Methods for particle size distribution, micromorphology and chemistry

Our aim was to try a new approach to investigate aeolian particle granulometry and micromorphology in a sandy-loessial area. We used such parameters as: total aeolian deposition, particle perimeter (P), particle area (A), long (L_1) and short (L_2) particle axes, particle roughness ($R_1 = 4\pi A/P^2$) and particle sphericity ($R_2 = L_1/L_2$). These parameters were analyzed on spatial and temporal scales.

The size distribution of the particles on the filters was analyzed with a computerized scanning electron microscope (SEM), which facilitated investigation of the micromorphological and chemical characteristics of the particles (Figure 14.2A, B, C, and D).

Our estimation of the particle size distribution was based on a planar projection of the particles. The value of the particle area (in μm^2), was defined as the surface of the particle enclosed within the projected border, i.e., the perimeter, (in µm). This must be compared to the classic size parameter, i.e., the diameter of the smallest circle enclosing the whole plane projection of the particle (Alshibli et al., 2004).

Using an SEM instrument (type JSM 5410 JEOL), a series of parameters were measured on a large number of particles: the projected surface area (A, in μm^2), the projected perimeter (P in µm) and the projected long (L_1) and short (L_2) axes (in µm). From these values two dimensionless ratios (projected particle roughness R_1 and projected particle sphericity R_2) were computed for a large number of particles.

The first parameter, R_1, was defined as $R_1 = 4\pi A/P^2$. This parameter characterizes the irregularity of the contour of the particle, i.e., the roughness of the particle surface, as compared to the smoothness of a perfectly spherical surface (Alshibli et al., 2004; Donat and Ruck, 2004; Vanderstraeten et al., 2006a, 2006b).

The second parameter, R_2, refers to the elongation of the particle and corresponds to the ratio of the projected major axis, L_1, divided by the minor axis, L_2, of the smallest ellipse enclosing the planar projection of the particle.

The interest in introducing R_1 and R_2 comes from the fact that they are largely uncorrelated parameters. In other words, there is no obvious correlation between the elongation and the roughness of the particles and, therefore, these parameters provide independent and complementary information on the morphology of the particles.

Figure 14.2. SEM (Scanning Electron Microscope) photographs of aeolian particles (see Table 14.4 for size distribution, micromorphology and chemistry details for each particle)

To determine the value of the projected area (as particle size) and the form of the particles (especially the very small ones), the particles were analyzed at various resolutions (magnitudes). The micromorphology of the airborne particles was analyzed by computerized SEM investigation with a Soft Imaging System (SIS) (digital solutions for imaging and microscopy). From the data obtained by the SEM, the values for R_1 and R_2 could then be calculated (Table 14.4). The chemical compositions of the particles, also gained from the SEM, are given as percent of the total weight of the elements presented (Table 14.4).

Among the four types of dust collectors tested—wet-flat (water), dry-contoured (marbles), dry-flat and moist-flat—the water collector received the greatest quantity of both deposited and accumulated sediment. Deposition on the marble collector is comparable to that on the water collector, but as the former is more sensitive to post-depositional erosion, accumulation on it is markedly lower. The effect of post-depositional erosion is even more clearly observable in the case of the other collector types tested: in general, deposition was lower on the moist-flat collector than on the dry-flat collector, but because the latter is extremely sensitive to erosion, accumulation on it was always smaller (Goossens and Offer, 1994).

The type of collector used in a particular situation depends largely on the aim of the measurements. If, for example, the chemical composition of the sediment deposited is to be determined, no water (or moisture) collectors can be used because the water will react with the sediment. If, on the other hand, the total amount of sedimentation (or accumulation) is of primary concern, water will give the best results, provided algal development and insect penetration can be kept to a minimum. Careful consideration of the sampling method should, therefore, be made in advance (Goossens and Offer, 1994). Only four types were tested in this study. Many other types can be used and future research is necessary to test them. At present, we consider the use of marble collectors to be most convenient, because of their simplicity, handiness and lack of maintenance, and we propose that this type should be used as a standard in future dust deposition (or accumulation) research (Goossens and Offer, 1994).

Table 14.4. SEM (Scanning Electron Microscope) analyses of the aeolian particle size distribution micromorphology and chemistry (see Figure 14.2)

Sample	Area	Perimeter	Area/ Perimeter	Long Distance	Short Distance	Long/ Short
	μm²	μm		μm	μm	
A	2871.48	250.07	11.48	66.42	57.53	1.15
B	16656.63	512.50	32.50	188.81	106.78	1.77
C	2709.16	201.37	13.45	67.93	50.25	1.35
D	934.82	108.38	8.63	34.43	34.43	1.00

Sample A		Sample B		Sample C		Sample D	
Element	%	Element	%	Element	%	Element	%
O	22.67	O	44.29	O	43.49	O	23.93
Si	5.01		29.11	Si	3.15	Si	3.74
Sn	19.75	Zr	26.59	S	3.80	Fe	72.33
Zn	14.63			Cl	5.91		
Pb	37.94			Ca	1.11		
				Cr	11.09		
				Fe	31.46		
Total	100.00	Total	100.00	Total	100.00	Total	100.00

Concerning **dust deposition**, the results of the Avdat experiment demonstrate that it is possible to simulate actual dust deposition in a particular area with a topographic scale-model experiment (Goossens and Offer, 1994). Secondly, the Avdat experiment demonstrates the order of magnitude of the gross sedimentation of atmospheric dust during a moderately to very dusty day in the northern Negev desert. Sedimentation is strongly influenced by topographic location (Goossens and Offer, 1988). And finally, the Avdat data can be used to calculate the deposition velocity of Avdat dust during a moderately to very dusty windstorm. The deposition velocity, an important parameter in theoretical dust deposition models, is usually noted as v_d and is defined by the following equation:

$$v_d = F/C \tag{1}$$

where F is the downward sediment flux and C the airborne sediment concentration. The reference height for C is usually 1–1.5 m for land surfaces, and 10–15 m for ocean surfaces. Particle deposition velocity is greater than or equal to gravitational settling velocity (v_t), which increases proportionally with particle density and the square of particle diameter (Goossens and Offer, 1990).

As for **dust erosion**, the wind tunnel and the field results show reasonable agreement. In a previous paper (Offer and Goossens, 1994), wind tunnel simulations of aeolian dust deposition on the same Avdat scale model were compared with field deposition data obtained during a full-scale dust storm. These simulations showed a 90% agreement with the field deposition data. However, wind tunnel simulations of aeolian dust deposition are conceptually different from wind tunnel simulations of aeolian dust erosion: local surface parameters (e.g., surface roughness) have a very strong impact on the erosion process.

The general sedimentation pattern of dust deposition is largely determined by the global airflow pattern over the region, which depends primarily on macro-topography. Thus, accurate simulations of aeolian **dust deposition** require a correct simulation of the general airflow field over the scale model, while a correct simulation of aeolian **dust erosion** at any particular point on the model also requires a correct simulation of surface roughness at that point. As the latter is much more difficult to realize, erosion simulations on topographic scale models can be expected to be less accurate than deposition simulations (Offer and Goossens, 1994). The Avdat experiment shows that wind-tunnel simulations of aeolian dust erosion on scale-models of a complex topography may reasonably predict the location of erosion-sensitive areas in a hilly landscape of nearly uniform surface roughness. When surface roughness is nearly uniform, aeolian erosion of (dry) dusty sediments is almost entirely determined by macro-topography, as was illustrated by the Avdat experiment (Offer and Goossens, 1994).

As for particle size distribution, micromorphology and chemistry, the difference between the specific soil-particle size-distribution for a given location and the aeolian material deposited in the collectors at that location might be explained by tropospheric wind activity and its interaction with the soil surface. The specific form of the collectors induced turbulent air movement and decreased the possibility of deflation of the relatively larger particles.

In our estimation, one of the most important parameters for understanding the characteristics of aeolian particle micromorphology and atmospheric particle dynamics is **particle area**. We analyzed this parameter from two different aspects—spatial and temporal. The size differential may be explained by terrain morphology and by the overland water runoff that easily transports smaller particles downstream.

The *R1* micromorphological parameter ($4\pi A/P^2$) describes the roughness of the particles. Particles with the same projected area may have perimeter lengths (hence shapes) that are very different.

The soil surface may be covered with annual vegetation during specific seasons that might decrease wind velocity close to the ground. This may affect the particle configuration, far from the spherical shape, as a result of decrease in wind erosion, which causes shortening of their time-movement in the atmosphere.

The classic method for determining particle size (Alshibli et al., 2004) does not facilitate investigation of particle roughness (R_1) or sphericity (R_2). Our method offers possible advantages in human-health fields by providing more detailed information on size distribution and shape (roughness and sphericity) of the particles that affect respiratory processes. This method also offers advantages in the field of mechanical engineering, providing greater understanding of the influences of airborne particles on turbines, motors, glasses and filters in corrosive processes. Furthermore, this method may contribute to the meteorological study of cloud formation by offering a greater understanding of how particle size distribution and shape affect condensation mechanisms and ice nuclei formation.

Atmospheric particle dynamics are a direct result of wind activity and ground surface features. Our comprehensive approach better facilitates detailed calculation of particle size distribution and particle roughness and sphericity for these purposes.

Notes:

1. Eli Zaady[a,c], David J. Eldridge[b] and Moshe Shachak[c]
   a Department of Natural Resources and Field Crops, Agriculture Research Organization, Ministry of Agriculture, Gilat Research Center, Israel.
   b Department of Land and Water Conservation, Center for Natural Resources, School of Geography, University of New South Wales, Sydney, Australia.
   c Mitrani Department for Desert Ecology, Blaustein Institutes for Desert Research, Ben-Gurion University of the Negev, Sde-Boker Campus, Israel.

2. By Z. Y. Offer[a] and E. Zaady[bc]
   a Department of Solar Energy and Environmental Physics, Blaustein Institutes for Desert Research, Ben-Gurion University of the Negev, Sde-Boker Campus, Israel.
   b Mitrani Department for Desert Ecology, Blaustein Institutes for Desert Research, Ben-Gurion University of the Negev, Sde-Boker campus, Israel
   c Department of Natural Resources and Field Crops, Gilat Research Center, Agriculture Research Organization, Mobile Post Negev, Israel.

References

Abrahams, A. D., Parsons, A. J., & Wainwright, J. (1994) Resistance to overland flow on semiarid grassland and shrubland hillslopes, Walnut Gulch, southern Arizona. *Journal of Hydrology*, 156:431–446.

Alshibli, K. A., Asce, M., Alsaleh, M. I. & Asce, A. M. (2004) Characterizing surface roughness and shape of sands using digital microscopy. *Journal of Computing in Civil Engineering*, 18:36–45.

Bailey, D., Mazurak, A. P., & Rosowski, J. R. (1973) Aggregation of soil particles by algae. *Journal of Phycology*, 9:99–101.

Belnap, J. (1995) Surface disturbances: Their role in accelerating desertification. *Environmental Monitoring and Assessment*, 37:39–57.

Belnap, J. (2006) The potential roles of biological soil crusts in dryland hydrologic cycles. *Hydrological Processes*, 20:3159–3178.

Belnap, J., & Gardner, J. S. (1993) Soil microstructure in soils of the Colorado Basin: The role of cyanobacterium *Microcoleus vaginatus*. *Great Basin Naturalist*, 53:40–47.

Belnap, J., & Gillette, D. A. (1998) Vulnerability of desert biogenic soil crusts to wind erosion: The influences of crust development, soil texture, and disturbance. *Journal of Arid Environments*, 39:133–142.

Bertocchi, C., Navarini, I., & Cesaro, A. (1990) Polysaccharides from cyanobacteria (review paper). *Carbohydrate Polymers*, 12:127–153.

Beven, K., & Germann, P. (1982) Macropores and waterflow in soils. *Water Resources and Research*, 18:1311–1325.

Brotherson, J. D., & Rushforth, S. R. (1983) Influence of cryptogamic crusts on moisture relationships of a soil in Navajo National Monument, Arizona. *Great Basin Naturalist*, 43:73–79.

Bruins, H. J. (1990) The impact of man and climate on the central Negev and northeastern Sinai deserts during the Late Holocene. In S. Bottema, G. Entjes-Nieborg, & W. Van Zeist (Eds.), *Man's role in the shaping of the Eastern Mediterranean landscape* (pp. 87–99). Rotterdam: Balkema.

De-Philipis, R., Margheri, M. C., Pelosi, E., & Ventura, S. (1993) Exopolysaccharide production by a unicellular cyanobacterium isolated from a hypersaline habitat. *Journal of Applied Phycology*, 5:387–394.

Donat, J. & Ruck, B. (2004) Simulated ground deposition of fine airborne particles in an array of idealized tree crowns. *Boundary-Layer Meteorology*, 93:469–492.

Dregne, H. E. (1986) Desertification of arid lands. In F. El-Baz & M. H. A. Hassan (Eds.), *Physics of desertification*. Dordtrecht, the Netherlands: Martinus, Nijhoff.

Dunne, T., Zhang, W., & Aubry, B. F. (1991) Effects of rainfall, vegetation and microtopography on infiltration and runoff. *Water Resources and Research*, 27:2271–2285.

Eldridge, D. J. (1993) Cryptogams, vascular plants, and soil hydrological relations: Some preliminary results from the semiarid woodlands of eastern Australia. *Great Basin Naturalist*, 53:48–58.

Eldridge, D. J., Zaady, E., & Shachak, M. (2000) Infiltration through three contrasting soil crusts in patterned Negev Desert landscapes. *Catena*, 40:323–336.

Eldridge, D. J., Zaady, E., & Shachak, M. (2002) Microphytic crusts, shrub patches and water harvesting in the Negev desert: The shikim system. *Landscape Ecology*, 17:587–597.

Esser, U. (1989) Zum Stickstoff-Haushalt arider Hangökosysteme im nördlichen Negev-Hochland, Israel und den Auswirkungen der "Hang-Minicatchment-Technologie" auf Stickstoffumsatze und -vorräte. Dissertation, Institute of Geography, Münster, Germany.

Evenari, M. (1985) The desert environment. In M. Evenari, I. Noy-Meir, & D. W. Goodall (Eds.), *Hot deserts and arid shrublands, ecosystems of the world* (Vol. 12, pp. 1–22). Amsterdam: Elsevier.

Feinbrun-Dothan N. & Danin, A. (1991) *Analytical flora of Eretz-Israel*. Jerusalem: Cana Publishers.

Fletcher, J. E., & Martin, W. P. (1948) Some effects of algae and molds in the rain-crust of desert soils. *Ecology*, 29:95–100.

Friedmann, E. I., & Galun, M. (1974) Desert algae, lichens and fungi. In G. W. Brown (Ed.), *Desert biology* (Vol. 2, pp. 165–212). London: Academic Press.

Garner, W. & Steinberger, Y. (1989) A proposed mechanisms for the formation of 'fertile islands' in desert ecosystems. *Journal of Arid Environment*, 16:257–262.

Garcia-Pichel, F. & Pringault, O. (2001) Cyanobacteria track water in desert soils. *Nature*, 413:380–381.

Gerson, R., Amit, R., & Grossman, S. (1985) *Dust availability in desert terrains: A study in the deserts of Israel and the Sinai*. Institute of Earth Sciences, The Hebrew University of Jerusalem, 220 pp.

Gillette, D. A. & Dobrowlski, J. P. (1993) Soil crust formation by dust deposition at Shaartuz, Tadzhik, S.S.R. *Atmospheric Environment*, 27:2519–2525.

Goodman, G. T., Inskip, M. J., Smith, S., Parry, G. D. R., & Burton, M. A. S. (1979) The use of moss-bags in aerosol monitoring. In C. Morales (Ed.), *Saharan dust: Mobilization, transport, deposition* (pp. 211–232). New York: John Wiley.

Goossens, D., & Offer, Z. (1988) Loess erosion and loess deposition in the Negev Desert: Theoretical modeling and wind tunnel simulation. The Jacob Blaustein Institute for Desert Research, Desert Meteorology Papers, Ser. A, No. 13, 65 pp.

Goossens, D., & Offer, Z. Y. (1990) A wind-tunnel simulation and field verification of desert dust deposition (Avdat Experimental Station, Negev desert). *Sedimentology*, 37:7–22.

Goossens, D., & Offer, Z. Y. (1994) An evaluation of the efficiency of some aeolian dust collectors. *Soil Technology*, 7:25–35.

Gosz, J. R., French, C., & Sprott, P. (2000) *The international long term ecological research network*. Albuquerque, NM: University of New Mexico.

Goudie, A. S., & Middleton, N. J. (1992) The changing frequency of dust storms through time. *Climatic Change*, 20:197–225.

Graetz, R. D. (1991) Desertification: A tale of two feedbacks. In E. Medina, D. W. Schindler, E. D. Schulze, B. H. Walker, & H. A. Mooney (Eds.), *Ecosystem experiments, SCOPE 45*. (pp. 59–88). New York: Wiley.

Greene, R. S. B., Chartres, C. J., & Hodgkinson, K. C. (1990) The effects of fire on the soil in a degraded semi-arid woodland. I. Cryptogamam cover and physical and micromorphological properties. *Australian Journal of Soil Research*, 28:755–777.

Johansen, J. R. (1993) Minireview: Cryptogamic crusts of semiarid and arid lands of North America. *Journal of Phycology*, 29:140–147.

Jungerious, P. D., & Van der Meulen, F. (1989) The development of dune blowouts, as measured with erosion pins and sequential air photos. *Catena*, 16:469–476.

Kidron, G. J., & Yair, A. (1997) Rainfall-runoff relationship over encrusted dune surfaces, Nizzana, western Negev, Israel. *Earth Surface Processes and Landforms*, 22:1169–1184.

Köhler, A., & Fleck, U. W. (1963) Untersuchungen zur Festlegung eines Standardmessgerätes für Staubniederschlag. TH Darmstadt, Institut für Meteorolgie. Abschlussbericht No. J (84) 2.

Lange, O. L., Kidron, G. J., Büdel, B., Mayer, A., Killian, E., & Abeliovich, A. (1992) Taxonomic composition and photosynthetic characteristics of the 'biological soil crusts' covering sand dunes in the western Negev Desert. *Functional Ecology*, 6:519–527.

Ludwig, J., & Tongway, D. T. (1995) Spatial organization of landscapes and its function in semi arid woodlands, Australia. *Landscape Ecology*, 10: 51–63.

Ludwig, J., Tongway, D. T., Freudenberger, D., Noble, J., & Hodgkinson, K. (1997) *Landscape ecology: Function and management*. Melbourne: CSIRO Publishing.

Maestre, F. T., Huesca, M., Zaady, E., Bautista, S., & Cortina, J. (2002) Infiltration, penetration resistance and microphytic crust composition in contrasted microsites within a Mediterranean semi-arid steppe. *Soil Biology and Biochemistry*, 34:895–898.

Mehta, V. B., & Vaidya, B. S. (1978) Cellular and extracellular polysaccharides of the blue-green alga Nostoc. *Journal of Experimental Botany*, 29(113):1423–1430.

Metting, B. (1981) The systematics and ecology of soil algae. *Botanical Review*, 47:195-312.

Metting, B., & Rayburn, W. R. (1983) The influence of microalgal conditions on selected Washington soils: An empirical study. *Soil Science Society American Journal*, 42:682–675.

Moore, P. D. (1998) Life in the upper crust. *Nature*, 393:419–420.

Morin, J., & Cluff, C. B. (1980) Runoff calculation on semi-arid watersheds using infiltration equations determined from rotodisk rainulator. *Water Resources Research*, 16:1080–1093.

Noy-Meir, I. (1985) Desert ecosystem structure and function. In M. Evenari, I. Noy-Meir, & D. W. Goodall (Eds.), *Hot deserts and arid shrublands. Ecosystems of the world.* (Vol. 12, pp. 93–103). Amsterdam: Elsevier.

Offer, Z. Y., & Goossens, D. (1994) The use of topography scale models in predicting aeolian dust erosion in a hilly area: Field verification of a wind-tunnel experiment. *Catena*, 22:249–263.

Offer, Z. Y. & Goossens, D. (2001a) Ten years of aeolian dust dynamics in a desert region (Negev desert, Israel): Analysis of airborne dust concentration, dust accumulation and the high-magnitude dust events. *Journal of Arid Environments*, 47:211–249.

Offer, Z. Y. & Goossens, D. (2001b) Airborne particle accumulation and composition at different locations in the northern Negev Desert. *Annals of Geomorphology*, 45:101–120.

Plasschaert, T. (1985) Analyse van de windwerking op een kleine kustduinstrook. M.Sc. Thesis, K. U. Leuven, 247 pp.

Pluis, J. L. A. (1992) Relationships between deflation and near surface wind velocity in a coastal dune blowout. *Earth Surface Processes and Landforms*, 17:663–673.

Ralph, M. O. & Barrett, C. F. (1976) A wind tunnel study of the efficiency of the deposit gauges—interim report. Warren Spring Laboratory. L.R. 235 (AP), 17 pp.

Rheis, M. C. & Kihl, R. (1995) Dust deposition in southern Nevada and California, 1984-1989: Relations to climate, source area, and source lithology. *Journal of Geophysical Research*, 100:8893–8918.

Rostango, C. M. (1989) Infiltration and sediment production as affected by soil surface conditions in a shrubland of Patagonia, Argentina. *Journal of Range Management*, 42:382–385.

Schlesinger, W. H., Reynolds, J. F., Cunningham, G. L., Huenneke, L. F., Jarrell, W. M., Virginia, R. A., & Whitford, W. G. (1990) Biogenic feedbacks in global desertification. *Science*, 247:1043-1048.

Scott, G. A. M. (1982) Desert bryophytes. In A. J. E. Smith (Ed.), *Bryophyte ecology.* London: Chapman and Hall.

Shachak, M., Sachs, M., & Moshe, I. (1998) Ecosystem management of desertified shrublands in Israel. *Ecosystems*, 1:475–483.

Skärby, L. (1977) *Correlation of moss analysis to direct measurement by deposit gauge of cadmium, lead and copper.* Swedish Water and Air Pollution Research Laboratory, Publication B377a, 24 pp.

Skujins, J. (1984) Microbial ecology of desert soils. In C. C. Marshall (Ed.), *Advances in Microbial Biology* (Vol. 7., pp. 49–91). New York: Plenum Press.

Sokal, R. R., & Rohlf, F. J. (1995) *Biometry* (3rd ed.). San Francisco, CA: Freeman and Company.

Steen, B. (1979) Techniques for measuring dry deposition. Summary of WMO Expert Meeting on dry deposition, April 18-22, 1977, Gothenburg. In C. Morales (Ed.), *Saharan dust: Mobilization, transport, deposition*, (pp. 287–289). New York: John Wiley.

Stern, A., Gradus, Y., Meir, A., Krakover, S., & Tsoar, H. (1986) *Atlas of the Negev.* Beer-Sheva, Israel: Ben-Gurion University of the Negev.

Teomim, N. (1990) *Soil survey of the Sayeret Shaked Park.* Gilat, Israel: Jewish National Fund.

Thomas, D. S. G. (1995) Desertification: Causes and Processes. In *Encyclopedia of environmental biology* (Vol. 1, pp. 463–473). London: Academic Press.

UNEP – United Nations Environment Programme (1991) Status of desertification and implementation of the United Nations Plan of Action to Combat Desertification (p. 77). Nairobi, Kenya UNEP.

Vanderstraeten, P., Lénelle, Y., Meurrens, A., Carati, D., Brenig, L., & Offer, Z. Y. (2006a) Airborne particle granulometry and micromorphology during working and non-working days in the Brussels environment. *Air Pollution XIV, WITPress*, 86:287–295.

Vanderstraeten, P., Lénelle, Y., Meurrens, A., Carati, D., Brenig, L., & Offer, Z. Y. (2006b) Temporal variations of airborne particles concentration in the Brussels environment. *Environmental monitoring and assessment.* DOI 10.1007/s10661-006-9531-z.

West, N. E. (1989) Spatial pattern: Functional interactions in shrub dominated plant communities. In C. M. McKell (Ed.), *The biology and utilization of shrubs* (pp. 283–305). London: Academic Press.

West, N. E. (1990) Structure and function of microphytic soil crusts in wild land ecosystems of arid to semi-arid regions. *Advances in Ecological Research*, 20:179–223.

Yair, A. (1990) Runoff generation in a sandy area—the Nizzana sands, Western Negev, Israel. *Earth Surface Processes and Landforms*, 15:597–609.

Zaady, E., Groffman, P., & Shachak, M. (1996) Litter as a regulator of nitrogen and carbon dynamics in macrophytic patches in Negev desert soils. *Soil Biology and Biochemistry*, 28:39–46.

Zaady, E., Gutterman, Y., & Boeken, B. (1997) The germination effects of cyanobacterial soil crust on mucilaginous seeds of three desert plants: *Plantago coronopus, Reboudia pinnata* and *Carrichtera annua. Plant and Soil*, 190:247–252.

Zaady, E., Kuhn, U., Wilske, B., Sandoval-Soto, L., & Kesselmeier, J. (2000) Patterns of CO_2 exchange in biogenic soil crusts of successional age. *Soil Biology and Biochemistry*, 32:959–966.

Zaady, E., Levacov, R., & Shachak, M. (2004) Application of the herbicide, Simazine, and its effect on soil surface parameters and vegetation in a patchy desert landscape. *Arid Land Research and Management*, 18:397–410.

Zaady, E., & Shachak, M. (1994) Microphytic soil crust and ecosystem leakage in the Negev desert. *American Journal of Botany*, 81:109.

Zaady, E., Shachak, M., Moshe, Y. (2001) Ecological Approach for Afforestation in Arid Regions of the Northern Negev Desert, Israel. In D. Vajpeyi (Ed.), *Deforestation, environment and sustainable development: A comparative analysis* (pp. 219–238). Westport, CT, USA: Greenwood Publishing Group.

Chapter 15

Safety Management in the Manufacture and Supply of Chemicals

Robert Gilmour[1]

The chemical industry has several voluntary programs to promote the continuous improvement of health, safety and environment management in the industry. The main program, established by the International Council of Chemical Associations, is known as "The Responsible Care® Global Charter." It has several elements, one of which has the appropriate name of "Product Stewardship."

These days, Responsible Care has been adopted by most countries worldwide and has an international logo (Figure 15.1). Separate companies or organizations can become members if they agree to uphold the principles of Responsible Care, which entitles them to display the logo on their literature and communications. Participating chemical companies have upgraded their programs from a voluntary initiative to a fully integrated program, because they simply could not otherwise survive today without doing so.

Figure 15.1. Responsible Care® logo

Participating companies must be more open and communicate, i.e., they must transfer information on sensitive issues concerning their products and operations. The Responsible Care program must be part of their business plan and policies and Responsible Care procedures must be in place. This means that all health, safety and environmental problems must be addressed. When the required standards and performances have been achieved, they must be maintained.

The whole process has gradually converted from a reactive to a proactive management system. In this way all the standards set by the authorities for air emissions, liquid effluent discharges, solid wastes, ground pollution and noise must be achieved or, ideally, be substantially bettered. Risk-evaluation and assessment studies must be in place to predict emergency situations, and responses must be planned to mitigate an incident in the shortest possible time. Responses must be practiced and drilled to show that they will work when put to the test.

Practically speaking, this means that process upsets that have the potential to cause damage must be identified in advance. This refers, in particular, to fire, explosion or toxic release, which could have adverse affects both inside and outside the plant limits. In other words, companies are not only dealing with the protection of their workers, but also their neighbors, or—more specifically—the general public.

Unfortunately, the chemical industry has a bad name, due to several catastrophic accidents that caused deaths and injuries among the general public. Fortunately, however, these accidents have been well documented and much can be learned from them to prevent similar catastrophes from occurring. The best known, perhaps, are the Flixborough fire and explosion in the United Kingdom in 1974, and the Bhopal, India toxic-gas release in 1984 that killed thousands.

The main lessons learned (among many other recurring lessons) are that one should minimize the inventory of hazardous chemicals used in the manufacturing process, that hazardous plants must not be built close to populated areas, and that process changes must be studied and approved. A well-known safety guru, Trevor Kletz, has written many books and articles on learning from accidents. Concerning the Flixborough accident, for instance, there are three basic theories about how that accident was initiated. Kletz (2007) pointed out that it is important to remember that if these theories are plausible, then steps need to be taken to prevent them in the future.

Fortunately, safety and risk assessment has crept slowly, but surely, into university curricula in recent years and many universities offer first and second-degree courses on this topic. However, there is still a shortage of skilled professionals in the field. This is mainly because companies are only just realizing that when experienced personnel leave a company, they take their know-how with them. Old books, technical articles, calculations and data are safeguarded on the shelf or under lock and key in desks. Companies are just waking up to the fact that all this knowledge must be harnessed for their intranet sites, so it can be referred to when needed. Safety and risk data may be critical to saving lives, preventing injuries and avoiding damage to equipment. There is no point in learning these lessons the hard way time after time.

At a chemical plant there is a great deal of know-how to be learned and data to be found. One might be tempted to say that the Internet, with all of its various search engines, can guide the seeker to the source of knowledge within seconds. However, not all information is free and getting to the bottom of a difficult problem may require having a credit card handy or having a healthy budget available from the company for technical publications. Many companies just pay a consultant to sort it out for them, but this is also costly and dampens the enthusiasm for understanding the nature of the problem. Only if the technical aspects of the problem are defined within the scope of a work document and only if the consultant's report is read thoroughly (rather than just accepting the bottom line) will know-how actually be transferred.

Product stewardship

To illustrate the methods employed and the effectiveness of technology transfer in the field of Chemical Safety, I would like to concentrate on the elemental program mentioned above—Product Stewardship—which endeavors to ensure that all occupational health, safety and environmental aspects are fully assessed during the stages of research and development, plant design, manufacturing, supply and use of hazardous chemicals during the life of the product. Some refer to this as being a "cradle-to-grave" stewardship of the product (see Figure 15.2). It is a continuous cycle because after the first introduction of the chemical to the market, there may be a need to change the process to solve a production problem or to improve the quality of the product; there may be a change in the environmental regulations that also require process changes; there may be a demand for a new packaging size or material that requires changes, or customers may have run into processing difficulties in their plant that require the manufacturers' expertise.

Rather than trying to cover the whole of the circular path, let us concentrate on the responsibility involved in transport and storage of the supplied chemical, to ensure that the customer knows how to handle it. Note that the customer can either be an end user, or an intermediate user, whereby your product is a raw material for further processing. In the latter case one is dealing with another chemical company or some allied industry, such as pharmaceuticals, plastics, biocides, and so on. Your customers would therefore be similar to your own company and have professional staff to deal with safety and environmental matters. However they may not be so familiar with your material and may need guidance for storage of containers and before unloading the chemical into their system. In this case, you would be sending all the documentation to them so they can assess the risks in their operation. One document that all chemical suppliers are legally bound to provide is a Material Safety Data Sheet or MSDS. This provides physical and chemical data, toxicology data, basic storage recommendations, basic emergency response advice, and various regulatory information for shipping and trucking the material. It is a document written by professionals for use by professionals.

Figure 15.2. A Product Stewardship cycle

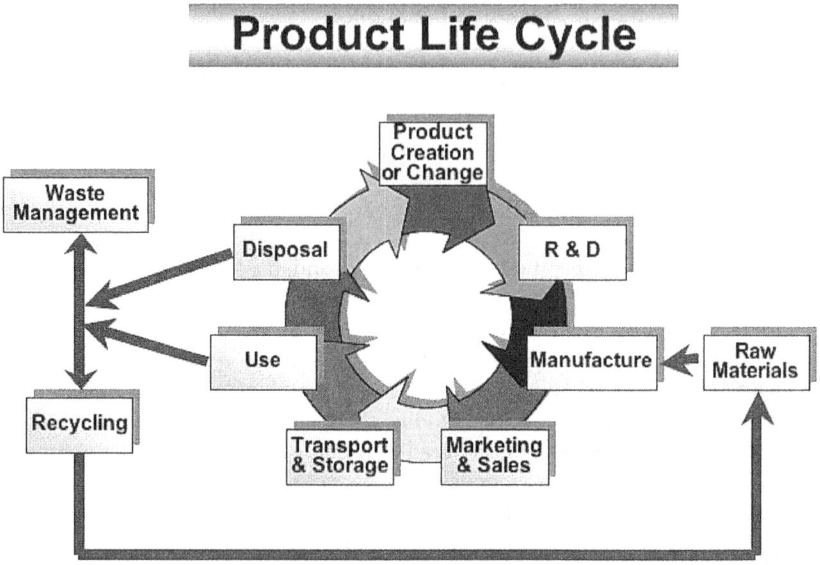

Thus, one important item of know-how that has to be transferred and documented involves training customers who are newly receiving hazardous chemicals. This should include all logistics, production and maintenance staff; i.e., all personnel who could possibly be exposed to the chemical. For highly hazardous chemicals, delivery of the material itself should never be made until after transfer of this know-how has been completed.

We have so far only considered the customer. Yet, whether we are responsible for transport and storage of our product or not, we must ensure that this part of the logistics supply chain is also included in our Product Stewardship program. When one is supplying hazardous chemicals to end users, the need for Product Stewardship is even greater because many of the personnel in the supply chain are not at the same technical level. The supply chain could include warehouses, distributors, wholesalers, or contractors applying the chemicals.

No chemical company wants a significant incident to occur in the logistics supply chain, even with a competitor's material. It only takes one accident to bring the unfavorable attention of the general public to your product and maybe even your company. The press will make maximum negative use of the event. After all, it is good business for them but awful business for you. A significant accident in the logistics supply chain can bring extensive court cases, fines, regulation changes, supply hold-ups, and costs beyond all expectations. You may have a sound insurance policy or a good lawyer, but at the end of the day it is a big blow to your reputation and heads will roll.

One may ask the question: What can I do about shipping, trucking, and storage when these activities are not under my direct control? The answer is that although the control of their operations is not being directly controlled by you, know-how must be transferred to personnel along the logistics supply chain. You must select good partners who will uphold good procedures and know exactly what to do in an emergency. They should be trained and procedures drilled. All the information may not be in their heads, but it must be made available to them for their use. Experts need to be brought to the accident scene quickly to mitigate an incident before it gets out of hand.

Companies that supply chemicals on a global basis need to coordinate these activities via emergency centers that have the necessary contacts. The delivery notes and packaging labels must contain 24-hour contact numbers. Supplying to Asia-Pacific countries often involves language problems and therefore bilingual local response services are essential when coordinating response activities. Training material and safety information should be translated into the local language. It may be advisable to have spare parts for bulk packaging kept at strategic locations along the supply route in case such items as valves and sealing materials need to be replaced. Some highly hazardous materials, such as chlorine and anhydrous ammonia, have special sealing kits for the dedicated containers used for their transport. Local emergency teams will need to bring protective clothing and neutralizing materials to the accident scene for quick and effective response.

Technology transfer occurs initially in the classroom where the hazards of a particular chemical are presented, various scenarios are developed and the correct response procedures are discussed. As chemical spills can often result in exposure to hazards by the general public, evacuation procedures need to be considered. The "hot" zone must be established and all people trapped therein must be either moved out or instructed to shelter in place, until the danger has passed. This decision needs be made by the scene commander, who is supported by the technical staff of the team. These must be able to evaluate a cloud dispersion to determine if it is a toxic release or a flammable cloud that may ignite.

No training is complete without an emergency drill in the field that runs through one or two accident scenarios. This would involve securing the scene, alerting response personnel and equipment, evacuating exposed personnel who cannot be part of the response activities, attending to and evacuating the injured, neutralizing spills, fighting fires, and cleaning up the area. Key communications must also be tested. These procedures must be practiced once or twice a year so the response team will remain familiar with all response activities. Often the company spokesman is also tested on his ability to deal with the local press and with TV crews.

Should a large-scale incident occur, back up must be arranged to support the team's activities for 24 hours or more. Local emergency services (police, firebrigade, paramedics and civil defense teams) often participate in these drills.

When incidents occur in isolated or inaccessible locations, the use of helicopters may be required. Also, large mechanical equipment may need to be hired to lift containers, trucks or rail cars.

The first step in carrying out a Product Stewardship program is to prepare the training kit that will be presented to the various groups along the logistics supply chain, i.e., port emergency teams, drivers, transportation and logistics personnel, warehouse personnel, and the customer's technical staff. One must bear in mind that training will take one to three hours out of someone's busy day, so the presentations must be interesting to hold the audience's attention—because so much information needs to be transferred in those few hours. The audiences are often people who are not used to going to intensive courses, so the sessions must involve a lot of visual presentations and discussions to arouse the interest of the listeners.

The first step is to explain who you are and what your company does and where you come from (particularly if you are in a foreign country). It is also interesting to know who your audience is, so ask them to introduce themselves. This often breaks the ice and improves the atmosphere for transfer of information. Of course, one can always start with a PowerPoint presentation, just like everyone else, but always be mindful of the rules of an effective presentation, e.g., not too much information on one slide, large clear text and pleasing color schemes. It is often good to start with an explanation of what Product Stewardship is and what you are trying to achieve. Emphasize that incidents and accidents with your product are very rare, but that it is still essential that they know what to do and who to contact in the unlikely event that an incident occurs.

Once this has been introduced, start to summarize the properties of your product—it is a brown fuming liquid that boils at … and has a density of …. Its hazards are that it is a toxic substance and is corrosive to most metals, and burns the skin, etc. Then take them through some of the regulatory material. It has a United Nations number (this will identify the substance anywhere in the world to the local emergency services). Tell them about the packaging you use: bulk (isotanks), intermediate or small packaging (drums or sacks) and what placards and labels are regulatory when transporting the substance. Talk a little about occupational health and what the acceptable and dangerous concentrations are, what personal protection equipment must be worn when handling the packaging or when dealing with an emergency. This is a slow build up to what happens if you have a spill and the emergency plan has to be implemented.

This is where participation of the audience is essential to understanding the procedures, which, when written on paper are totally boring, but, when put into practice develop into a drama. So, ask someone to sketch the layout of the site on the white board. Where is the chemical stored? Where is the perimeter fence? What is the prevailing wind direction? Where are the offices? Where is the nearest residential area? Where is the entrance to the site? How far is it to the fire station, the nearest hospital or clinic? Where is the personnel protection equipment, neutralizing materials, water supply, etc.?

Next, a scenario is developed together with the group. For example, there is a loss of containment from a 15-ton isotank (a hole in the tank) and a pool of toxic liquid has started to evaporate and form a toxic cloud. What should be done first? Do you know whom to call? People must be evacuated. How? Are there injured? How does the emergency team approach the scene? How is the area secured? Who is going to be in charge of the response activities? Who is going to talk to the press, TV, etc.? Do they know what to say? What are the neutralization procedures? How can the leak be stopped?

Anyone in your audience may be an important member of the team, passive or active. The procedures have therefore to be tested for various scenarios. Often problems are discovered in one or more aspects of the response activities and corrective action taken.

At the end of the frontal presentation and discussion, it is often good to show a video of response activities where all the procedures you have presented are repeated and are absorbed by the audience. This should be followed by a practice drill where everyone has a "role" to play.

There are several practical lessons that need to be mentioned time and time again in order to prevent injury or death when responding to emergencies. The main point that should be made is that a responder should not consider himself or herself a hero. The last person a scene commander needs on a team is an impulsive person who acts without thinking or consulting with his or her colleagues beforehand. For instance, if someone is lying on the ground, apparently unconscious, there is a reason for it. He could have been injured physically by a blow or a fall. But if there are toxic vapors or gases involved, he may have been exposed to them and was not protected. The first impulse is to rush in and save him, but anyone doing this without a respirator and suitable clothing may also become a victim, resulting in additional injury (or worse!) to deal with.

Confined spaces are typical places where negligent behavior could have lethal consequences. If someone has to enter a drain, a pit or a processing tank, there may be insufficient oxygen inside to work safely. One must first test the atmosphere and decide if self-contained breathing apparatus is needed. Unfortunately a large number of accidents occur, where people have lost their lives; because they did not, or were not taught to, follow these entry procedures.

Another killer is the power supply to equipment. When someone needs to work on mechanical equipment for maintenance or in an emergency, the power supply must be isolated; otherwise someone may unknowingly start the equipment while a colleague is working on it. There are strict procedures for isolating, locking and tagging starters for equipment, and the key to the lock should be in the worker's pocket. If there are two workers that have to work on the same equipment, then two locks are placed on the starter unit, so that the equipment can only be operated when both locks have been removed.

These are just two of the safety procedures that have to be a part of daily life in the chemical and allied industries. Safety procedures must be known, understood and practiced before anyone is permitted to work on a chemical plant.

Returning to the stewardship aspects of supplying a chemical product to the market, one is legally committed to supplying a user label on the packaging. This enables the customer to review the safety and environmental aspects of the material and to make a risk assessment of his own. However there is no guarantee that he will do this and therefore highly hazardous chemicals must only be sold to a new customer after he has received a formal transfer of the know-how from the supplier, if Responsible Care principles are to be applied.

A recent incident, involving the sale of a chemical to a large refinery, is a good example of this. The refinery employed a service company to supply this chemical and to design the equipment. The system was relatively simple. The chemical needed be loaded into a vessel, which was connected to the water supply of a cooling tower. The water passed over a bed of this chemical and the reactive ingredients in it were slowly released into the feed water as it dissolved the material. These additives prevented algae or other biological growths from forming in the cooling tower. When all of the chemical has been dissolved, the flow is stopped and a fresh batch of material is added. The safety instructions clearly stated that the water flow should be started immediately (or at least within an hour or two). If there is a problem with the water supply, then an alternative water source should be connected. Unfortunately, the refinery employed a sub-contractor to refill the unit. Apparently, the sub-contractor had not been properly trained, because, contrary to instructions, the water flow was turned off and left off for 24 hours. The vessel became pressurized due to gassing inside and the filling lid blew off. Fortunately, no one was injured. Subsequently, the chemical manufacturer put together a training package and presented the safety instructions to all the staff involved. The refinery operating personnel, the designer, and the contractor were included so that everyone understood the correct operating and re-filling procedures. Additional safeguards were added. It was lack of transfer of critical know-how to a critical employee that failed to prevent the incident from occurring. The conclusion drawn by the chemical manufacturer following this incident was that either the refinery operates as prescribed, or the supply of this particular chemical will be stopped.

Product Stewardship sometimes requires hard decisions, but company reputation and the right to bear the Responsible Care logo on the company correspondence must be maintained.

One of the most important databases that must be available is the information on each chemical that is being used in a chemical plant. Even a medium-sized chemical manufacturer could have hundreds of different chemicals present at the plant. Typically, this includes products, raw materials, intermediate chemicals, laboratory chemicals, cleaning materials, boiler-treatment chemicals, etc. Each chemical will have an MSDS (see above); but, as mentioned, this is a long document and is intended for the professional staff. It is best to issue a simple, one page summary to enable the plant operators, maintenance staff and all those that could use or be exposed to the substance, to be aware of the hazards involved, the safety equipment to be used and other information that has to be accessed quickly. These should include the storage conditions, the instructions to be followed to clean up spills and fight fires that involve the specific chemical.

Again, there have to be formal training sessions to ensure that the employee knows how to identify the material, how to use the material, and what to do in an emergency.

Chemicals that are hazardous must be stored and used according to the instructions. Six basic hazard classes are relevant to chemical plants:

- Class 2: Gases that are flammable, pressurized or toxic;
- Class 3: Liquids that are flammable;
- Class 4: Solids that are flammable, hazardous when wet or spontaneously combustible;
- Class 5: Oxidizers;
- Class 6: Toxic chemicals,
- Class 8: Corrosive substances (See Figure 15.3).

Figure 15.3. Some international hazard placards, from the Globally Harmonized System (GHS) of classification and labeling of chemicals

Although these chemicals are supplied in reliable, approved packaging, the risk of an incident occurring can be minimized by keeping incompatible chemicals apart. As an obvious example: acidic materials should not be stored or transported with alkaline materials. There are many other incompatibilities that are not so obvious and must be identified.

The know-how, technology, and procedures for handling hazardous chemicals must be readily available to all those in the logistics supply chain and at plants storing, transporting and handling them. In view of the sheer volume of data that may be needed, a computerized system via the Internet or local company intranet, sometimes referred to as knowledge management, is essential. But this is not enough. Management of a Product Stewardship program must also involve regular visits to service providers along the logistics supply chain and to customers. Key locations, including ports, warehouses, transportation companies, and customer facilities, must be audited to ensure that they are complying with all safety recommendations and that arrangements are in hand to deal with emergencies. Training must be carried out and repeated to ensure that all the data and procedures are understood. Those involved should be encouraged to report accidents, incidents, and near misses so that they can be fully investigated and can be prevented in the future.

International companies must ensure that all the technology, know-how, and documentation is available in the appropriate language. The MSDS, the user labels, safety instructions, training kits and other critical information must be in the local language of the user. Even in the United States, Spanish is required in certain areas. Quite often, a plant operator, a maintenance worker, a port employee, or a warehouse manager may not have the ability or the patience to read English documents. Often an interpreter is useful to have in training sessions. Once the importance of preserving the safety and health of those directly involved in handling the chemicals has been understood by those likely to be exposed, good working practices will be established.

Unfortunately, there are still a significant number of accidents during the handling, storage and shipment of hazardous chemicals. However, professional responders and technicians, armed with the knowledge and the equipment, can mitigate the incident to keep injuries and damage to a minimum. Good Product Stewardship is the key to maintaining a high state of alertness and a high level of professional skill in the field.

Note

1. Since graduating in chemical engineering from London University in 1962, Bob Gilmour has been employed in industry in the UK, The Netherlands, and Israel. He has worked as a commissioning engineer and project manager for investments in the chemical, steel, and mining industries both in engineering offices and directly for manufacturing plants. He moved to Israel in October 1983 and worked for 6 years in an engineering office designing and constructing plants for the plastic and chemical industry. For the past 18 years he has worked for a large chemical manufacturer as a safety management consultant. His responsibilities include risk assessment studies, leading safety studies with the design teams, and advising on process plant safety. With this background, he has traveled the world to train, audit and advise on transportation, storage, and use of hazardous chemicals.

Reference

Kletz, T. (2007) Another look at Flixborough. *The Chemical Engineer* (UK). May 2007.

Chapter 16

Pastoral Nomads: Static Versus Nomadic Pastoralism

Gideon M. Kressel and Hendrik J. Bruins,
in honor of the late Ya'acov Orev

Introduction

Studying and comparing migrations of pastoral nomads from aridity-stricken areas across fertile valleys provides vital insight into issues that have divided "the desert from the sown" for millennia. Dealings between pastoral nomads and farmers, based as they are on the exchange of goods and services, have long been affected by the vicissitudes of climate and fluctuations in the fertility of soil and herds. Due to the superior political organization of the pastoral nomads—which has guaranteed tribal cohesion—and due to the particular nature of their capital (mobile herds), it has been easier for them, as compared to the farmers, to amass and wield power, enabling them to tilt the balance of exchange between them, hence to be recipients of farmer goods and services, as compared to providers of livestock products.

Where a third party (i.e., a kingdom and more recently a state) has come into being, it has tended to protect the interests of the farmers—the food producers and the more submissive and compliant tax-paying clients vis-à-vis the pastoralists—and has also supervised dealings between farmers and pastoralists. Through the ages the balance of power wielded by the three parties has occasionally shifted, with the pastoralists or organs of the state occasionally gaining the upper hand (Kressel, 2003; Ch. 1).

Three main phases can be discerned in the varying degrees of state dominance: a totally weak state, a state of intermediate power (neither weak nor strong) and a totally strong state. This third and last phase has evolved since the 19th century. The strength of nation states has grown while the relative freedom afforded to pastoralists to make a living has diminished, accordingly. Shepherds who migrated between winter pastures in the desert and summer pastures in the vicinity of perennial water sources in sown districts were also able to serve as mercenaries, guard caravan routes, work in remote wilderness locations, extort money from occasional travelers and from farmers and smuggle forbidden goods —until these activities were curtailed by the state. Consequently, many former pastoralists have renounced livestock breeding and have left the desert for good.

As more pastoralists quit herding in steppe zones, one finds them in growing numbers settling at the outskirts of towns. This article presents the outcome of studies of current, past and likely future exchanges between these three constituents: i.e., farmers, pastoralists and state organs. With an aim to returning inhabitants to forsaken arid lands-cum-foraging lands, this article proposes to offer the former pastoralists squatting at the outskirts of towns a better life as herders, predicated upon a viable symbiosis between the two rural factions, i.e., the farmers and the pastoralists. The chain of events proposed here includes:

1. State investment in agro-forestry and amelioration of forage capacity in semi-arid lands—in coordination with the pastoralists;
2. upgrading of pastoralists' herds and livestock production;
3. arranging symbiotic dealings between farmers and pastoralists—for the direct and mutual benefit of both factions;
4. and finally, enlargement of the Gross National Product (GNP)—for the benefit of society as a whole and the organs of state in the process.

Research projects and their potential applications

On studying the enduring practices common to pastoral nomads and determining which of these will not be constructive in the coming years, plans of action to cope with their unwanted effects were developed at the Social Studies Center of the Jacob Blaustein Institute for Desert Research at Sde Boker, Israel, (by the authors, together with the late Ya'acov Orev). Complementing the efforts invested by Evenari, Shanan and Tadmor (1958, 1971) in harvesting rainwater for agriculture, we worked on use of rainwater for amelioration of pasture flora, (e.g., annual grasses, perennial bushes and edible parts of forest trees), to provide forage for livestock and to solve problems for breeders of sheep, goats and camels.

Orev's handbook (1986, 1987) followed the Evenari team publications of 1958 and 1971. Based on three years of fieldwork in Niger, this handbook was directed mainly at countries of Sub-Saharan Africa and has been accepted by many African countries. Applicable to similar conditions elsewhere, the Israel Land Administration approved Orev's handbook—in essence, though not in practice—regarding the Negev Bedouin. During the 1990s, Orev's material served the Israeli delegation to the International Center for Agricultural Research in the Dry Areas (ICARDA) workshop groups, addressing the future of shepherding in Middle East deserts and semi-desert zones.

Led by Bruins, a joint Kenyan, Israeli and Dutch project was carried out during 1994–1999, sponsored and administered by the Netherlands Israel Research Program (NIRP). Its main objective was to study the durability of skilled Turkana and Maasai pastoralists of Kenya, and the Negev Bedouin, particularly under natural disaster conditions. The results were summed up in a research report and published in 2003 (Bruins et al., 2003). The main intent of this project was to offer input and formulate policy recommendations for the development of integrated drought contingency planning.

Some cultural contingencies of Middle East economies, with particular regard to shepherding, were summed up by Kressel (1998). This knowledge (translated into Chinese and reprinted in 2000) was transferred to Northwest University in Xi'an, China in 1999 and to the Bharatia Agro-Industries Foundation (BAIF) in Pune, India in 2001. Although the work-plan was acceptable in essence to scholars in China, because pastoral nomads comprise distinct ethnic minority groups in that country, the Chinese Government viewed it as "meddling" in their political affairs and did not accept it. India, on the other hand, favorably considered its practical application, despite its many logistic difficulties.

A Better plan of action: the Middle East model

Almost all potential desert rangeland in Israel (receiving <100 mm annual rainfall) and most of the steppe rangeland (receiving 100–200 mm annual rainfall) is under state control. Except for small experimental farms, none of these lands are managed as rangeland proper. Yattir Forest, covering approximately 10,000 acres of the northern Negev region of Israel, began as an agro-forestry project and has, however, yielded an unplanned benefit for shepherding: the provision of annual pasturage for ruminants. During the months of January to April 2000, 4,500 head of sheep and goats belonging to the Negev Bedouin benefited from foraging in Yattir forest. During the spring of 2001, that figure rose to 7,500 head and by 2007, to 12,000 head.

This ongoing development project also provides firewood to hundreds of neighboring families—villagers in an area extending from the suburbs of Hebron to the center of the Negev. Agro-forestry supplements the devastated range during the winter and spring. During the summer, the herds graze on stubble after cereal crops have been harvested. Both of these solutions to the problem of providing food for animals are, however, improvised; there is no overall plan for the maintenance and development of herds. Herding entrepreneurs who own the largest herds (200 head and more), resort more and more to feeding the animals in pens with purchased feed, even when pasture is available. Thanks to the farming sector in Israel, it is cheaper to pay for feed and to feed livestock *in situ* than to pay for the labor required to tend the grazing herd. Shortage of skilled and responsible shepherds is ubiquitously evident.

Preservation of shepherding—albeit a modernized version of the traditional occupation—is possible, despite the challenges that lie head. This is the task we take upon ourselves as we search for ways to facilitate shepherding in years to come.

In 1950, the feasibility of developing the Negev's potential pasturelands, including the Negev Highlands (our focal point of departure) were assessed by botanists appointed by Israel's newly formed Ministry of Agriculture (Seligman, Tadmor and Raz, 1962; Shanan, 1987, 1992, 1998). Despite their optimism and recommendations for projects to improve provision of forage, not much has been achieved since that time. Even though the land available for herding has increased, as has the size of the herds, this land is used less and less by the herders because the provision of forage has drastically decreased after years of drought and overgrazing; and the number of shepherds who can take the animals to pasture and tend them has declined. Encouraging investment to ameliorate deteriorating rangelands for the benefit of shepherding requires solutions to three basic problems:

1. Long-term leasing of tenure rights over rangelands by tribal herds leads to the phenomenon known as "the tragedy of the commons" (Hardin, 1968; Hardin and Baden, 1977; McCay and Acheson, 1990). In effect, this means that pastoralists are unlikely to graze their herds judiciously, since they stand to gain in the short-term, and the cost is passed on to society as a whole. In such a situation, even if a few individuals were to act responsibly to restrict harm done to plants, there would be no general change in the quality of rangelands. Recognition of open-access rights of grazing on tribal grazing land (the *dîra*) has traditionally acknowledged that tribal families have the right to share in the take but, so far, no noticeable effort has been seen to share in putting back—no tribal provision that all should invest equally in the effort to ameliorate their common rangeland forage.
2. The pool of potential shepherds is shrinking fast since the children of herders are acquiring better educations and other job opportunities are opening up for them.
3. Many tribesmen are increasingly turning to *in situ* feeding of the herds; i.e., employing a stationary (non-nomadic) method. As already noted, food is bought for the herds, which are fed in pens, even if grazing land is available in the vicinity, when hay is cheaper than herders' rent. Invitations to pastoral nomad herders to graze their livestock in the spring pastures of Yattir Forest are answered only to a limited extent, largely due to the preoccupation of herd owners who lack a solution to the question of who will take their sheep up into the forest and watch them while they are grazing "up there." The future of skilled shepherding depends on bettering its image, and this will not be feasible until the methods and tools of the trade are modernized so as to improve the living standards of shepherds.

When rangelands have been destroyed and no attempt is made to rehabilitate them, the desert takes over. Arresting the spread of desertification has thus far proved an overwhelming task for both the tribesmen who forage on wasteland located within the common land, and for reckless entrepreneurs, acting out of personal interests. In the State of Israel, the forestation department of the Jewish National Fund (JNF), backed by the Israeli government, has been the only agent to plant forests on desert land. In other Middle East countries, the situation is almost the same—only the state, helped by non-government organizations (NGOs) from Western countries, has exerted a visible impact on the desert landscape.

Initial development projects, led by the state and enforced by agricultural legislation throughout the Middle East, usually delegate responsibility for rangelands to the state, notwithstanding tribal claims to grazing rights; e.g., Jordan Agricultural Law No. 20 of 1973; the range-inventory project in Tunisia 1989 pertaining to all rangeland areas of the country, comprising ~30% of all Tunisian lands; the Matruh Resource Management Project (MRMP) in Egypt aimed at encouraging rangeland development in the northwest coastal area of Egypt (ICARDA, 2000). By so doing, control of rangeland encroachment and limits on degradation and soil erosion can be exercised by the state, and negative processes can be reversed. The "sense of tragedy" pertaining to common rangelands has become a telling point in the argument that governments must play a stronger role in dealing with environmental problems.

On the other hand, some observers have advocated laissez-faire policies that encourage privatization by having the state leave the role of development to individual developers. Private groups, however, are often reluctant to engage in a war against desertification or to ameliorate vegetation in fallow lands. For this reason, private investment remains negligible. Why should one strive to change the "natural" course of events, i.e., reverse land deterioration? Because this tragedy surpasses individual prosperity and impinges on environmental, economic, and societal well-being. However, since no specific group can be sure that it will profit in the future, not even if land is ameliorated, no one feels compelled to try.

In whose interest? Who will benefit?

Some wasteland areas in developing countries—where, in the past, shepherds led their herds to seasonal forage—are increasingly being taken over now by agro-forestry and national parks. Meant to benefit tourists, mainly those from foreign countries, governments leave the indigenous herders out of their thinking (Chatty, 1996). Development projects are frequently funded by Western "green" concerns. After green spots within arid areas are developed, local agencies endeavor to "preserve" them—to provide tangible evidence that the money has been well-invested—by often making them out of bounds for herders. This attitude arouses resentment towards development agencies mainly during droughts that exacerbate the herders need for these enclosed green areas. Tracts of land for amelioration, on which the herders can pasture their herds unmolested, could mollify attitudes—hence prevent conflicts—while measures must also be taken to enforce laws against trespassers (people and livestock) on nature reserves.

The worsening state of pastures and the near-extinction of certain types of vegetation is a cause for concern not only among herd breeders, who make a living from wilderness flora, but also among "green-minded" individuals, ideological circles, and political groups in the West, since the dwindling of species of flora indigenous to arid zones is a fact, and botanists are alarmed at the irreversible lost of germplasm and the threat presented to bio-diversity and a balanced state of nature. These bodies have launched the current campaign for the preservation of species of flora in several Middle East countries. The emphasis on germplasm and the survival of all botanical species, however, are not necessarily compatible objectives, and at times appear to conflict with solutions offered to relieve herders' problems. Introduction of foreign, nutritive species of flora from other arid lands (outside the Middle East), to provide feed for indigenous stock, is often ruled out lest the local bio-diversity be altered.

Prime concern for the flora, on the one hand, and for the herders and their livestock, on the other, calls for compromise in the form of new initiatives to encourage a symbiosis of objectives. Many observers throughout the Middle East hold the pastoral nomads responsible for the deterioration of the vegetation of the region through over-grazing. Others, who assume the ability of traditional societies to modernize, support efforts to integrate agro-forestry projects into monitored grazing, aiming to reinforce forest maintenance. This objective could be achieved by training pastoralists to farm on their own or to use the pasturelands in a more judicious fashion, as has been tried in the past (for herders moving across the borders of Syria, Jordan, Saudi Arabia and the Gulf Emirates in the 19th and 20th century, hence attempts to settle them, see Bocco, Jaubert and Métral, 1993). Moreover, shepherding, duly monitored by experts, could enrich the savanna; e.g., grazing could clear away winter (wild) annuals before they dry out in the months without rain—thus limiting the risk of fire—and the droppings from the herd could provide fertilizer for the trees. In other words, although forestry projects deny the pastoral nomads free access to land, pastoral nomads would, in the long run, be the first to benefit from improved pasturage. Grand forestry plans sponsored by states to catch runoff water by terracing slopes would, in the long run, also relieve the botanists' concern for wild plants since this would promote the growth of indigenous species carried by the wind alongside newly-introduced nursery plants on the slopes. The JNF Negev forests exemplify this scenario in an impressive manner (Perevolotsky and Landau, 1992).

By shifting attention to sustainable ecological solutions, grand forestry projects may benefit national communities by providing answers to ecological and human problems. Savanna belts improve the quality of the air the town dwellers breathe and, at the same time, provide attractive resorts for tourists. Solutions must also be provided to the herders, to enable them to make a living from their livestock. Efficient (not old-fashioned) shepherding necessitates a more economic use of areas that are marginal for cultivation purposes. Rangelands are not only a cheap way to supply ruminant feed, but are also a habitat for wildlife in danger of extinction. Disappearance of savanna belts—wherever this is allowed to occur—paves the way for encroachment by the desert. Moreover, the condition of the soil and vegetation on which scarce precipitation falls has a major influence on the amount and quantity of water available for human uses. Above all, proper management of desert frontiers (e.g., meadows) helps accommodate bio-diversity, and ensures retention of that land as a significant natural resource.

Shepherding, economy and society

Without state intervention, areas of the Middle East that were in use as common (tribal) property will decrease at an alarming rate. In Turkey, for example, the total area estimated as rangeland amounted to 12.4 million hectares in 1991, about a quarter of the 44.2 million hectares of rangeland in 1940 (see Firincioglu, 2000). Regarding the pastoral-to-agricultural transition in Syria since 1860, see Lewis (1987), Jaubert (1993), Khalaf (1993) and Métral (1993); for similar reports from Jordan see Lewis (1987), Tell (1993) and Lancaster and Lancaster (1993); regarding Saudi Arabia, see Fabietti (1993) and Cole and Altorki (1993). The main factor responsible for this great reduction is conversion of fallow/meadow areas to agricultural land, primarily for the cultivation of cereals. This is the consequence of privatization and of the immense population growth in the region (Firincioglu, 2000, pp. 2–3). Plowing and sowing, even if for a negligible yield, symbolize personal entitlement to land and a change in the possession patterns of territories, formerly considered to be common or tribal lands, which have now been subdivided into arable plots (Kressel, Ben-David and Abu-Rabi'a, 1991).

Even a cursory survey of the region's history and of the present situation suffices to establish the fact that only states and their organs (and not private concerns) have been able to implement new initiatives with any degree of success in past efforts to check the advance of the desert. For instance, state development agencies have in the past encouraged their pastoralist tribal constituency to forsake its *dîra* tradition, and to endorse *ifraaz* (subdivision of their commons into private plots), with expectations, from the viewpoint of the state, that taxation revenues from farming products are greater than from livestock products (Kressel and Ben-David, 1995 and 1996). But those empowered to carry out dry cultivation on virgin soil in such cases, i.e., the new and unskilled farmers, do not feel bound by ecological principles and social norms; rather they will try to maximize short-term gains, which will inevitably lead to over-exploitation of the resource and to its depletion. Therefore, government policy has tended to shift during the last decades towards giving agro-forestry a chance. Despite their importance, rangelands, when compared to other agricultural areas, have received much less attention. This can be accounted for by the Middle East attitude entertained towards regions in which farming is marginal and towards the inhabitants of these districts, the pastoralists, who are also regarded as marginal.

As a result of decades of mismanagement, overstocking and overgrazing, forage provision has steadily diminished. States initially failed to act, and when they did enter the picture, the trend throughout the region was to act alone, giving little thought to the affected population. From the examples provided by the Faysaliyah and Wadi Mujib range projects in Jordan, the Matruh development project in Egypt and the Yattir forest project in Israel, one senses the wish of these states, first and foremost, to bring green to otherwise brown terrain. This is in accordance with the wishes of NGO donors, who willingly invest in such projects. They all contemplate control over improved lands for farming while hardly ever addressing the needs of the herders. Only later do they attend to the needs of the pastoral nomads and their herds and promote amelioration of forage yields in fallow and meadow areas.

The MRMP in Egypt allotted developed plots—formerly used as common tribal areas—to individual families to develop, and, by so doing, weakened the traditional authority of the tribe. In addition, the MRMP confronted the problem of army officers commanding nearby military bases, who put pressure upon lands within the Selected Range Management Area (SRMA). Military personnel employed people to carry out cereal farming on their behalf on rangeland originally meant to serve the Awlad 'Ali pastoral nomads who inhabit the Western Desert of Egypt. Aided by the introduction of tractors, they plowed and marginally cultivated wheat and barley on the best land and, consequently, increased pressure on the remaining SRMA plots. Cereals (unlike plantations or forests) are annuals and therefore suited to army officers who are frequently rotated and moved from one part of the country to another.

The Israeli authorities did not subdivide the Yattir forest for use and did not reassign land ownership. Doing so would not only be harmful to the forest, but would also imply giving in to Arab demands for the entire land and losing a part of the Jewish State. They acknowledge the usufruct rights of neighboring pastoral nomad families and allow them access to spring forage in the forest pastures, when available, and allow them to gather trimmed branches for firewood all year long. The development authorities retain full control of the forest and monitor the pastoral nomads as they reap the forest's benefits, free-of-charge. As a matter of policy, care for the flora must come first; only later does care of the herds follow. Preparation of rangelands takes perseverance, and years are needed for these efforts to yield substantial forage.

Jordan's two impressive range projects have not yet (by 2003) reached maturity and so have not provided pasture for any tribal group or private initiatives, but the development authorities are already preparing for the coming phase, in which collaboration with families of herders will follow as a matter of course. Collaboration with the tribes, however, may not work at first, because the tribes will tend to assume that this is not in their best interest. Nonetheless, collaboration will pay in the long run—for both the development agencies and for the local pastoral nomad communities. After a number of years, when the forest can provide fodder and firewood, the pastoral nomads will see that the success of these projects are indeed in the best interest of the majority.

Although agro-forestry projects can hardly be profitable in the Middle East, the return on the effort and investment is one of future benefit—by preventing the spread of desertification. Therefore, costs to rectify the situation should not be cut. Like the historic networks of canals that restrained and redirected the inundating Tigris and Euphrates when their banks were overrun, or the breakwater dikes along Europe's northwest coast, which protect inland fields, the forestry plots of Middle East states have the potential to prevent the desert from overrunning tilled fields.

Everywhere in the Middle East, private concerns or tribes are vying for possession of desert territories in order to control mineral deposits or traffic routes. Rarely do people, as once in the past, fight for rangelands for their botanical potential in order to graze herds more efficiently. The simple explanation for this is that only now—for the first time—is there a possibility of ameliorating the range through intervention by NGOs and state organs.

Rights of usufruct: integrating shepherding into modern agriculture

After conceding that governments—and not private or communal concerns—must take the initiative and lead the campaign to reinvigorate the dwindling flora along desert frontiers, the question arises as to how to implement this objective in a way that will encourage the inhabitants of these regions to actively collaborate and thereby guarantee success. A precondition of the first order is to secure the right of usufruct for the herders, i.e., to ensure their right to use of the provisions of the project. The authorities in charge of development and the pastoralist groups—the users—or lawyers who represent the two sides, should jointly meet with an advisory board that will establish how ameliorated range plots will be used in years of abundant rainfall—and in years of drought. When droughts occur, state organs should facilitate herding on stubble fields in the fertile plains (in Israel, these are located in the northern Negev, and in the center and the north of the country) and facilitate use of silage and food-industry waste to feed herds in distressed districts, or they should import hay and concentrated feed from abroad. Herding along the desert frontier would then potentially become part of the contemporary agricultural scene and herders would be instructed and supported by experts acting on behalf of state ministries.

New efforts are seen throughout the Middle East to restructure agriculture and implement policies that are more market-oriented. This contrasts with the plan, mentioned above, that would relegate the initiative exclusively to the state in areas of agro-forestry, improvement of rangelands and provision of alternative feed for herds, when needed—a plan that would cost the state and the NGOs substantial sums of money. Economists, influenced by the spirit of capitalism in the West, have encouraged involvement and participation of stockholding farmers in the making of rangeland policy, while preferring not to involve herders; they consider the pastoral nomads-cum-tribesmen to be—by their nature—unsuitable as members of a company because stockholders should be individuals and not agnatic (combatant) groups (Meyer, 1958; Ch. 7, 8). This may explain why the trend to consult with and include pastoral nomad users of rangelands, among other potential beneficiaries of development projects, has been modest thus far.

States are apparently able to protect rangelands and the herds grazing on them against raiders and rustlers. In doing so, however, does this diminish the need for tribal entities? Livestock have always been privately owned and tribes have always been the corporate entity that guarded their "walking property" before states and police forces came into being. Since time immemorial, it has not been the private possession of herds, but rather the need to guard them that has forged tribal entities. Would state-provided security of livestock property eliminate the tribe? Probably not. As long as the tribe serves as the central pillar of micro-social order and as long as social hierarchy ostensibly depends upon it, Middle East communities will retain this entity, which provides a sense of structure and order alongside all super-structural organs (Kressel, 1992, 1996). Clearly, this means that agro-forestry planning should endeavor to include tribal leaders in the planting and guarding of forage plots.

Which pro-active initiatives should the states of the Middle East carry out to combat spreading aridity and promote shepherding? We suggest the following:

- Range development areas must have persons in charge. They would be responsible for summoning the herds to pasture, notifying the herders of the foraging boundaries and monitoring the shepherding to give the vegetation, bushes, and trees a chance to rejuvenate.
- Ways to include herders as participants in inspecting the forest range must be carefully formulated. The herders' mindset must be modified to convince them of the long-term importance of conservation over short-term economic gains. This change in priority of concern depends on management information systems (MIS) specifically directed at herder self-education, to enable them and motivate them to keep pace with the general effort.
- Alternative round-table councils—patterned neither after Western-style stockholder associations nor after the *sheikhdom*-and-tribal-bond associations of the Middle East—are badly needed now. Lack of a tradition among rangeland managers to include legitimate range users in open public meetings, and an initially deep-rooted custom of agnatic ascription of council constituents by tribal stockholders committed only to their own group of agnates (Kressel, 1998), may spoil the accountability that should be conveyed by the rangeland authority to the stockholder teams. This calls for the participation of judicious minds—experts who can design straightforward modes of operation.
- Strategies to minimize the risks of interminable disputes between the foresters and the shepherds over range management policy should entail methods that circumvent open public debates. Exchanges of opinion regarding preferred lines of action and uses of agro-forestry require strict confinement to joint public concerns and cost and benefit considerations (i.e., they must be free of interference from inter- and intra-tribal affairs). This implies that inculcation of range management principles must come first.

If non-sustainable resource practices are condoned, (e.g., allowing overgrazing in a drought year when insufficient rain limits the amount of available pasture), the principle that equal forage must be provided for all becomes compromised. A feasible method for enabling a *Naturvolk* to share in accountability for state-managed commons implies that pastoral nomads are acquainted with the requirements for successful implementation of development policy. Other preparations must include (after King, 2000):

- Mutual recognition by owner (state) and user (mainly pastoral nomads) of long-term lease of land-use tenure rights over rangelands.
- Mutual recognition of principles regarding inclusion in, and conditions for, membership in leasehold groups, devoid of tribal (agnatic) favoritism.

- Coordination between farming and shepherding communities—to the benefit of both groups—requiring liaison offices to receive requests of farmers seeking the aid of herds to clean out the fields or of herders in search of foraging on farm surplus, thus delivering a much-needed service to both groups.
- Improved road systems to facilitate movement of animals to and from stubble fields and to the verge of plantations in sown districts during periods of rangeland stress.
- Agreement about transfer and inheritance of membership rights.
- Consent upon beneficiary/obligatory principles of allocating rights and commitment to accountability, in order to sustain the resource and to collect and deposit money in a fund during years of plentiful rain to provide feed during years of drought.
- A mechanism must be devised to prevent/solve domestic disputes and to divide up and fairly allocate pasture plots on the basis of herd needs, as opposed to consideration of agnatic strengths. There could be conflicting interests within the community of rangeland users, and, therefore, they may not be able to reach a consensus as to the sustainable uses of the (rangeland) resources and the amounts to which each is entitled. Agreement upon stock ratios (de-stocking if needed) and formulae for lease of pasture under various rainfall and agro-ecological conditions can be another stumbling block. If decision-making power about the size of herds were to be transferred to tribal leaders, then this would imply government approval of an inequitable distribution of power and opportunity in intra-tribal affairs. This could result in a widening social gap within the tribe as the rich and powerful become richer and more powerful.
- State ownership of rangelands must be unmistakably felt, seen and reaffirmed, while secure long-term (10, 49, 98 years) user rights over pasture must be guaranteed within the legal system. This implies that the leases of rangeland must be reconfirmed by legal decrees. Once guaranteed by jurisprudence, oversight by the legislature will counterbalance arbitrary shifts in jurisdiction that may, for instance, deprive eligible families access to pastures.
- Auxiliary measures that help better the forage are an inadequate solution to secure the future of shepherding. Additional stewardship practices are required to keep the shepherds' community life intact and improve their quality of living. This includes medical care, both for humans and animals, and measures to reduce the fear of natural disasters, such as plagues, floods, and droughts, so that herders will not consider leaving for good. An improved road system to connect deep desert sites with sown districts would improve access to distant pasture plots, hence, provide the range authorities with a vital tool to alleviate the stress occasioned by droughts, as they do for peasant communities during time of drought.
- Any measure to augment security for animals grazing on public meadows should be welcomed because it helps herd owners save on manpower. Portable, easily dismantled fencing is needed to confine the herding to separate plots, and forage on rangeland or at the boundaries of forest development projects should be cleared. Efficient herding has the potential to encourage the Jewish sector in Israel to breed herding stock as well and return to rangeland- and forest-grazing, rather than feeding livestock in pens.

- Wherever intruding herds jeopardize plants, "cut-and-carry" methods may offer a solution. Tree trimmings brought out of the forest could serve as feed. Once the capacity of the forest to provide green fodder is recognized, the choices of species to be planted in arid and semi-arid zones should be re-evaluated. For example, planting of pines, eucalyptus and tamarisk could be limited and other species that ruminants favor, such as acacia, could have greater representation[1]. Obviously, herd owners and nutrition experts would be provided with input regarding the best mix of species to be planted in the forest. The kinds of trees required would have to be grown in the forestry tree nurseries—perhaps also including types not grown at present.[2]
- With improvement of herding conditions, herd composition (sheep, goats, camels) could be readjusted. For example, goats—once notorious for the damage they cause to natural forests—and camels—a danger to plantations—would not present such a threat; their merits could be reconsidered, their assignments programmed anew, and their number increased (Perevolotsky, 1991a, 1991b).
- Housing conditions for shepherds must be improved to enhance the status and future of shepherding. Stone-age dwellings and goat-hair tents should be relegated to hosting guests and serving as exhibits in museums. Modern housing that provides better shelter and comfort is desired by the new generation of pastoral nomads.
- Accommodate the establishment of villages for herd-owners near range-development projects to ensure availability of forage for consumption during the springtime.
- Increase the variety of employment options, other than herding. Supply these villages with electricity, running water and all other modern amenities so that the younger generation will wish to stay.
- A vital organizational change is needed in the set-up of livestock markets where small ruminants and their products are sold, in order to increase the success of shepherding[3]. Demand for sheep and goat milk and cheese will rise once people are educated about their superior value (in contrast to cows' milk). Demand for Awassi sheep wool and goat hair may exceed its present level if research and subsequent funding increase their modern uses, thus opening additional markets. Likewise, mutton consumption cannot develop without massive efforts to change the consumption patterns of Jews, who prefer chicken, while Middle East consumption patterns evidently favor mutton (Kressel and Ben-David, 1995, 1996; Kressel, 2003, Ch. 8).[4]

Fitting the Middle-East model to Kenya's Kajiado and Turkana Districts

Our study aimed to find solutions to problems that pastoral nomads in Kenya face at times of drought and then propose auxiliary lines-of-action for households and governments to take to diminish pastoral nomad vulnerability. We paid particular attention to rainwater harvesting to mitigate the effect of forced migrations. Harvesting water that would otherwise run off entails constructing bulwarks to divert streams onto fields and plantations or into storage reservoirs (ponds or subterranean cisterns). Installation of "family-sized" micro-catchment devices requires a sedentary frame of mind, i.e., a determination to stay put, and hence to monitor the earthworks and mend them and improve their operation, as needed. Stability must precede family investment in real estate, and intra-tribal agreements must precede the construction of family micro-catchment systems and the planting of trees before families can settle on land that had previously been used as common pasture.

Furthermore, "state-sized" macro rainwater catchment installations require broad (tribal) consent because rainwater-harvesting installations cover wide areas of grazing land. The role of the state in Kenya regarding drought relief is as important there as it is in the Middle East. Last-resort famine relief requires an adequate food supply, gained through sound national enterprise. But additional, more profound measures are needed to preempt the spread of hunger. While a great many of the pastoralists in the Middle East are ethnically and linguistically Arabs, the pastoral nomads of Kenya differ in ethnic origin and language of communication. In the case of the Turkana, inter-tribal peace would help with restocking herds. Deterring armed tribal raids across and within its national borders; initiating livestock vaccination programs to reduce incidences of disease; establishing efficiently organized livestock markets; facilitating herder-farmer exchanges and improving symbiotic relationship patterns between them—these are all within the purview and responsibility of the state.

Modern farmers barter fodder, hay and use of stubble fields for grazing with the pastoral nomads in exchange for surplus husbandry products. When these resources are in short supply, the role played by the state is enhanced. Investing in water catchments; encouraging transition from strict husbandry to dryland-farming; monitoring use of agro-forestry plots; coordinating pastoral use of stubble fields—all these require the state to create networks of intermediary officers who collect information on potential surpluses among farmers and pastoralists and facilitate encounters between the two parties for bargaining and exchange of their surplus products. The state must constantly maintain order by monitoring transactions between farmers and pastoralists at markets lest they, as dissimilar as they can be, deceive or abuse one another.

Fitting the Middle-East model to India's Rajasthan and Gujarat States

Analogous to the Middle East "desert and sown" relationship, the herder-farmer relationship of the Indian subcontinent manifests points of similarity yet, nonetheless, substantial dissimilarity. During the spring of 2003, Kressel spent a month as a guest of the Bharatiya Agro-Industries Foundation (BAIF), a development/research foundation committed to sustainable rural development, food security and clean environment, in Pune, Maharashtra, India. Together with his hosts, he visited the headquarters of the Gujarat Institute of Desert Ecology (GUIDE), an organization that aims to preserve the Banni Grasslands in Bhuj, Kachchh, Gujarat, and the headquarters of the Lokhit Pashu Palak Sansthan (LPPS), an organization that aims to promote the breeding of camels in Sadri, Rajasthan. The three aforementioned organizations share a commitment to bettering the future of India's rural communities by improving the quality of their production, which would, in turn, raise their standard of living and improve their quality of life.

The arid zones of India are small compared to the vast rural areas that spread out beyond them. Likewise, pastoral nomads in India constituent a small, varied and largely innocuous group—much as they do in the Middle East. India's deserts do not harbor grand cohorts of pastoral nomads. Those that the author met along the roads at the end of March—leading thousands of head of livestock to temple shelters—belonged to several ethnic groups, including the Rabari in Gujarat and the Raika in Rajasthan. The steady incursion of pastoral nomad tribes into sedentary farming zones presently occurring in the Middle East—thanks to a single shared mother tongue (Arabic)—could not occur in India. Hindu pastoral nomads must always negotiate for favorable assistance from sedentary farmers for pastoral relief for their herds. Due to the scarcity of telephones and electronic devices, most contacts are oral. pastoral nomads who come begging to barter are welcome.

In India, the state provides far greater assistance to sedentary farmers than to pastoral nomads. Relief for pastoral nomads—or better, for their flocks—comes from NGOs who finance the collecting and storing of forest leaves in temples during the two to three months (March to June) that precede the monsoon rains, to prevent death from hunger. Symbiotic cooperation between sedentary farmers and pastoral nomads must be encouraged and facilitated by a third party capable of employing modern means of communication to invite the herders to bring their livestock to forage on fields that need clearing out after the harvest. This third party would then organize stubble field grazing, monitor adherence to agreements, preserve natural forests and invest in agro-forestry for the years to come.

Conclusion

A major reason for the decline in shepherding in the Middle East—and in Israel in particular—is modern economic development (mostly petroleum-stimulated growth). The countries of this region can now supply their needs and purchase livestock goods abroad. Therefore, shepherding, which only provides a modest source of income, is neglected. This reason for the decline in pastoralism has yet to receive due attention the way range destruction due to drought, over-grazing, and urbanization of herders has. As pastoral nomad power dwindles, Middle East countries fail to attend to the needs of this constituency; e.g., their need for potable water. These countries import livestock and livestock products at increasingly rapid rates to keep up with the local demand. Moreover, in Israel, the Arab-Jewish conflict over the land and the impact of development projects on the Negev (mainly for the benefit of the Jewish population), have resulted in a further reduction in herding.

Efforts in Israel to induce the settlement of nomads have proceeded on a much larger scale than those in other Middle East countries. Nearly all the former pastoralists in the Galilee and more than half of the former pastoralists in the Negev dwell in townships, in concrete homes. Most pastoralists in the Middle East no longer wish to return to their situation of a century ago, but rather opt for the comforts of urban life. One may contend that a nomadic society is transitional; i.e., a society constantly evolving toward a settled existence. Pastoralists can thus be visualized as a group always seeking incentives to settle. Their overall state could be assessed as one of dynamic equilibrium: one pole drawing them toward continued pastoral herding of livestock, the other drawing them toward a sedentary existence combined with continued raising of livestock using modern *in situ* methods—or even toward alternative employment opportunities. The crucial question is whether settling into a sedentary existence inevitably eliminates this balance and brings shepherding to an end. Can a way be found to maintain shepherding in concert with new economic factors and changing ecological conditions?

The answer could be to have a smaller population of herders keeping smaller-sized herds than in the past. Could a decline in numbers provide a new self-sustaining model—that of settled (non-nomadic) herding or of semi-pastoral herding (shepherding along with another source of income)? If male pastoralists look for employment elsewhere, will the traditional pastoral system become unworkable? The increased contribution from female labor in the herding economy does not make up for the loss of contribution from males. Moreover, as the wealth of the family increases, women, too, will quit this occupation, thus jeopardizing its existence. As the amount of stock raised on the range diminishes, the few capital-intensive, livestock-raising farms (where animals are kept and fed in pens) will take up the slack and supply local market demands, thus competing with imported livestock and further depriving the local breeders of their livelihood. Therefore, Alternative "C"—the essence of which is that national societies replenish destroyed pasture with the aid of international associations, so that herder families can flourish—remains the last and the sole solution available to prevent unwanted developments, such as the raising of animals only in pens.

One must consider the advantages of monitored shepherding, for the pastoral nomad herds and also for the environmental and agro-forestry development projects. The more limited grazing patterns that result from farm-based herds will leave wide areas of rangeland untouched. Even when animals are occasionally transported by vehicle to graze in more remote areas, this will, in the long run, still entail neglect of a major asset—the natural range. Forest terraces—designed to slow down runoff water so it may reach the roots of plants—require large-scale infrastructure work, which tribes of former herders, not to mention single families of herders, cannot afford. It is up to states, augmented by NGO efforts, to carry out this work, rendering agro-forestry viable.

Small family-tended herds with limited transhumance movements reduce the former political function of the tribe—one of which was to defend the livestock property of its affiliated families. On the positive side, however, regarding the foraging capacity and sustainability of tribal *dîra* ranges, the tribe still retains a shared interest in maintaining their rejuvenation potential. Without a tribal focus on control, the enterprise—including forage production—would be at risk. Family herds grazing locally are liable to exacerbate the problems of over-grazing. Moreover, the long-term viability of sedentary animal production based on provision of fodder increases the demand for drinking water. At present, water tables in Israel, and the Middle East as a whole, are continually being lowered. Formerly nomadic families squatting in pasture areas drill their own water wells, the cumulative effect of which is to lower the water-table, resulting in a local water supply that will be depleted in the future. The general public, led by the state, must take the initiative to supply drinking water, thus providing possible sustenance to village life in arid zones, inhabited by herders and their herds. The state must take the initiative to supply the required drinking water; an undertaking that is otherwise too large either for individuals or for a concerted tribal initiative to accomplish. The state organs are called upon here to increase their involvement by reclaiming ruined rangelands and lending a helping hand to communities of *in situ* pastoralists.

Notes

1 The once pastoral nomads of the Negev Highlands refuse to use the tamarisk either for fodder or firewood. However, the Senussi tribes of Menzel el Habib in Tunisia do make use of the tamarisk—an indicator of a cultural variant.

2. Government development agencies—as opposed to agrarian communities—may view the merits of pine forests differently. For an example from Asturias in southern Spain, see Fernandez (1990, pp. 272–3).

3 Concerning the trade in livestock in the Be'er-Sheva market and the eastern deserts of Saudi Arabia and Jordan during the Ottoman and British periods, see Kressel and Ben-David (1995, 1996).

4. The price differential of meat for the Israeli consumer is tied to ritual factors. Meat consumers in Israel, including the Arab sector, predominantly favor kosher over non-kosher meat. In essence, whereas Jews are concerned about ensuring proper slaughtering procedures (e.g., slaughtering, not strangling the animal, and letting the blood flow out), the Arab sector demands principally the exclusion of pork, which is automatically and strictly ensured by Jewish Orthodox practice. The Orthodox religious establishment sends its employees to slaughterhouses to inspect a wide range of factors relating to the animals' hygiene and feeding practices. *Kashrut* supervision affects pricing in a variety of ways: payment of inspector's salaries, fees payable by the meat supplier in return for the rabbinate's stamp, and the additional work entailed by the practice of porging sinews from the hind legs of ruminants. As far as mutton is concerned, these constraints render it five times as expensive as mutton slaughtered without ritual observances (see Kressel (2003, Ch. 8).

References

Bocco, R., Jaubert, R., & Métral, F. (Eds.) (1993) *Steppes d'Arabies: Etats, pasteurs, agriculteurs et commerçants; le devenir des zones sèches*. Paris: Presses Universitaires de France et Genève.

Bruins, H. J., Akonga, J. J., Rutten, M. E. M., & Kressel, G. M. (2003) *Drought planning and rainwater harvesting for arid-zone pastoralists: The Turkana and Maasai (Kenya) and the Negev Bedouin (Israel).*NIRP Research for Policy Series 17. Groningen, the Netherlands.

Chatty, D. (1996) *Mobile pastoralists: Development planning and social change in Oman*. New York: Columbia University Press.

Cole, D. P., & Altorki, S. (1993) Commerce et production dans le nord de l'Arabie central: changement et développement à 'Unayzah. In R. Bocco, R. Jaubert, and F. Métral (Eds.), *Steppes d'Arabies: Etats, pasteurs, agriculteurs et commerçants; le devenir des zones sèches*. Paris: Presses Universitaires de France et Genève.

Evenari, M., Shanan, L., & Tadmor, N. (1958) The ancient desert agriculture of the Negev. III. Early beginnings. *Israel Exploration Journal*, 8:231–268

Evenari, M., Shanan, L., & Tadmor, N. (1971) *The Negev: The challenge of a desert*. Cambridge, MA: Harvard University Press.

Fabietti, U. (1993) Politiques étatiques et adaptations bédouines. In R. Bocco, R. Jaubert, & F. Métral (Eds.), *Steppes d'Arabies: Etats, pasteurs, agriculteurs et commerçants; le devenir des zones sèches*. Paris: Presses Universitaires de France et Genève.

Fernandez, J. W. (1990) The call to the commons: Decline and recommitment in Asturias, Spain. In B. J. McCay, & J. M. Acheson (Eds.), *The question of the Commons: The culture and ecology of communal resources* (pp. 226–289). Tucson: The University of Arizona Press.

Firincioglu, H. K. (2000) Range and management in Turkey. In J. Sabet, (Ed.), *Proceedings: First Regional Seminar on Rangeland Strategies in Jordan*. Amman. ICARDA, pp. 1–12.

Hardin, G. (1968) The tragedy of the commons. *Science*, 162:1243–48.

Hardin, G., & Baden, J. (Eds.) (1977) *Managing the commons*. San Francisco: Freeman.

ICARDA (2000). *Proceedings: First Regional Seminar on Rangeland Strategies in Jordan*. J. Sabet, (Ed.). Amman, May 2–4, 2000. International Center for Agricultural Research in the Dry Areas (ICARDA).

Jaubert, R. (1993) Evolution des systèmes agro-pastoraux et politiques de développement des régions sèches de Syrie. In R. Bocco, R. Jaubert, & F. Métral, (Eds.), *Steppes d'Arabies: Etats, Pasteurs, Agriculteurs et Commerçants; le Devenir des Zones Sèche*. Paris: Presses Universitaires de France et Genève.

Khalaf, S. (1993) Cheikhs, paysans et membres du Parti Ba'th: Changements politiques en Syrie du Nord. In R. Bocco, R. Jaubert, & F. Métral (Eds.), *Steppes d'Arabies: Etats, pasteurs, agriculteurs et commerçants; le devenir des zones sèches*. Paris: Presses Universitaires de France et Genève.

King, D. J. (2000) In J. Sabet, (Ed.), *Proceedings: First Regional Seminar on Rangeland Strategies in Jordan*. Amman. ICARDA

Kressel, G. M. (1992) *Descent through males*. Wiesbaden: Otto Harrassowitz.

Kressel, G. M. (1996) *Ascendancy through aggression*. Wiesbaden: Otto Harrassowitz.

Kressel, G. M. (1998). The Arab world's travails: The desert burden. *Middle East Quarterly*, 5(1): 49–57. Reappeared 2000, in Chinese translation of the *Middle Eastern Studies*, Xi'an University, 36:68–73.

Kressel, G. M. (2003). *Ascendancy through aggression*. Wiesbaden: Otto Harrassowitz.

Kressel, G. M., & Ben David, J. (1995). The Bedouin market—Corner stone for the founding of Be'er Sheva: Bedouin traditions about the development of the Negev capital in the Ottoman Period. *Nomadic Peoples*, *36/37*, 119-144.

Kressel, G. M., & Ben David, J. (1996). The Bedouin market—The axis around which Beersheba developed in the British mandatory period. *Nomadic Peoples*, *39*, 3-28.

Kressel, G. M., Ben-David, J., & Abu-Rabi'a, K. (1991) Changes in the land usage by the Negev Bedouin since the mid-19th century: The intra-tribal perspective. *Nomadic Peoples*, 28:28–55.

Lancaster, W. & Lancaster, F. (1993) Sécheresse et stratégies de reconversion économique chez les Bédouin de Jordanie. In R. Bocco, R. Jaubert, & F. Métral (Eds.), *Steppes d'Arabies: Etats, pasteurs, agriculteurs et commerçants; le devenir des zones sèches*. Paris: Presses Universitaires de France et Genève.

Lewis, N. (1987) *Nomads and settlers in Syria and Jordan, 1800-1950*. Cambridge, UK: Cambridge University Press.

McCay, B. J., & Acheson, J. M. (Eds.) (1990) *The question of the Commons: The culture and ecology of communal resources*. Tucson: The University of Arizona Press.

Métral, F. (1993) Elevage et agriculture dans l'oasis de Sukhné (Syrie): gestion des risques par les commerçants-entrepreneurs. In R. Bocco, R. Jaubert, & F. Métral (Eds.), *Steppes d'Arabies: Etats, pasteurs, agriculteurs et commerçants; le devenir des zones sèches*. Paris: Presses Universitaires de France et Genève.

Meyer, A. J. (1958) *Middle Eastern capitalism*. Cambridge, MA: Harvard University Press.

Orev, Y. (1986) *A practical handbook on desert range improvement techniques*. Geneva: Interagency Group on Agricultural Biometeorology.

Orev, Y. (1987) *Manuel pratique pour l'amélioration des pâturages en zones semi-arides*. Genève: Centre Technique de Coopération Agricole et Rurale.

Perevolotsky, A. (1991a) Rehabilitation of the black goat. *Journal of Arid Environments*, 13:153–161.

Perevolotsky, A. (1991b) Goats or scapegoats—Overgrazing controversy in Piura, Peru. *Small Ruminant Research*, 6:199–215.

Perevolotsky, A. & Landau, Y. S. (1992) Droits pastoraux en Israël: Perspectives historique et écologique sur le statut des terres de parcours. In A. Bourbouze, & R. Rubino (Eds.) (1992) *Terres collectives en Méditerranée: Histoire, législation, usages et modes d'utilisation par les animaux* (117–136). Rome: Villa d'Agri, FAO.

Seligman, N. G., Tadmor, N. H., & Raz, Z. (1962) Survey of natural pasturelands in the Central Negev. Jerusalem, The Volcanic Center. *Land Preservation Division Bulletin*, 67. (in Hebrew).

Shanan, L. (1987) The impact of irrigation. In M. G. Wolman & F. G. A. Fournier (Eds.) *SCOPE 32: Land transformation in agriculture*, New York: John Wiley & Sons. pp. 115–131.

Shanan, L. (1992) Planning and management of irrigation systems in developing countries. *Agricultural Water Management*. 22(1–2): 1–234.

Shanan, L. (1998) Irrigation development: Proactive planning and interactive management. In H. J. Bruins, & H. Lithwick (Eds.) *The arid frontier: Interactive management of environment and development.* Dordrecht, Netherlands: Kluwer Academic Publishers, p. 251–276.

Tell, T. (1993) Paysans, nomades et Etat en Jordanie orientale rural (1920–1989). In R. Bocco, R. Jaubert, & F. Métral (Eds.), Steppes d'Arabies: Etats, pasteurs, agriculteurs et commerçants; le devenir des zones sèches. Paris: Presses Universitaires de France et Genève.

Chapter 17

Conflict Resolution: A Gandhian and Quaker Approach*

A. Paul Hare

The model that I use for conflict resolution is based on the practice of the *Shanti Sena* (Peace Brigade) in India, which in turn is based on the approach developed by Gandhi. One of the leaders of the movement, whom I have worked with, was Narayan Desai, whose father was Gandhi's secretary and who grew up in Gandhi's *ashrams* (intentional communities). Gandhi's way of life included three main aspects: (1) living in an *ashram* without class distinctions, where cooperation was practiced and decisions were made by the process of reaching consensus; (2) nonviolently resisting aspects of structural or physical violence in a society with "truth force" (*satyagraha*) and bringing members of opposing sides together to reach a new definition of the situation; and (3) improving the structure of society through "constructive programs," such as using homespun material or introducing cottage industries (Desai, 1977, 1999; see also Hare, 1968). I also use my experience as a Quaker: using consensus in reaching decisions and having a commitment to nonviolent solutions to conflicts.

For this chapter I draw on my experiences in India during the 1960s through the 1990s, from inter-communal conflicts, and from visits to instances of "constructive programs." Prior to these, I had participated in various "Marches on Washington" for civil rights in the U.S.A.—including hearing Dr. Martin Luther King, Jr. give his "I have a dream" speech—taken part in a "sit-in" at a restaurant, and founded several "Fair Housing" committees that promoted equal op-

---

* From: Chapter 4, pages 52 through 62, "Informal mediation by private individuals" by A. Paul Hare in Jacob Bercovitch and Jeffery Z. Rubin (Eds.), *Mediation in international relations*, 1992, St. Martins Press, Inc. Reproduced with permission of Palgrave Macmillan.

portunities for buying real estate in a community. While teaching at Haverford College in the 1960s, I established the Center for Nonviolent Conflict Resolution with a major grant from the National Institute of Mental Health, where research on nonviolent action was carried out over a period of several years. I was assisted by the late Charles "Charlie" Walker, a Quaker activist, who introduced me to Narayan Desai and who later joined me on other projects. Charlie was responsible for recording the guidelines for nonviolent action and the activities of marshals and observers at Kent State University that we later used in other parts of the world: in Curaçao, Cyprus, India, and South Africa. Possibly influenced by my experience as editor of the newspaper for the high school I attended in Washington, D. C., I have left a "paper trail" of books, articles, and other publications that provide the details of these experiences.

I became familiar with the Quaker method of reaching decisions using consensus while a student at Swarthmore College, which had a Meeting House on campus, and later joined the Cambridge Meeting while a post-doctoral student and research assistant at Harvard University. For a few years I was also on the Board at Pendle Hill, a Quaker Center near Philadelphia. For a year, in 1961–62, I observed the "transfer of technology" as the Deputy Representative for the U.S. Peace Corps in the Philippines, where volunteers were teaching English in schools. During my interview with Sargent Shriver, then Head of the Peace Corps, he was impressed with my experience as an Infantry Rifle Platoon Leader in Europe during World War II, which he considered to be suitable experience in "administration" for a Peace Corps staff member. In the periods when I was actively involved in the nonviolent movement, my status as a College or University Professor appeared to make it easier for me to talk with persons in authority.

Applications of nonviolent approach

My experience with the Gandhian and Quaker approaches to nonviolence—and the similar approach derived from the work of Martin Luther King, Jr.—was to adapt aspects of their approaches and apply these to the "culture" and the "situation" in which the conflict occurred. I was familiar with the influence of situational variables from my study for a Ph.D. in sociology at the University of Chicago. The more usual approach by a person going into an area of conflict was to use some "package" involving group discussion, such as those for "brainstorming," by withholding criticism so that more, possibly creative, ideas may be heard (Osborn, 1963), or holding a problem-solving workshop with representatives of the various sides involved (Doob, 1974, 1978; Kelman, 1992), or "paying attention to the other" (Bar-On and Grinberg, 1997), or using some method to enhance some characteristics of one of the sides involved, such as providing achievement motivation (McClelland, 1953; see also Berlew and LeClere, 1974). My own approach has been to apply whatever aspects of the Ghandian/Quaker/King method seem appropriate and not follow a fixed program, which I happen to have in my pocket, for transfer of technology.

A major example of an application with this approach derives from Kent State University, where members of a research team went to do research after an incident in the 1960s, during which four students were killed by members of the Ohio National Guard, who had been brought in to quell a disturbance. We stayed on to provide support and training in nonviolent methods of protest for students who acted as marshals, and I was given a teaching position at Kent State University to provide a seminar on nonviolence (Hare, 1977). A second example derives from the island of Curaçao in the Caribbean, where I had been asked to study the "attitudes of the people" who had taken part in a riot that brought down the government after police had intervened at a strike at the main oil processing plant. On the bases of responses to interviews with a small cross-section of residents of Curaçao, I proposed a Summer Institute to review the problems that had led to the riot. The Institute was funded and two sessions were held. Although we had proposed to continue the Institute the following summer, it was not necessary since the major issues had been solved (Hare, Carney and Ovsiew, 1977). A third example is the intervention in South Africa (Hare, 1983) where I helped students conduct a major demonstration at a funeral without violence. The last example, on Cyprus (Hare and Wilkinson, 1977; Hare, 1992) is given in some detail as a reprint of a published article. As noted, additional experience was gained in India by working with Narayan Desai and members of the *Shanti Sena* and with the group of African Americans who immigrated to Israel to form the Hebrew Israelite Community (Hare, 1998).

Cyprus

Here I reprint a chapter from a collection by Bercovitch and Rubin's *Mediation in International Relations* on "Informal mediation by private individuals" (Hare, 1992). The selection provides an indication of some of the characteristics of informal mediation by private individuals in conflicts between nations, or in conflicts within a nation where external involvement may be present. I draw mainly on my experiences as the coordinator of a third party team that mediated between the Turkish and Greek sides on Cyprus from 1972 until the invasion of the Turkish army in 1974.

Following the initiative of the International Peace Academy, we were working in parallel with the United Nations Forces to help resettle the Turkish villagers who had become refugees during the "time of troubles" in 1963 (Hare and Wilkinson, 1977; Hare, 1984; see also a collection of documents from the project in the Peace Collection, Swarthmore College library). However, the experience had many similarities to my experience in mediation in inter-group relations in communities or larger social entities in the United States (Hare, 1977), Curaçao (Hare, Carney and Ovsiew, 1977), and South Africa (Hare, 1976, 1983). The same principles of informal, private mediation seem to apply, no matter what the system level.

In these mediation efforts I was quite consciously applying the Quaker approach to reaching consensus as a way of finding solutions to conflicts to which all parties would agree. As background for this report I read the accounts of several other Quaker mediators who also used a "low profile" approach to mediation, and for contrast, accounts of mediation by Michael Scott and Armand Hammer who were "high profile" in their activities (Vining, 1958; Warren, 1964; Considine, 1975; Yarrow, 1978; Hare and Blumberg, 1980; Lampen, 1983; Ullmann, 1984; Curle, 1986; Fisher, 1987; Hammer, 1987; Lee, 1987; Pickvance, 1988; Van der Merwe, 1988). Although there are differences in the goals of mediation and the extent of the use of economic or other resources among the different cases, the main points in my experience are common to all of these efforts.

Some characteristics of the mediation process occur in all types of mediation. Others take the form of a set of choices that the mediator may take or may be imposed by the situation. The general characteristics of mediation that provide a context for the present description are given by Young (1967, 1972), Bercovitch (1989), Rubin (1981), Zartman and Berman (1982), and Fisher (1972, 1983).

Some characteristics of private mediation

The phases in the development of any social process, including mediation, can be described in terms of dramaturgical analysis (Hare and Blumberg, 1988). In the dramaturgical view, the four phases of action are, first providing a definition of the situation in the form of an *actable idea*, second, arranging the *staging* by providing the action area and the resources, third, recruiting the actors (if they are not already in place) and training them for their *roles*, and fourth, the *enactment* (in this case carrying out the mediation). There may be one or more instances of enactment followed by a final phase in which the *new meanings* are assessed for the actors and the audience.

Rubin's (1981, pp. 7–19) sets of continua by which he derives a "partial listing of third party roles" can be placed in the order in which they would appear according to dramaturgical theory. One of the decisions a mediator must make is whether or not to initiate mediation or to seek an invitation to mediate. In the cases I have reviewed, and in my own experience—with the exception of Curaçao—none of the private mediators was formally invited, although they all found ways to make themselves acceptable. In Rubin's terms, the private mediators would also be classed as *conflict managing*, *advisory* (since the individuals have no formal authority to direct), and *temporary*. They all intended to be relationship facilitating, impartial, informal, and of course, individual (although in several instances a team was involved). Rubin's examples of informal mediators are mainly persons delegated by some formal government body to sound out the opposing side, and thus are generally not impartial. In contrast, the private mediator combines informal status with impartiality.

A characteristic of the formal mediator that the private individual does not possess is the power of economic or political backing. Based on his comparison of examples of international dispute mediation, Bercovitch (1989, p. 246) finds the manipulation strategies such as "leverage, resources, power, and influence are at the heart of successful mediation in international relations." This leaves the informal mediator with two other strategies: communication and formulation. Fortunately, according to Bercovitch's account, these methods are effective at least 30 per cent of the time, although manipulation is effective over 50 per cent of the time.

An indication that the third party can be a collective and not just one individual is emphasized by Bercovitch (1984), who suggests that the third party should be a group of professionally qualified persons who work as a team, since no one person could be expected to have all the necessary skills. Bercovitch goes on to recommend that the team be drawn from an organization with an "impartial and trustworthy reputation." His example is the Quakers.

Quaker conciliation

In his description of Quaker conciliation, Yarrow (1978) cited two characteristics that are shared with many other private individual mediators: namely, a *humanitarian approach* and *confidentiality*. However he also indicated two characteristics that are less likely to be widely held, one is *pacifism* and the other is the use of the *consensus* method of problem solving, coupled with support for an individual "concern." The Quaker's pacifist stance is a signal to both sides that the mediator will not intentionally support violence from either side.

The Quaker use of consensus in reaching decisions rather than majority vote is based on the assumption that "there is that of God in every person." Thus each individual is credited with some insight into a problem and each individual's view should ideally be represented in the solution to the problem (Hare, 1973). The term "concern" is used when a person has a religious leading to pursue some humanitarian activity or to act as a mediator in a crisis situation. This "concern" may be tested in consultation with members of the local Quaker meeting, or it may find support from some Quaker organization. Will Warren's activity in Northern Ireland is an example (Lampen, 1983). The consensus method is also used by the Gandhians in India.

The Cyprus Resettlement Project

Armed with letters of introduction from members of the International Peace Academy, who had served with United Nations Forces and knew members of the UN Forces and civilians on Cyprus, the first team of five persons visited Cyprus in August 1972. In third-party negotiations of this sort the principal resources are the talents and values of the project personnel. From the first we knew that even our exploratory team had to represent the commitment to nonviolence and the kinds of problem solving skills that we planned to use throughout the project if our third-party services were to be accepted by the Greek and Turkish sides. In the first team we also needed persons who could help present the project to outside agencies, such as the American Friends Service Committee, that we hoped would become sponsors and also help with the recruiting of other volunteers, especially from the *Shanti Sena* (Gandhian Peace Brigade) in India. Thus it was valuable to have the Peace Secretary from the American Friends Service Committee and the Secretary of the *Shanti Sena* in the team. We also had persons who were young, old, male, female, peace activists, and social scientists. Subsequent teams were composed similarly with volunteers from India, England, South Africa, and the United States. On three occasions I was the only member of the team on Cyprus.

In Cyprus the first team held interviews with representatives of the United Nations and Greek and Turkish officials. Mr. Osorio-Tafall, United Nations Representative on Cyprus, suggested that we work on psychological problems that divide people rather than on problems requiring only technical assistance. In particular, he suggested that the resettlement of Turkish displaced persons who had left their homes at the "time of troubles" in 1963 was a humanitarian problem that might provide a focus for the work.

Officials on the Greek side were already committed to the principle of resettlement and had built houses in some villages. On the Turkish side, the officials warmly recalled the work of a group of British volunteers who had rebuilt houses in a village some years before. However, at this time no progress had been made toward resettlement for some years. Persons on the Greek and Turkish side gave different reasons as to why the Turkish villagers had not returned. Representatives of the United Nations had tried for ten years to activate the resettlement process. They agreed that a private initiative might be helpful. Although the Turkish side was considered a part of Cyprus at that time and had not yet declared itself an independent state, all political arrangements with the Turkish side were handled by the Greek Ministry of Foreign Affairs. It is my experience that, in general, individuals or community members are unwilling to let a private third party intervene unless they have given up on the problem or a crisis situation indicates that they may be unable to solve the problem by themselves.

When Makarios, the Greek President, discovered that Narayan Desai, one of the leaders of the Gandhian movement in India, was a member of the first team, he asked for a special interview. He told us that he had unveiled a bust of Gandhi in Nicosia a few years before and that Gandhi had been one of the major influences in his own life, the other being his mother. Thus our nonviolent approach was clear to the Greek side from the beginning.

Later, when members of the third, and largest, team began to arrive, I took a small group to meet the Turkish Vice President Denktash. Two of the men from India were wearing their traditional dress and one woman had recently returned from a reconciliation project in Northern Ireland. Denktash came around from behind his desk to admire the group. "Oh what a lovely sight," he remarked. He was indicating that he was pleased that people with experience in nonviolence had come from all over the world to show interest in the problems of the Turkish Cypriots. In these days it may be difficult to find a conflict between two nations that does not involve the interests of many more nations. The presence of an international team symbolized the concern of persons from other parts of the world that a solution be found.

Throughout the project I held the title of "professor." However, I was not an official representative of any college or university, nor any other organization since our project was an *ad hoc* group. My professional status never seemed to be a handicap in talking to officials of various ranks, possibly because it suggested a nonpolitical orientation. Although one does not have to be an academic to engage in third party activity, I would suggest that one should be able to provide better facts or a better theory (preferably both), if one is to find a solution that has not been apparent to the parties in conflict. It is also helpful to have some affiliation with an institution to which grants can be made on a tax-deductible basis. In this case the funds that were eventually received from the Lilly Endowment were given to Haverford College to administer.

A second team of five persons worked on Cyprus during the month of July 1973. Initially they gathered more detailed information about conditions on Cyprus and the resettlement problem. Reports were written giving the positions of the Greek side, the Turkish side, and the United Nations. These reports were given to each party to check the accuracy of the team's understanding of the problem. The Greek side and the United Nations approved the first draft. Mr. Denktash suggested that we add several additional pages outlining the Turkish position. We did so. Copies of the final versions were sent to all three parties. We did not attempt to use these documents as a basis for negotiation, rather as a means of letting each party know that we understood its position. We knew that each side would tend to present only the facts that would support its position. If a solution could be found using only the facts as stated, the parties would presumably have found it some time before. Current information about the houses in the villages and the persons who wished to return was needed. The sides had not been in communication about the issue for some time. We therefore continued to collect information from the Greek and Turkish sides separately since they had not yet agreed to meet for face-to-face discussion.

In the Cyprus case, and in other projects in which I have been involved, it would have served little purpose to have kept a team or even an individual on Cyprus when it was necessary for one of the sides to carry out some commitment. Several months were necessary for Turkish social workers to contact the Turkish families who had left their homes to determine who would be willing to return. An exception to this is when one or more individuals actually move to the community or country in question and engage in some form of employment between crises or while waiting for further developments. This is the approach I prefer. However, if one is involved in international disputes, one must be prepared to move around.

In September 1973, I visited India to recruit persons from the *Shanti Sena* to be members of the third team. In November the third and largest team arrived to work for three months on Cyprus. This was the period of the most extensive work on the project. There were teams of three or four persons living in each of four villages in areas where resettlement would take place. Each team collected information from villagers in its area and tried to find ways to increase the involvement of villagers in the resettlement process. Members drawn from each team formed a negotiating team to meet with representatives of the Ministry of Foreign Affairs, on the Greek side, and the Turkish Cypriot leadership.

During the visit of the third team the officials of the Greek side told us that one of our earlier reports on conditions in one of the formerly mixed villages, which we had intended for internal use, was biased in favor of the Turkish side. This was one of several occasions when our attention was called directly to the sensitivity of each side.

Looking back over some of our Cyprus documents some years later, I found it hard to see why certain wording would be offensive to one side or the other. We needed to learn the history of the use of various words and phrases and attempt to use only terms that were mutually acceptable. This applies to terms in this account, such as "time of troubles" and the references to the Greek and Turkish "sides."

Eventually we came to realize that the Greek side was very hesitant to do anything, such as be represented on a joint committee, that might recognize the Turkish side as having separate status, and that the Turkish side was equally hesitant to engage in any activity that did not grant them appropriate recognition. The issue of mutual recognition of basic identities would seem to be present in all conflict situations.

By this time we knew enough about the situation to be able to make suggestions to the two sides concerning moves that might facilitate the process of resettlement. Although the United Nations had some 3,000 men stationed on the island at any one time, only seven of them were concerned with community affairs, and only part of their time was spent on issues of reconciliation. Thus a relatively small team, of only five persons for most of the visits, by working full time on a single issue could accumulate a great deal of information in a relatively short period. It also helped to have access to persons at all system levels, from villagers, though *mukhtars* (village heads), through regional officials, to Denktash and Makarios. In this way we could search for solutions that would be acceptable at all levels.

It is quite in keeping with the method of arriving at decisions by consensus for the private mediator to offer suggestions concerning points on which the two sides might agree. In the more formal use of consensus, when all parties take part in the discussion, it is the responsibility of everyone, and especially the person chairing the meeting, to frame statements that all those present would be willing to endorse. In the Cyprus case, when representatives of the two sides were not yet meeting face-to-face, the private mediator was the only one who had a detailed sense of the concerns of both sides.

Our relationship with the United Nations Forces was unofficial. At first I did not understand that this would be the case. I approached the Chief UN Political Officer. I noted that we were on Cyprus to work along the lines of the UN mandate and in cooperation with them. I asked what our relationship would be. "You have no relationship," was the reply. Thinking that I had misunderstood, I asked again: "What is our relationship to the UN?" Once more he said: "You have no relationship." It finally dawned on me. "You mean we have no (official) relationship," I said. "Yes," he said. He then handed me his personal card with his business and home phone numbers. "Call me day or night," he said, "whenever you have anything that we should know."

It was obvious, then, that the UN could have no responsibility for whatever we might say or do. If we succeeded in bringing the Greek and Turkish sides together on the resettlement issue, they would be pleased. If we failed, they could not help us. This, of course, is common to all private third party activity. No official or organizational pressure can be brought to bear on the parties involved. If the "good offices" become unacceptable to any of the parties, the "third party" is out of business.

In December 1973 a front page story in the *Cyprus Mail*, headed "Further aid to displaced Turks," announced the fact that the government of Cyprus would repair houses of the Turkish Cypriots who had left their homes ten years before during the "time of troubles." The next day the paper quoted Mr. Osorio-Tafall, the United Nations Special Representative on Cyprus, as saying that this step by the government would "create better conditions for the furtherance of a solution" to the Cyprus problem. In press releases, both the Greek and Turkish sides noted the contribution of the Cyprus Resettlement Project in helping to facilitate the process of resettlement.

In January, after many consultations on both sides of the Green Line, officials of both sides met face-to-face with members of the project and a representative of the United Nations to discuss the next steps in the resettlement process. There is no written statement describing the purpose of the meeting, since we were unable to find wording for a statement that was acceptable to both sides.

In mid-April the fourth team of four persons arrived to review the progress toward resettlement and to decide whether or not any further service by the project seemed desirable. They found that a new political development had brought the inter-communal talks to a halt and that the government had not yet voted money to begin reconstruction in the villages. However, it appeared that the political situation might be "clarified" by the end of May, allowing the work to proceed.

As an interim project, to assure some movement towards resettlement and as a sign of the goodwill of the citizens of both sides, the team proposed that a four-week work camp, with participants from the youth of both sides, be held in one of the villages, to rebuild Turkish homes. The work camp was in progress on 15 July when a *coup* brought a change in the Greek government and finally came to a halt with the invasion of the Turkish Army on 20 July. As a result of the military and political events, there were many more refugees, now mostly Greek Cypriots, and there was no need to continue our particular project.

Was the venture on Cyprus a success or a failure? Certainly we did not succeed in uniting the two sides of Cyprus. We could argue that to deal satisfactory with a political coup and a military invasion required resources well beyond those of our third-party team. However the enthusiasm with which the Greek and Turkish sides supported out efforts, including the work camp during which Greek and Turkish youth repaired Turkish homes, suggests that our intervention was a timely activity as a step toward possible reconciliation.

Some highlights of private mediation

As regards resources, in several instances the Quaker mediators could turn to the American and British Friends Service Committees to provide food, clothing, and other forms of relief if this was necessary. On occasion they might intersperse their period of active mediation with a period of supervision of relief. In my own case, even without any official connection with one of the Service Committees, I have found persons in different parts of the world who recall receiving aid from a Quaker feeding program when they were young or receiving relief in some other form. They seemed much more willing to cooperate in a project even though there was no expectation of additional resources at that time. However, the ability to supply a variety of resources was prominent in the activities of other private mediators, such as Armand Hammer.

The principal characteristic of private "unofficial" mediation is that it is "unofficial." In most of the cases of private mediation that were reviewed for this analysis, one of the parties was the "official" government of the country and the other was a breakaway group that did not yet have equal official status. In any formal negotiation, the side with the formal status would have to formally recognize the other side. When a private mediator is involved, the higher status side can avoid giving official status to the negotiations to the other side (Van der Merwe, 1988).

A second important characteristic of private mediators is they are not responsible for implementing the policy of a particular government or organization, such as the United Nations. In the case of the Cyprus Resettlement Project it was quite clear that the private team was able to take initiatives that the UN officials could not, since the UN officials were bound by their mandate.

A third characteristic of many of the interventions by private mediators is that they usually receive no publicity. However, Hammer's high profile "shuttle diplomacy," with frequent press interviews, indicates that other approaches are also possible.

In addition, the private mediator must be aware that there are many "players" involved in any conflict and that private efforts are, at best, only one influence among many. It is also important to know when to leave. In the Cyprus case the situation had changed so dramatically that it was obvious to all that the third party could no longer be useful unless some new definition of the situation could be agreed upon.

Conclusions

Informal mediation by private individuals in international relations has much in common with more formal and official mediation. Ideally all mediators, either as individuals or as teams, have the necessary intellectual and social skills to assist in finding creative solutions to social problems. All forms of mediation involve a range of activities with increasing degrees of involvement, from providing a communication channel to being a full participant in the problem-solving process.

There are some disadvantages to being a private individual, mainly in the amount of resources and political power that one is able to bring to bear. However, this can be a hidden advantage in that it moves the mediator to seek solutions that are based on a new definition of the situation rather than trade-off for some economic or power gain. In principle, solutions based on an exchange of resources or the use of power are likely to be less permanent than those that involve higher levels of creativity, such as those based on mutual recognition or a common set of values (Hare, 1982, Ch. 10). The process of reaching decisions by consensus has the possibility of leading to more creative solutions. Here, Quaker mediators have experience, but the method can be learned and practiced by anyone (Hare, 1982, Ch. 9).

The advantages of being a private mediator include the fact that an official government may avoid—at least temporarily—giving formal recognition to an adversary group; and the mediator does not need to stay within any governmental or organizational mandate in searching for a solution. Private individuals acting as mediators may lack resources or "leverage," but they do possess freedom of movement and latitude of decision that is often denied to their official counterparts. In all the instances of individual mediation reviewed for this paper, many formal mediation efforts were also in progress. Still, private initiatives offer considerable advantages in protracted conflicts, or in the preparation phase for mediation in a conflict to warrant adding them to the list of mediation attempts in any conflict in international relations. Private individuals acting as mediators may not always achieve a substantial agreement, but they can certainly pave the way for others to do so.

References

Bar-On, D., & Grinberg, L. (1997) *The center for dialogue between populations in conflict*. Be'er Sheva: Ben-Gurion University, Center for Dialogue between Populations in Conflict.

Bercovitch, J. (1984) In *Social conflict and third parties* (p. 31). Boulder, CO: Westview Press.

Bercovitch, J. (1989) International dispute mediation: A comparative empirical analysis. In K. Kressel & D. G. Pruitt (Eds.), *Mediation research: The process and effectiveness of third party intervention* (pp. 284–99). San Francisco: Jossey-Bass.

Berlew, D. E., & LeClere, W. E. (1974) Social intervention in Curacao: A case study. *The Journal of Applied Behavioral Science*, 10(1): 29–52.

Considine, B. (1975) *The remarkable life of Dr. Armand Hammer*. New York: Harper and Row.

Curle, A. (1986) *In the middle: Non-official mediation in violent situations*. Leamington Spa: Berg Publishers Ltd.

Desai, N. (1977) Intervention in riots in India. In A. P. Hare & H. H. Blumberg (Eds.), *Liberation without violence: A third party approach*. Totowa, NJ: Rowman & Littlefield.

Desai, N. (1999) *My Gandhi*. Ahmedabad: Navjivan Publishing House.

Doob, L. W. (1974) A Cyprus workshop: An exercise in intervention methodology. *Journal of Social Psychology*, 94(2) December: 161–178.

Doob, L. W. (1978) A Cyprus workshop: Intervention methodology during a continuing crisis. *Journal of Social Psychology*, 98(1) February: 143-144.

Fisher, D. (1987) Afghan settlement may be near, Hammer says. *Los Angeles Times*, 15, October, Part I, p.14.

Fisher, R. J. (1972) Third party consultation: A method for the study and resolution of conflict. *Journal of Conflict Resolution*, 16:67–94.

Fisher, R. J. (1983) Third party consultation as a method of inter-group conflict resolution: A review of studies. *Journal of Conflict Resolution*, 27:301–34.

Hammer, A., with Lyndon, N. (1987) *Hammer*. New York: G. P. Putnam and Sons.

Hare, A. P. (1968) Introduction to theories of nonviolence. In A. P. Hare & H. H. Blumberg (Eds.). *Nonviolent direct action* (pp. 1–30). Washington, D.C.: Corpus Books.

Hare, A. P. (1973), Group decisions by consensus: Reaching unity in the Society of Friends. *Sociological Inquiry*, 43:75–84.

Hare, A. P. (1976) Riot control: A third party approach. *South African Outlook*, 106 (1264, September): 134–135.

Hare, A. P. (1977) Applying the third party approach. In A. P. Hare & H. H. Blumberg (Eds.), *Liberation without violence* (pp. 265–287). London: Rex Collings.

Hare, A. P. (1982) *Creativity in small groups*. Beverly Hills, CA: Sage Publications.

Hare, A. P. (Ed.) (1983) *The struggle for democracy in South Africa: Conflict and conflict resolution*. Cape Town: Centre for Intergroup Studies, University of Cape Town.

Hare, A. P. (Ed.) (1984) Cyprus Resettlement Project: An instance of international peacemaking. Be'er Sheva, Israel: Department of Behavioral Sciences, Ben-Gurion University.

Hare, A. P. (1992) Informal mediation by private individuals. In J. Z. Rubin & J. Bercovitch (Eds.), *Mediation in international relations* (pp. 52–63). New York: Macmillan.

Hare, A. P. (1998) The community as mediator. In A. P. Hare (Ed.), *The Hebrew Israelite community* (pp. 81–96). Lanham, MD: University Press of America.

Hare, A. P., & Blumberg, H. H. (Eds.) (1980) *A search for peace and justice: Reflections of Michael Scott*. London: Rex Collings.

Hare, A. P., & Blumberg, H. H. (1988) *Dramaturgical analysis of social interaction*. New York: Praeger.

Hare, A. P., Carney, F., & Ovsiew, F. (1977) Youth responds to crisis: Curacao. In A. P. Hare & H. H. Blumberg (Eds.), *Liberation without violence* (pp. 220–238). London: Rex Collings.

Hare, A. P., & Wilkinson, E. (1977) Cyprus: Conflict and its resolution, in A. P. Hare & H. H. Blumberg (Eds.) *Liberation without violence* (pp. 239–247), London: Rex Collings.

Kelman, H. C. (1992) Informal mediation by the scholar/practitioner. In J. Bercovitch & J. Z. Rubin (Eds.), *Mediation in international relations* (pp. 64–96). Houndmills, UK: The Macmillan Press.

Lampen, J. (Ed.) (1983) *Will Warren: A scrapbook*. London: Quaker Home Service.

Lee, G. (1987) Hammer has Afghan peace role: Shuttle diplomacy on Moscow's behalf. *Washington Post*, 17 October, A 15.

McClelland, D. C. (1953) *The achieving society*. New York: Appleton-Century-Crofts.

Osborn, A. F. (1963) *Applied imagination: Principles and procedures of creative problem solving*. (Third revised edition). New York: Charles Scribner's Sons.

Pickvance, T. J. (1988) Third party mediation in national minority disputes: Some lessons from the South Tyrol problem (pp. 131–146). In C. R. Mitchell & K. Web (Eds.), *New approaches to international mediation.* New York: Greenwood Press.

Rubin, J. Z. (Ed.) (1981) *Dynamics of third party intervention: Kissinger in the Middle East.* New York: Praeger.

Ullmann, R. K. (1984) *The dilemmas of a reconciler.* London: Quaker Peace and Service.

Van der Merwe, H. W. (1988) South African initiatives: Contrasting opinions in the mediation process (pp. 180–194). In C. R. Mitchell & K. Web (Eds.), *New approaches to international mediation.* New York: Greenwood Press.

Vining, E. G. (1958) *Friend of life: The biography of Rufus M. Jones.* Philadelphia, PA: J. B. Lippincott.

Warren, R. L. (1964) The conflict intersystem and the change agent. *Journal of Conflict Resolution, 8,* 231–241.

Yarrow, C. H. M. (1978) *Quaker experience in international conciliation.* New Haven, CT: Yale University Press.

Young, O. R. (1967) *The intermediaries: Third parties in international crisis.* Princeton, NJ: Princeton University Press.

Young, O. R. (1972) Intermediaries: Additional thoughts on third parties. *Journal of Conflict Resolution,* 16:51–65.

Zartman, W. I., & Berman, M. R. (1982) *The practical negotiator.* New Haven: Yale University Press.

Name Index

Subject Index